GREENHOUSE GAS BALAN
BIOENERGY SYSTEMS

GREENHOUSE GAS BALANCES OF BIOENERGY SYSTEMS

Edited by

Patricia Thornley

Paul Adams

ELSEVIER

ACADEMIC PRESS

An imprint of Elsevier

Academic Press is an imprint of Elsevier
125 London Wall, London EC2Y 5AS, United Kingdom
525 B Street, Suite 1800, San Diego, CA 92101-4495, United States
50 Hampshire Street, 5th Floor, Cambridge, MA 02139, United States
The Boulevard, Langford Lane, Kidlington, Oxford OX5 1GB, United Kingdom

Library of Congress Cataloging-in-Publication Data

A catalog record for this book is available from the Library of Congress

British Library Cataloguing-in-Publication Data

A catalogue record for this book is available from the British Library

ISBN: 978-0-08-101036-5

Publisher: Joe Hayton
Acquisition Editor: Raquel Zanol
Editorial Project Manager: Mariana L. Kuhl
Production Project Manager: Sruthi Satheesh
Cover Designer: Miles Hitchen

Typeset by SPi Global, India

Contents

11. Combustion of Energy Crops for District Heating

PAUL GILBERT

12. Production of Wood Pellets from Waste Wood

CARLY WHITTAKER

13. Second-Generation Ethanol from Lignocellulose

SOPHIE PARSONS, MARCELLE C. McMANUS, CAROLINE M. TAYLOR

14. Electricity From North American Forest Residues

MIRJAM RÖDER

15. Agricultural Lessons

IAN SHIELD

16. Engineering Lessons—Using Engineering Design to Minimise GHG Emissions From Bioenergy Production

PAUL ADAMS

17. Environmental Lessons—Making Bioenergy System Decisions That Benefit the Environment

PATRICIA THORNLEY

18. Policy Lessons: The Role of Policy Regimes in Maximising GHG Savings in Bioenergy Systems

PATRICIA THORNLEY, PAUL ADAMS

19. Outlook–for Low Carbon Bioenergy

PATRICIA THORNLEY, PAUL ADAMS

Contributors

Paul Adams University of Bath, Bath, United Kingdom

Tony Bridgwater Aston University, Birmingham, United Kingdom

Laura Craggs University of Manchester, Manchester, United Kingdom

Lorenzo Di Lucia Imperial College London, London, United Kingdom

Julia Drewer NERC Centre for Ecology & Hydrology, Penicuik, United Kingdom

Paul Gilbert University of Manchester, Manchester, United Kingdom

Zoe M. Harris Imperial College London, London, United Kingdom

David Howard NERC Centre for Ecology & Hydrology, Lancaster, United Kingdom

Amanda Lea-Langton University of Manchester, Manchester, United Kingdom

Marcelle C. McManus University of Bath, Bath, United Kingdom

Niall McNamara NERC Centre for Ecology & Hydrology, Lancaster, United Kingdom

Suzanne Milner University of Southampton, Southampton, United Kingdom

Sophie Parsons University of Bath, Bath, United Kingdom

Mirjam Röder The University of Manchester, Manchester, United Kingdom

Andrew Ross University of Leeds, Leeds, United Kingdom

Ian Shield Rothamsted Research, Harpenden, United Kingdom

Raphael Slade Imperial College London, London, United Kingdom

Caroline M. Taylor EarthShift Global, Kittery, ME, United States

Gail Taylor University of Southampton, Southampton, United Kingdom

Patricia Thornley University of Manchester, Manchester, United Kingdom

Ian Watson University of Glasgow, Glasgow, United Kingdom

Carly Whittaker Rothamsted Research, Harpenden, United Kingdom

CHAPTER

1

Sustainable Greenhouse Gas Reductions From Bioenergy Systems—Climate Change: A Bioenergy Driver and Constraint

Laura Craggs, Paul Gilbert
University of Manchester, Manchester, United Kingdom

1.1 INTRODUCTION

In response to the increasing evidence of human impact on the climate, the International Panel on Climate Change (IPCC) was created in 1988 with the intention of stabilising global greenhouse gas (GHG) emissions in the atmosphere [1]. Even with the increased attention on GHG emissions and climate change over the last 30 years, anthropogenic GHG emissions continue to increase and in 2010 they stood at 49 $GtCO_2$ per year [1]. The observed impact of climate change can be seen through occurrences of extreme weather, with melting ice caps in the arctic, rising sea levels, severe flooding, and increased rain levels [2].

Climate change theory states that an increased abundance of GHGs in the atmosphere traps reflections of solar radiation from the earth's surface, keeping this radiation within the earth's atmosphere and causing a warming effect [3]. The major anthropogenic GHGs are carbon dioxide (CO_2), nitrous oxides (NO_x), and methane (CH_4), with CO_2 accounting for 60% of the observed global warming [1,4]. The balancing movement of carbon between the atmosphere and land storage is an important component of climate change. Large volumes of carbon are frequently exchanged between the land and the atmosphere, through biological, chemical, geological, and physical processes, highlighting the importance of balance, as atmospheric CO_2 will only be stable when these processes are in equilibrium [5]. Certain human activities can alter this equilibrium, for example, deforestation, which reduces the carbon stored on land, and burning fossil fuels, which increases the carbon released into the atmosphere [6].

Greenhouse Gas Balances of Bioenergy Systems
https://doi.org/10.1016/B978-0-08-101036-5.00001-X

Although burning fossil fuels has smaller net GHG emissions than most natural carbon exchanges, the key difference is the absence of any counteracting storage of CO_2; meaning this is a one-way movement and all fossil fuel burning contributes to carbon in the atmosphere [5]. Seventy-eight percent of the increase in GHG emissions between 1970 and 2010 were from fossil fuel combustion and industrial processes, highlighting the important role humans play in the climate change process [1]. This chapter provides an understanding behind the urgency for reducing greenhouse gas emissions and the potential role of bioenergy in avoiding the very real threat of an increased global temperature.

1.2 SCALE OF THE GLOBAL CHALLENGE

Since 1971, cumulative emissions from burning fossil fuels have increased from 13,995 million tonnes to 32,189 million tonnes of CO_2 in 2013 [7]. This trend is projected to continue into the future, due to increasing populations and economic development of countries such as China and India creating an increased energy demand [1]. A continued growth of CO_2 emissions will lead to long-term and irreversible impacts on the climate to the detriment of the environment and human health. With no mitigating actions against these increasing emissions, the temperature increase compared to pre-industrial levels is expected to be 4°C by 2100 [2].

In a 4°C warmer future, sea levels will rise dangerously, cities will be sub-merged, food security will be at risk from reduced crop productivity, and society will experience extreme weather patterns, including forest fires, violent storms, and devastating droughts [2,8]. In a 2°C warmer world, these risks are still apparent but with less severe and life-threatening impacts [9]. Evidence suggests that society's previous emissions are so significant that a 1.5°C temperature increase from pre-industrial levels can no longer be avoided. This means that urgent and dramatic action must be taken to ensure that we keep the global temperature below the dangerous levels of 2°C [2].

1.3 CLIMATE POLICY OBJECTIVES

Following building evidence and increasing concern around the effects of climate change, a global commitment has been made through the highly anticipated Paris agreements (COP 21), to keep global temperature increases 'well below 2°C' and to explore the potential for 1.5°C [10]. COP 21 also led to an agreement that, from 2020, richer countries will spend money helping lower income countries either mitigate climate change or try to adapt to the impacts of rising temperatures, such as rising sea levels [9]. This is a key change from previous agreements, as historically commitments have been made only by developed countries. This commitment also highlights the fact that climate change can no longer be fully avoided and measures must be taken to mitigate the inevitable impacts.

Following the Paris agreement to limit global warming, the 195 countries of the United Nations Framework Committee on Climate Change (UNFCCC) laid out their most ambitious targets through Intended Nationally Determined Contributions (INDC's), which are declarations of their best efforts on emissions reduction [11]. There is concern among scientists that the INDC's as set out will not provide the emissions reductions necessary to keep warming below 2°C [9,11]. Moving forward, countries need to set ambitious emissions reduction targets and effective policy frameworks in order to meet global goals and avoid the consequences of a 2°C global temperature increase.

There are multiple ways in which emissions reductions can be achieved, which essentially fit into two categories; reducing emissions of carbon to the atmosphere and increasing the carbon stored on land [10]. Methods of increasing carbon sinks include afforestation and improving forest management. Reducing emissions involves decarbonising energy generation by moving to renewable sources, reducing energy demand, and reducing fossil fuel use in other industries, such as construction or transport [1]. Ambitions to tackle climate change are global, but each country declares their contribution to meeting these goals through their INDCs.

In addition to INDCs, there are global mechanisms to encourage emissions reductions. The Emissions Trading scheme under the Kyoto Protocol is a market-based approach to mitigating emissions; the largest of these schemes is the EU Emissions Trading Scheme, which sets a cap on total emissions and allows companies to trade allowances for emissions between businesses [12]. Another policy mechanism is REDD+, first created in 2007 and later modified; REDD+ intends to maximise carbon sinks on land through minimising deforestation and degradation of forest land, although implementation of REDD+ in developing countries is currently relatively slow [13]. The purpose of these mechanisms is to address the issues of climate change from a range of perspectives to provide a suite of measures that span multiple sectors and nations.

1.4 THE ROLE OF THE ENERGY SECTOR

Almost 70% of GHG emissions from human activity are related to energy production, and with energy demand projected to increase in the future, the energy sector is a key area of interest [14]. However, provisional numbers suggest that 35 billion tonnes of CO_2 were emitted in 2015, which is relatively consistent with emissions in the previous 2 years, suggesting that CO_2 emissions are beginning to plateau [7]. This stabilisation of CO_2 emissions has happened alongside economic growth, and therefore, has an important implication that our decarbonisation does not have to require the slowing down of economic development [7]. This potential stabilisation could be attributed to improved energy efficiency or an increase in energy generation from renewable sources; however, in 2013, the three major types of energy generation in total global primary energy supply were still oil (31%), coal (29%), and gas (21.4%) [15]. If society is to reach net zero CO_2 emissions between 2050 and 2100, it will need to rapidly increase the uptake of renewable energy, so that all generation from fossil fuels is replaced with a renewable alternative.

Renewable energy has great potential for reducing CO_2 emissions, but there are challenges associated with large-scale development of renewables. The potential for renewables is limited by land and technical constraints and energy is usually required to convert renewable energy potential into usable energy. These energy requirements can make renewable energy more costly and reduce the available useful energy [16], but it is clear from the science that emissions from fossil fuel use need to be eliminated, so these challenges should be an area of priority.

1.5 GLOBAL RENEWABLE ENERGY TARGETS

From mid-2015, 164 countries had at least one type of renewable energy target, compared to 43 countries in 2005, highlighting the significant development over these 10 years [17]. These 'targets' can range from declarations and plans, to legally binding targets for the uptake of renewable energy [17]. Fig. 1.1 shows the total EU target for renewable energy generation is 20%, but there are significant targets also being set in developing countries: Nicaragua, for example, is aiming to get more than 90% of their energy from renewable sources by 2020 [18].

India and China are among the major GHG emitters globally and, in 2013, they had a joint contribution of 35% to global GHG emissions [19]. India is now the fourth largest user of energy in the world, and although renewable energy currently only accounts for only 12.5% of total installed capacity, a number of policies have been implemented to promote renewable energy [20]. China has also committed to a target of 15% of primary energy consumption from renewables by 2020 and a medium target of 30% by 2030 [19].

1.6 RENEWABLE ENERGY TARGETS FOR EUROPE

The current target for renewable energy generation in the EU is 20% of the total energy generation by 2020, apportioned across EU countries. The share of the target for each country is based on a standard increase from the amount of renewable energy produced in 2005 and also takes into account the wealth of each country, so wealthier countries take on a more challenging target [21].

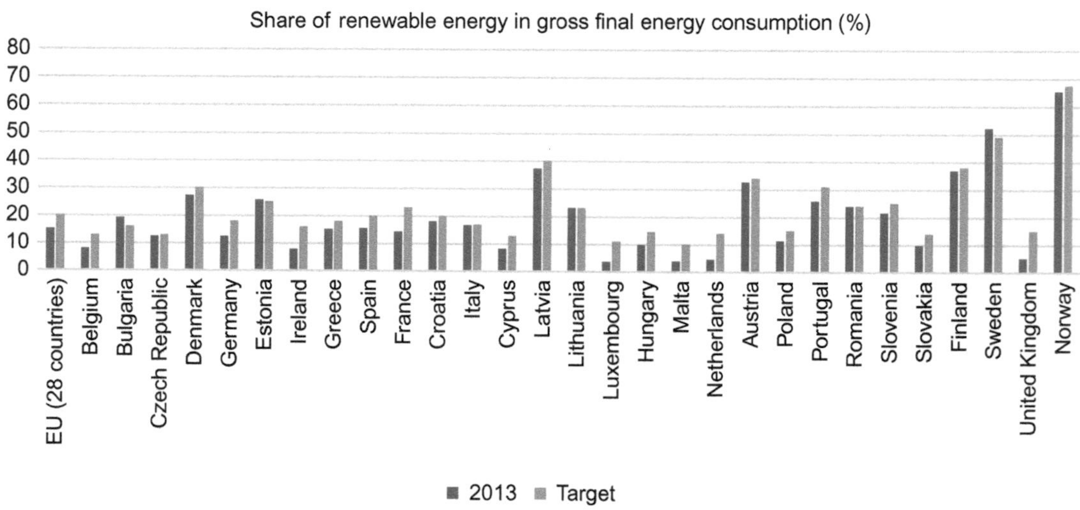

FIG. 1.1 Graph comparing the 2013 share of renewable energy for each EU member state to the 2020 target. *Data from http://ec.europa.eu/eurostat/data/database; http://ec.europa.eu/eurostat/statistics-explained/index.php/Renewable_energy_statistics.*

Fig. 1.1 shows that the EU is making good progress towards the 2020 targets, with 15% of all energy generation now from renewable sources, but it needs to look into the future in order to avoid a 2°C temperature increase. Beyond 2020, the EU must look to dramatically increase the proportion of renewables in energy generation, from 20% to 100%, in order to meet the goals of the Paris agreement. In Europe, bioenergy provides almost two-thirds of total renewable energy generation, with the remainder from hydropower, wind, solar, geothermal, and tidal energy [22].

1.7 BIOENERGY

The definition of 'biomass' is any plant-based organic matter and bioenergy is the energy derived from this organic matter. The underlying principle behind using biomass fuels is that the CO_2 released in combustion is the same amount which is absorbed from the atmosphere as the plant grows, thus theoretically it is carbon-neutral, provided the biomass is regrown and does not drive wider change in land use [23]. The balance of CO_2 from uptake and combustion is discussed in more detail in subsequent chapters.

Bioenergy is energy created from the combustion of biomass and has always been an important source of fuel, widely used for heating before the industrial revolution and still used in developing countries today [24]. Bioenergy is currently thought to provide over 10% of energy globally, with nearly two thirds of this used in developing countries compared to developed countries, where there is a stronger reliance on fossil fuels [4,25]. Outside of the traditional use of bioenergy in developing countries, more modern uses of bioenergy in developed countries have also increased, as a requirement for countries to move to renewable energy has led to a significant increase in bioenergy use [26]. In order to move towards a fully decarbonised energy sector, renewable energy technologies must be deployed on a larger scale and must be competitive with fossil fuel generation. To be a realistic competitor to generation from fossil fuels, more modern applications of bioenergy are required to harness the energy from biomass in a cost-effective and efficient way.

Bioenergy is the most widely applied renewable energy source available and is currently the only technology available for transport fuels [27]. As we strive towards zero GHG emissions in this century, it is likely that we will see a dramatic increase in the uptake of bioenergy to displace carbon-intensive energy generation. Future levels of bioenergy use are expected to increase significantly, with predicted deployment levels in 2050 of between 100 and 300 EJ, made up of both traditional wood fuel use and modern uses of biomass [28].

As of September 2016, there are 104 policy support mechanisms for bioenergy globally, across 56 different countries [29]. These policy mechanisms include laws and plans for the uptake of renewable energy generally, increasing renewable fuels used in transport and bioenergy for heating and electricity. Across Europe, there are clear plans for the deployment of bioenergy as a renewable energy source, but across the globe, countries are planning increased bioenergy use, including an intention in the United States for an annual consumption of 36 billion gallons of biofuels each year by 2022 [30]. Other countries also have strong potential for bioenergy use, such as Brazil, where biomass currently produces 9.7% of all electricity generation [31]. The intended use of bioenergy for different types of energy within Europe can be seen in Fig. 1.2.

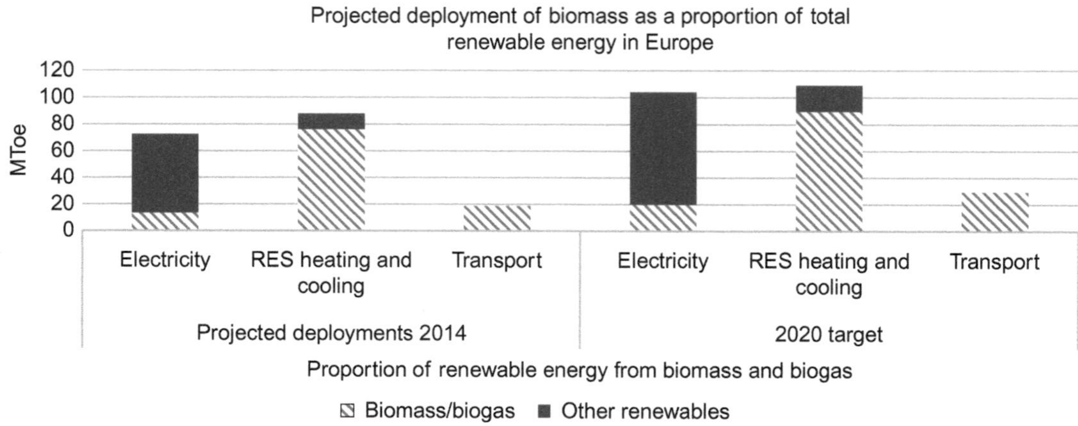

FIG. 1.2 Proportion of renewable energy targets to be met using biomass or biogas. *Data from European Commission. Renewable energy progress report. Brussels 15.6.2015. http://eur-lex.europa.eu/resource.html?uri=cellar:4f8722ce-1347-11e5-8817-01aa75ed71a1.0001.02/DOC_1&format=PDF.*

1.8 DELIVERING GREENHOUSE GAS REDUCTIONS FROM BIOENERGY

Expansion of bioenergy use will naturally be limited by the resources and land available for its production [32]. Any biomass can be converted into useful bioenergy, but technology must be applied to maximise the energy output. The emissions involved with these technologies and the emissions from converting and transporting the bioenergy to its end use must be calculated; these are known as the life cycle emissions. Total life cycle emissions must be significantly lower than the emissions from fossil fuel combustion for bioenergy to be a valuable form of renewable energy. Measuring these emissions is a mechanism for establishing the effectiveness of the bioenergy pathway in terms of GHG reductions when compared to its fossil fuel alternative. This is achieved via a Life Cycle Assessment (LCA) approach, which is explored further in Chapters 3 and 4. An LCA is dependent on a variety of factors, including the chosen conversion technology and transport methods [33]. The majority of LCA methodologies for bioenergy are based on the assumption that the biomass for energy use is carbon-neutral, as the CO_2 released in energy generation is equal to CO_2 taken up by the plant as it grows [33]. Fig. 1.3 depicts a highly simplified version of the terrestrial carbon cycle, highlighting the one-way carbon flow from fossil fuels, compared to the cyclical nature of biomass carbon cycling, as carbon is removed from the atmosphere through respiration.

Bioenergy may be applied to a number of different energy systems and products, which is one of the qualities which makes the development of bioenergy appealing for a renewable future. A paper by Thornley and Gilbert showed that bioenergy systems deliver 'substantial and cost-effective greenhouse gas reductions' [32]. It is difficult to compare greenhouse gas emissions savings from different systems, as the actual savings achieved is highly dependent on site and situation, but Table 1.1 highlights some of the potential greenhouse gas savings which could be achieved through the use of bioenergy in the place of fossil fuels. Table 1.1 also shows that bioenergy systems which have the highest GHG saving compared to the

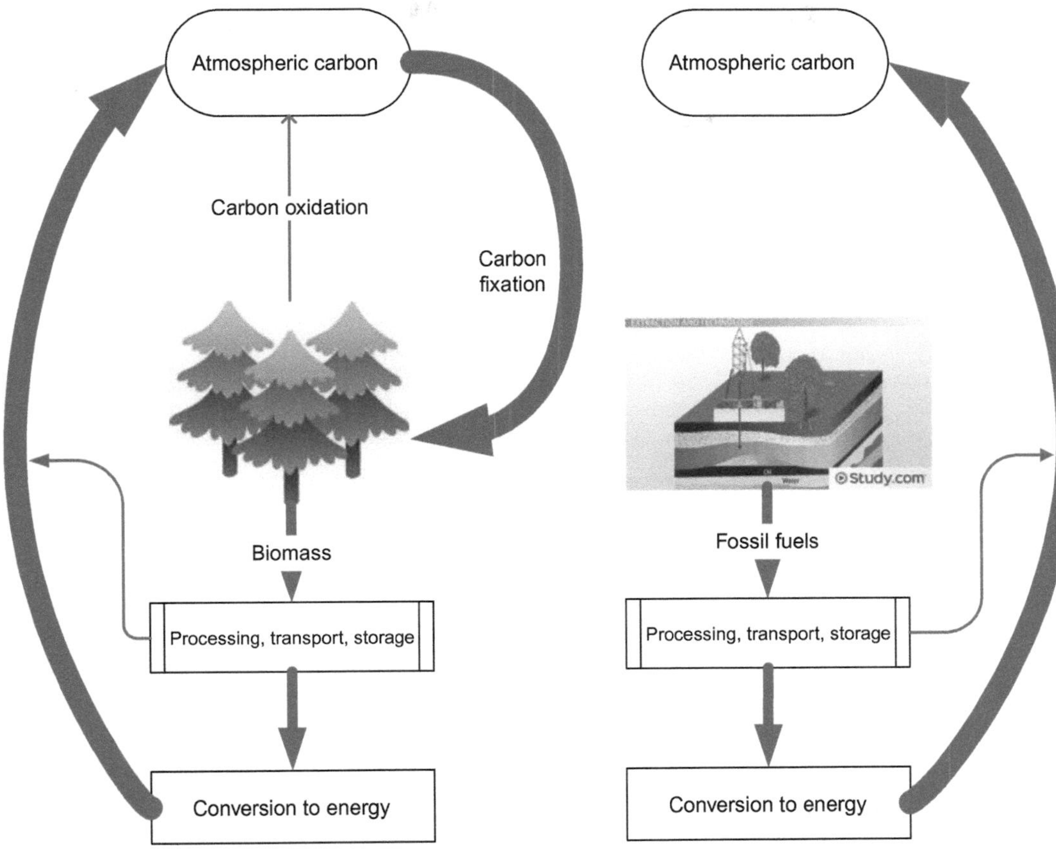

FIG. 1.3 Typical carbon flows for bioenergy and fossil fuel systems. *Modified from IEA Bioenergy task 38, Figure 1. http://www.task38.org/publications/task38_description_2013.pdf.*

energy generated do not necessarily show the highest GHG saving for each unit of biomass, suggesting that it is important to consider not only the emissions reductions of energy generation, but also the most resource-efficient uses of biomass.

1.9 IMPORTANCE OF MAINTAINING CARBON STOCKS

Biomass fuels are considered carbon-neutral because the CO_2 released when the biomass is combusted is equal to the CO_2 sequestered during the plant's growth [23]. This uptake of CO_2 means that plants are a net carbon sink, with an estimated 2000 and 3000 billion metric tonnes of carbon stored on Earth, for example, in forests [34]. However, changes to these carbon stocks can create emissions of their own, with land use change accounting for 17% of anthropogenic CO_2 emissions every year [35]. Since 1970, the cumulative emissions from land use change have increased by around 40%, which could be attributed to the growing demand for land, through urbanisation and agriculture to feed growing populations [1]. This shows

TABLE 1.1 Potential Greenhouse Gas Emissions Savings From Different Bioenergy Systems

Bioenergy System	Example Emissions Compared to Fossil Fuel Alternatives
Small-scale electricity from UK energy crops—gasification	Savings of 557 kgCO_2/MWh delivered. 90% GHG reductions relative to the reference case of UK grid average emissions Emissions reduction of 624 kgCO_2/tonne biomass.
Large-scale electricity systems from imported forest residues	Savings of 562 kgCO_2/MWh delivered 91% greenhouse gas reductions compared to the reference case of UK grid average emissions Emissions reduction of 897 kgCO_2/tonne biomass.
Imported pellets from forest products for small-scale domestic heat	Savings of 149 kgCO_2/MWh delivered 58% savings compared to the reference case of natural gas fired condensing boiler Emissions reduction of 538 kgCO_2/tonne biomass
Medium-scale wood chip district heat from energy crops	Savings of 225 kgCO_2/MWh delivered 94% savings against the reference case of a natural gas fired district heating system Emissions reduction of 1203 kgCO_2/tonne biomass
Ammonia from wood chip from imported forest products—large scale (through gasification)	Savings of 1317 kgCO_2/tonne 68% savings compared to the reference case Emissions reduction of 869 kgCO_2/tonne biomass
Biochar from wood chip from UK energy crop (medium scale)—slow pyrolysis and application of char to soil	Savings of 2264 kgCO_2/tonne Emissions reduction of 683 kgCO_2/tonne biomass

Based on Thornley P, Gilbert P, Shackley S, Hammond J. Maximizing the greenhouse gas reductions from biomass: the role of life cycle assessment. Biomass Bioenergy 2015;81:35–43.

the importance of maintaining carbon stocks, and importantly, ensuring that bioenergy does not encourage changes in land use.

Bioenergy is expected to play a significant role in meeting emission reduction targets globally and so it is important to quantify the potential resource availability [36]. There is significant debate on the potential supply of bioenergy globally, with the maximum global potential considered to be 1550 EJ each year, three times greater than current global energy supply [37]. However, there are also multiple studies which suggest the potential is significantly lower than this value, with one recent study suggesting the potential was between 61 and 161 EJ per year [38]. The greatest bioenergy potentials are thought to be in Latin America, China, and the United States, suggesting that as demand increases, trade will be necessary to ensure demands are met in an efficient and cost-effective way [39]. Global trade of bioenergy will allow more sustainable use of resources, but the transportation required will mean the life cycle emissions for the bioenergy will be higher and this should be monitored to ensure significant savings compared to fossil fuels are made.

Land use is strongly connected to climate and as populations continue to rise, urbanisation will likely increase, along with demand for agricultural land and bioenergy, all of which increase the demands on current land use, increasing the potential for tensions between demand for land for food or energy [40]. With greater demands on land, the CO_2 imbalance increases, so the potential impact of this additional demand must be mitigated through improving productivity on existing land or through the utilisation of degraded land to increase land carbon sequestration [40]. If the carbon storage on land is increased through improved practices or through using previously unused land and the resulting biomass is then used to displace fossil fuel combustion, this can have dual climate benefits, improving the carbon on land and avoiding CO_2 emissions to the atmosphere.

1.10 SUMMARY

The requirement to reduce emissions provides an incentive for the use of bioenergy, but an appropriate policy framework needs to be in place to ensure rapid and scalable uptake. Bioenergy, under the right circumstances, can be a strong driver for the mitigation of climate change, but controls are required as the extent to which bioenergy can be deployed sustainably is constrained by available land and sustainable practices. Effective policies for bioenergy should ensure that life cycle emissions from bioenergy provide significant savings against fossil fuels and that there are sufficient sustainable stocks to ensure GHG emissions are not created through land use change or clearing which reduces the carbon sinks on land.

References

[1] IPCC. In: Core Writing Team, Pachauri RK, Meyer LA, editors. Climate change 2014: synthesis report. Contribution of working groups I, II and III to the fifth assessment report of the Intergovernmental Panel on Climate Change. Geneva: IPCC; 2014. 151 pp.

[2] World Bank. Turn down the heat: confronting the new climate normal. Washington, DC: World Bank; 2014. License: Creative Commons Attribution—NonCommercial—NoDerivatives 3.0 IGO (CC BY-NC-ND 3.0 IGO).

[3] Klass DL. Biomass as an energy resource: concept and markets. In: Biomass for renewable energy, fuels, and chemicals. San Diego: Academic Press; 1998. p. 29–50 [chapter 2].

[4] Dhillon RS, von Wuehlisch G. Mitigation of global warming through renewable biomass. Biomass Bioenergy 2013;48:75–89.

[5] Schlesinger WH. The global carbon cycle and climate change. In: Howarth RB, editor. Perspectives on climate change: science, economics, politics, ethics (Advances in the Economics of Environmental Resources, volume 5). Emerald Group Publishing Limited; 2005. p. 31–53.

[6] IPCC. Climate change 2001: the scientific basis; 2001. Available from: http://www.grida.no/climate/ipcc_tar/wg1/pdf/TAR-02.pdf.

[7] IEA Report Highlights 2015. https://www.iea.org/publications/freepublications/publication/CO2EmissionsFromFuelCombustionHighlights2015.pdf.

[8] World Bank. Turn down the heat: why a 4 warmer world must be avoided. Washington, DC: World Bank; 2012.

[9] Watts G. Scientists welcome new global climate change pact. Lancet 2015;386(10012):2461–2.

[10] http://bigpicture.unfccc.int [Accessed 14 March 2016].

[11] Victor DG, Leape JP. Global climate agreement: after the talks. Nature 2015;527:439–41. https://doi.org/10.1038/527439a2015.

[12] http://ec.europa.eu/clima/policies/ets/index_en.htm [Accessed 21 March 2016].

[13] Fischer R, Hargita Y, Günter S. Insights from the ground level? A content analysis review of multi-national REDD+ studies since 2010. Forest Policy Econ 2016;1389-9341. 66:47–58.

[14] Höök M, Tang X. Depletion of fossil fuels and anthropogenic climate change—a review. Energy Policy 2013;52:797–809.

[15] IEA. Key world energy statistics; 2015. Available from: https://www.iea.org/publications/freepublications/publication/KeyWorld_Statistics_2015.pdf.

[16] Moriarty P, Honnery D. Can renewable energy power the future? Energy Policy 2016;0301-421593:3–7. https://doi.org/10.1016/j.enpol.2016.02.051. http://www.sciencedirect.com/science/article/pii/S030142151630088X.

[17] IRENA. Renewable energy target setting; June 2015. http://www.irena.org/DocumentDownloads/Publications/IRENA_RE_Target_Setting_2015.pdf.

[18] World Bank News. Nicaragua: a renewable energy paradise in Central America; October 26, 2013. http://www.worldbank.org/en/news/feature/2013/10/25/energias-renovables-nicaragua [Accessed 21 March 2016].

[19] Mittal S, Dai H, Fujimori S, Masui T. Bridging greenhouse gas emissions and renewable energy deployment target: comparative assessment of China and India. Appl Energy 2016;166:301–13.

[20] Tripathi L, Mishra AK, Dubey AK, Tripathi CB, Baredar P. Renewable energy: an overview on its contribution in current energy scenario of India. Renew Sustain Energy Rev 2016;60:226–33.

[21] Klinge Jacobsen H, Pade LL, Schröder ST, Kitzing L. Cooperation mechanisms to achieve EU renewable targets. Renew Energy 2014;63:345–52.

[22] http://ec.europa.eu/eurostat/data/database http://ec.europa.eu/eurostat/statistics-explained/index.php/Renewable_energy_statistics.

[23] Demirbas A. Combustion characteristics of different biomass fuels. Prog Energy Combust Sci 2004;30(2):219–30.

[24] McKendry P. Energy production from biomass (part 1): overview of biomass. Bioresour Technol 2002;83(1):37–46.

[25] Lamers P, Junginger M, Hamelinck C, Faaij A. Development in international solid biofuel trade—an analysis of volumes, policies and market factors. Renew Sustain Energy Rev 2012;16(5):3176–99.

[26] IEA. Renewables—bioenergy 2015. Available from: http://www.iea.org/topics/renewables/subtopics/bioenergy/.

[27] Anderson K. Talks in the city of light generate more heat. Nature 2015;528:437.

[28] Lamers P. Sustainable international bioenergy trade; 2013.

[29] Parameters: policy support, bioenergy, in force. http://www.iea.org/policiesandmeasures/renewableenergy/ [Accessed 09 September 2016].

[30] Hultman NE, Malone EL, Runci P, Carlock G, Anderson KL. Factors in low-carbon energy transformations: comparing nuclear and bioenergy in Brazil, Sweden and the United States. Energy Policy 2012;40:131–46.

[31] da Silva RC, de Marchi Neto I, Seifert SS. Electricity supply security and the future role of renewable energy sources in Brazil. Renew Sustain Energy Rev 2016;59:328–41.

[32] Thornley P, Gilbert P, Shackley S, Hammond J. Maximizing the greenhouse gas reductions from biomass: the role of life cycle assessment. Biomass Bioenergy 2015;81:35–43.

[33] Cherubini F, Bird ND, Cowie A, Jungmeier G, Schlamadinger B, Woess-Gallasch S. Energy- and greenhouse gas-based LCA of biofuel and bioenergy systems: key issues, ranges and recommendations. Resour Conserv Recycl 2009;53(8):434–47.

[34] IPCC. The physical science basis. Summary for policy makers; 2013.

[35] Ross CW, Grunwald S, Myers DB, Xiong X. Land use, land use change and soil carbon sequestration in the St. Johns River Basin, Florida, USA. Geoderma Reg 2016;7(1):19–28.

[36] Welfle A, Gilbert P, Thornley P. Securing a bioenergy future without imports. Energy Policy 2014;0301-421568:1–14. https://doi.org/10.1016/j.enpol.2013.11.079. http://www.sciencedirect.com/science/article/pii/S0301421513012093.

[37] Offernan R, Seidenberger T, Thran D, et al. Assessment of global bioenergy potentials. Mitig Adapt Strateg Glob Chang 2011;16:103–15.

[38] Searle SY, Malins CJ. Will energy crop yields meet expectations? Biomass Bioenergy 2014;65:3–12.

[39] International Energy Agency (IEA). Renewable energy outlook. In: Birol F, editor. World energy outlook 2012. Paris: International Energy Agency; 2012. p. 211–40.

[40] Lomax G, Helgeson ICF. The value of land restoration as a response to climate change. In: Land restoration. Boston: Academic Press; 2016. p. 235–45 [chapter 3.2].

Further Reading

[1] European Commission. Renewable energy progress report. Brussels 15.6.2015. http://eur-lex.europa.eu/resource.html?uri=cellar:4f8722ce-1347-11e5-8817-01aa75ed71a1.0001.02/DOC_1&format=PDF.

[2] IEA Bioenergy task 38, Figure 1. http://www.task38.org/publications/task38_description_2013.pdf.

CHAPTER

2

How Policy Makers Learned to Start Worrying and Fell Out of Love With Bioenergy

Raphael Slade, Lorenzo Di Lucia*, Paul Adams*[†]
[*]Imperial College London, London, United Kingdom [†]University of Bath, Bath, United Kingdom

Some problems are so complex that you have to be highly intelligent and well-informed just to be undecided about them. Laurence J. Peter

2.1 BIOENERGY AS A STRATEGIC TECHNOLOGY OPTION

Many elements of modern energy policy can be traced back to the political and institutional response to the 1970's oil crises and the growing awareness of the environmental impacts of the energy sector. One of the key institutional landmarks was the setting up in 1974 of the international energy agency (IEA) in response to the 1973 oil shock. Although the IEA was initially dedicated to responding to physical disruptions in oil supply, its mandate to enhance energy security led naturally to an assessment of alternative energy sources. Fuels from biomass were identified as one of the options most likely to achieve early commercial success, and a technology collaboration programme (IEA bioenergy) was set up in 1978 to support countries active in bioenergy research, development, and deployment [1].

In the 1970s and early 1980s, the role of environmental protection in energy policy had a relatively low profile, often being seen as a constraint rather than a necessity. This changed in 1985 when the IEA governing board—which comprises the energy ministers of member countries—adopted general principles that energy production, conversion, transport, and consumption should be carried out in an *'environmentally acceptable manner'*. Ministers also agreed to actively promote actions in their national energy policies which would, inter alia, enhance the development of new environmentally favourable energy technologies. In the late 1980s, concern about climate change was rising rapidly up the political agenda and the next IEA Ministerial policy statement, in 1989, went further still, stressing the need for *'integrated policies which further energy security, environmental protection and economic growth'* [1].

Greenhouse Gas Balances of Bioenergy Systems
https://doi.org/10.1016/B978-0-08-101036-5.00002-1

Another political and institutional landmark was the adoption of the United Nations Framework Convention on Climate Change (UNFCCC) at the Rio Earth Summit in 1992. This committed countries to stabilise greenhouse gas concentrations *'at a level that would prevent dangerous anthropogenic interference with the climate system'* [2]. It committed signatories to keep inventories of their greenhouse gas emissions and defined the GHG accounting frameworks they should use to report them. It also committed countries to report regularly on their climate change policies, many of which directly affected the energy sector.

These developments set the scene for what has become the key challenge for energy policy in many countries: how to balance the need for energy security, energy equity, and environmental sustainability; a problem that has come to be known as the energy trilemma [3]. Increasing the proportion of bioenergy in the global energy mix has emerged as one of the key technological responses to this challenge. Energy scenarios, such as those developed by the International Energy Agency (IEA) and on behalf of the Intergovernmental Panel on Climate Change (IPCC), indicate that bioenergy could make a major contribution to a future low-carbon energy system, potentially supplying 10%–50% of global primary energy by 2100 [4–7]. Most Integrated Assessment Model (IAM) results also show that imposing constraints on biomass supply would increase the cost of reducing global GHG emissions or necessitate reductions in energy consumption [7]. The same models show that limiting warming to 2°C or less would be virtually impossible if bioenergy was excluded from the mitigation toolkit [8].

Bioenergy has also come to be given a prominent role in many national energy strategies. By 2015, more than 60 countries had adopted policies designed to support bioenergy deployment including all European member states, the United States, Brazil, China, Japan, and Russia [9–11]. The impetus for these policies draws on a range of motivations: improving energy security, reducing GHG emissions, diversifying agricultural production, and stimulating rural development and job creation. Changing political priorities and improving scientific evidence, however, has demonstrably affected the weightings given to each of these motivations in policy reports and statements. Since 1997, when countries adopted the Kyoto protocol, the political importance given to sustainable carbon reductions has dramatically increased, but at the same time the consensus that increased bioenergy deployment will automatically or simplistically provide carbon reductions has been challenged, and a much more complex and contested picture has emerged.

The remainder of this chapter is presented as follow. First, we examine the extent to which GHG mitigation provided a rationale for policy makers to introduce bioenergy support mechanism and the backlash that followed. Second, we look at three key challenges to which policy makers have had to respond as market adoption increased: the shortcomings of carbon accounting frameworks, land-use change, and time critical changes in carbon stocks. Finally, we examine the governance challenges that are inherent in achieving large-scale carbon reductions through bioenergy deployment.

2.2 THE POLICY DRIVE FOR BIOENERGY

In the European Union, the coordinated promotion of renewable energy goes back to 1997 when a target (white paper) was adopted to increase the proportion of renewable energy from 5.2% of primary energy supply in 1995 to 12% by 2010 [12]. The benefits cited for the introduction of support policies included: security and diversification of energy supply, job creation, rural development, social and economic cohesion, and reducing carbon emissions. The white paper paved the way for Directives on electricity production from renewable sources in 2001 [13] and for the promotion of biofuels in 2003 [14].[1] Carbon emission reductions were simply assumed among other co-benefits of introducing biofuels, as the following excerpt from the 2003 Directive illustrates:

> In terms of environmental impact, biofuels are very attractive, emitting between 40 and 80% less in the way of greenhouse gases than other fossil fuels [... they] will also help to create jobs in rural areas and thus preserve the rural fabric by providing agriculture with new outlets.

Neither Directive included criteria for sourcing sustainable biomass, but the biofuel Directive did commit the EU commission to monitoring and reporting on the efficacy of biofuels in reducing carbon emissions and on the sustainability[2] of crops used for the production of biofuels [14].

Although an important milestone in the development of EU bioenergy policy, these Directives were relatively weak since targets were only indicative. The EC, however, considered progress to be 'sluggish' and, in 2004, committed to a plan [15] setting out a coordinated approach including precise legally binding targets and minimum sustainability standards [16]. This plan gave rise to the EU Climate and Energy Package, which was enacted by member states in 2009 to ensure that the EU met its climate and energy targets for the year 2020. These targets would be binding on member states and included a 20% cut in GHG emissions (from 1990 levels) and a 20% share of renewable energy sources (~244 Mtoe). Member states also had to adopt a National Renewable Energy Action Plan (NREAP) setting out their contribution to the overall EU target. Analysis of these NREAPs showed that member states expected the role of bioenergy in 2020 to be very substantial including: 19.9 Mtoe of bioelectricity (19% of target), 86.5 Mtoe bioenergy for heating and cooling (78% of target), and 29.2 Mtoe biofuels for transport (98% of target) [17]. One of the key pieces of legislation that was intended to drive renewable energy, and bioenergy, deployment was the Renewable Energy Directive (RED) [18].

The RED, however, did more than set out targets for member states' future share of renewable energy consumption.[3] It also set out rules for calculating the GHG impact of biofuels, bio liquids, and their fossil fuel comparators and specified a minimum set of sustainability criteria that member states would have to adopt. The RED was not comprehensive since it excluded from the accounting and reporting framework GHG impacts of solid and gaseous biomass fuels used for heating, electricity, and cooling. Nevertheless, the RED enshrined the

[1]The electricity directive committed member states to promoting renewable electricity and to reporting progress against indicative consumption targets differentiated by country. The biofuels directive committed member states to promote biofuels and report consumption volumes against uniform reference targets of 2% by 2005 and 5.75% by 2010.

[2]Sustainability was defined in terms of intensity of cultivation, crop rotation and use of pesticides.

[3]The RED 2020 targets were differentiated by member state, but included a minimum 10% renewable energy for transport across the Union.

principle that renewable energy—and bioenergy—deployment should not only contribute to carbon reduction targets, but that this contribution should be quantified and reported.

These policy developments in the EU illustrate the changing political priorities given to GHG mitigation in energy policy in one region, but there are parallels in the way policy has evolved in other parts of the world. In the United States, policies supporting biofuel deployment were originally strongly driven by the perceived need to diversify energy supplies and support the agricultural sector. These policies were subsequently modified, in the light of evidence that GHG savings from some biofuels were lower than expected, to include specific quotas for fuels with better overall GHG performance [19]. In Brazil, biofuels have been strongly promoted since the Pró-Álcool programme was adopted in 1975. This programme aimed to help Brazil achieve energy independence by replacing petroleum imports with domestic transport fuel (initially ethanol from sugar cane and later biodiesel from soy and electricity from sugarcane harvest residues (bagasse)). It has evolved to become one of the largest and most successful bioenergy programmes in the world, but is no longer thought of simply as a source of energy security and economic growth. Compulsory instruments and voluntary schemes have been introduced to improve environmental sustainability, reduce GHG emissions, and limit the expansion of the agricultural frontier towards fragile or valuable ecosystems [20].

GHG reductions have arguably never been the main driver for bioenergy policy, but the expectation that bioenergy can help deliver on GHG mitigation goals has historically been seen as an important co-benefit, and numerous scenarios and modelling exercises suggest that this will continue to be the case in the future.

2.3 MARKET UPTAKE AND THE BIOENERGY BACKLASH

As governments around the world put in place policies to promote renewables and bioenergy, the market response was remarkably rapid. Between year 2000 and 2013, bioelectricity production more than doubled from ~0.6 to 1.7 EJ. Biofuels for transport increased by more than a factor of seven from ~0.45 to 3.19 EJ produced on approximately 71 Mha of land (an area roughly twice the size of Germany). Direct heat from biomass including traditional biomass combustion for cooking increased from ~40 to 50 EJ. Overall, in 2013, bioenergy contributed ~57.7 EJ (10%) to global primary energy supply [21]. At the same time, the International Renewable Energy Agency (IRENA) estimated that, in 2015, of the ~8.1 million people working in the renewable energy sector, around 2.8 million were working in bioenergy. Measured simply in terms of the quantity of bioenergy produced, and jobs created, global bioenergy policy has been a success (Figs 2.1 and 2.2).

Yet, as efforts to accelerate the deployment of bioenergy gathered pace, the prospect of mobilising the large quantities of biomass required became increasingly controversial. Biomass availability tends to be intertwined with activity in other major economic sectors—agriculture, forestry, food processing, paper and pulp, building materials, etc. As feedstocks are diverted from established markets, some impact on these sectors is almost inevitable [10]. The way in which land resources are used may also be changed, and many commentators predicted growing land and resource conflicts between bioenergy and food supply, water use, and biodiversity conservation. Their fear was that the benefits offered by increased bioenergy production

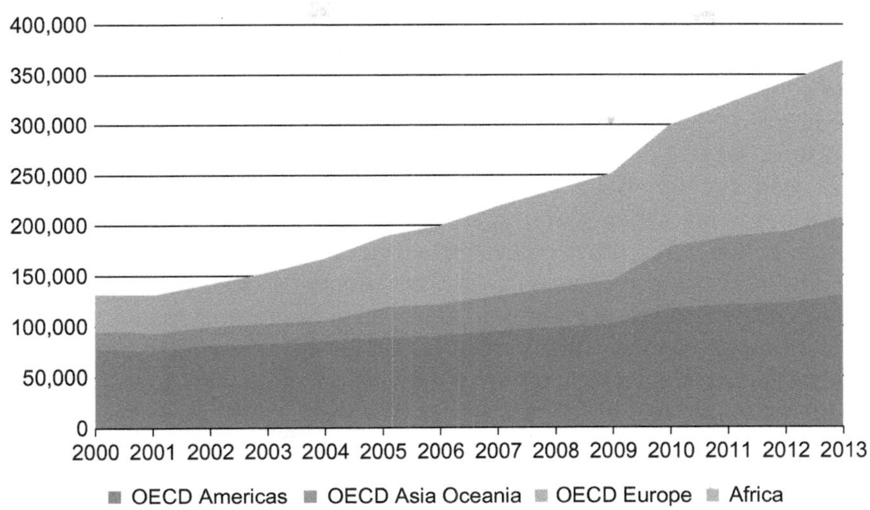

FIG. 2.1 Global generation of bioelectricity (GWh). *Data from IEA—World energy statistics.*

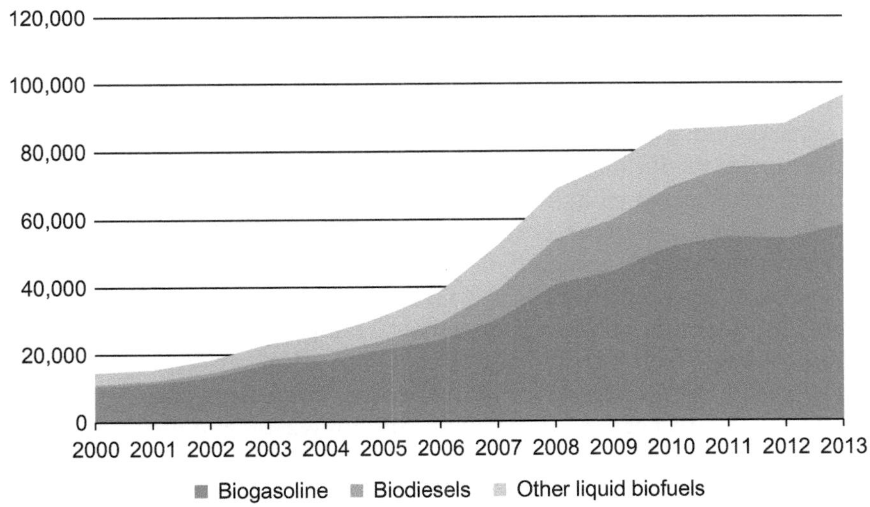

FIG. 2.2 Global biofuel production (kt). *Data from IEA—World energy statistics.*

could be rapidly outweighed by the penalties, and that increased deployment could exacerbate existing environmental problems. Sources of concern include both direct impacts, such as the effect of domestic stoves on urban air quality, and indirect impacts such as land-use change mediated through changing market prices [22–25].

Many different types of biomass may be used to produce bioenergy and there has been intense public and academic debate about resource availability and the merits and risks of dedicating particular resources to energy production both now and in the future [26,27].

Since 2000, however, bioenergy deployment has been underpinned by two key resources: forest biomass and agricultural land. Forest biomass has been used primarily to produce solid fuels (e.g. wood pellets) which can be combusted to produce electricity and heat. Agricultural land has been used to grow sugar, starch, and oil crops to produce liquid biofuels (and to lesser extent feedstocks for anaerobic digestion). How these resources are managed and used directly affects whether carbon emissions are increased or reduced.

Putting to one side the debates around energy security, welfare benefits, and competition between bioenergy, food, and other ecosystem services (these issues are critically important, but they are not the focus of this book), policy makers have faced three broad challenges to whether policies introduced to support bioenergy can genuinely contribute to climate change mitigation. The first challenge is that carbon accounting frameworks misrepresent the carbon saving benefits of bioenergy, potentially leading policy makers to support policies that have unintended and undesirable consequences. The second challenge is that increasing biomass production on agricultural land can directly, or indirectly, lead to increasing carbon emissions. The third challenge is that increased use of forest biomass does nothing to reduce emissions in the short term, but can only reduce carbon emissions in the distant future. The remainder of this chapter introduces each of these challenges and examines the issues this poses for policy and governance.

2.4 ACCOUNTING FOR CARBON

When plants photosynthesise, they capture carbon from the atmosphere. As they respire, when they are harvested and burnt, or when they die and decompose, this carbon is released back into the atmosphere. If biomass that would otherwise have decomposed is used to displace fossil fuels, then carbon emissions to the atmosphere can be avoided.

This is a simplified view of the carbon cycle that ignores a great many interactions between plant growth in the biosphere and global climate. For example, growing and decomposing biomass can produce long- and short-lived GHG (including carbon dioxide, methane, and nitrous oxide); burning biomass can release aerosols (sulphur dioxide, black carbon) that can have an overall cooling effect; and land-use changes can affect surface albedo and other physical properties [28,29]. Nevertheless, for annual crops and energy crops with short rotation cycles (assuming no indirect effects—see below), and for wastes and residues, the assumption that biomass combustion makes no net contribution to atmospheric carbon emissions is generally considered reasonable [24].

If, however, harvesting biomass for energy affects the stock of carbon stored in forests and soils or alters the fluxes between these carbon pools and the atmosphere, the simple assumption that emissions from biomass combustion are 'carbon neutral' could potentially be misleading [30–32]. A 2011 statement by the European Environment Agency (EEA) Scientific Committee makes this case, arguing that EU Directives[4] '*inaccurately assess the greenhouse gas consequences of different forms of bioenergy resulting in an "accounting error"*' with 'serious adverse consequences on a range of environmental concerns' [33].

[4]Renewable energy directive (2009/28/EC), fuel quality directive (2009/30/EC).

The accusation of an 'accounting error' undermines public support and threatens the credibility of policies designed to incentivise renewable energy and mitigate climate change. To understand this accusation, it is necessary to examine how emissions are reported.

National reporting guidelines for GHG inventories were decided under the UNFCCC and the 1997 Kyoto Protocol. Signatory countries agreed to report GHG emissions under different sectors including: Energy, Industrial Processes, Agriculture, and Land Use Land-Use Change and Forestry (LULUCF). These last two sectors were subsequently merged and reported as Agriculture Forestry and Other Land Use (AFOLU).[5] Biomass used to produce energy was classified as a Harvested Wood Product (HWP), resulting from Forest Management activities and reported as an emission in the AFOLU sector at the point of harvest. To avoid double counting, carbon emissions at the point of combustion were recorded as zero in the Energy sector.[6]

This approach is valid provided the land-use sector is fully reported and captures all carbon emissions. Considering the Energy sector in isolation, however, it appears that burning biomass makes no contribution to GHG emissions. The EU RED, EU emissions trading scheme (ETS) and the US Renewable Fuels Standard (US RFS2) adopt the convention that the direct emissions from biomass combustion are zero.

The AFOLU sector is one of the largest and most complex reporting sectors. It is also politically sensitive because it can be a source or sink of emissions and could affect how countries manage their natural resources.[7] When the Kyoto protocol was negotiated, political compromise resulted in mandatory reporting only being required for three categories of activity: deforestation, afforestation, and reforestation. Reporting emissions from Forest Management was optional for the first Kyoto commitment period (2008–12), but some countries, for example Australia, opted not to report [34]. The implications of this compromise become clearer with an example. The United Kingdom was one of the countries that elected to report Forest Management, and so biomass harvested from United Kingdom forests and used for energy would, at least in theory, have appeared in the United Kingdom's national inventory as a change in the carbon stock. But if the United Kingdom imported wood pellets from a non-reporting country like Australia[8] (or a country not signed up to the Kyoto protocol), the United Kingdom accounts would show that the emissions from burning the pellets were zero, but the change in carbon stock would not be reported.

Attempting to distinguish between anthropogenic and non-anthropogenic emissions from land-use is inherently complex, and the accounting rules for the first Kyoto commitment period had some significant limitations [35]. For instance, no distinction was made in cases where

[5]IPCC 2006 guidelines for national GHG inventories refer to Agriculture Forestry and Other Land Use (AFOLU). For brevity, we also refer to AFOLU in the text but readers should be aware that LULUCF is still a commonly used term in the literature.

[6]Direct methane and nitrous oxide emissions from biomass combustion for energy use would, however, be reported in the energy sector.

[7]The AFOLU sector is responsible for just under a quarter (~10–12 $GtCO_2$equiv./year) of anthropogenic GHG emissions, mainly from deforestation and agricultural emissions from livestock, soil and nutrient management [IPCC AR5].

[8]Australia chose not to report in the Forest Management sector.

carbon stocks might be reduced with a change of management, but where there was technically no change in land cover, e.g. between natural forest ecosystems and plantations or between primary forest and semi-natural forests logged for industrial wood production [34]. There are also significant limitations to the level of resolution that is achievable both spatially and temporally, as the following excerpt from the EU's 2013 decision on GHG accounting rules notes:

> Completing LULUCF accounts on an annual basis would make those accounts inaccurate and unreliable due to inter-annual fluctuations in emissions and removals, the frequent need to recalculate certain reported data, and the long time required for changed management practices in agriculture and forestry to have an effect on the quantity of carbon stored in vegetation and soils [36].

Nevertheless, it needs to be recognised that the accounting frameworks are not set in stone and continue to be revised with each round of climate negotiations. The UNFCCC Conference of the Parties held in Durban in December 2011, for instance, made accounting for emissions from forest management mandatory and also required accounting for conversion of natural forests to plantation forests [34]. Many alternatives have also been debated in the academic literature [31]. Incomplete implementation of the LULUCF requirements outlined in the Kyoto protocol might in some instances lead to the emission benefits from bioenergy being overstated [31]. But it is also important to recognise that GHG accounting frameworks represent a trade-off between efficacy and ease of implementation. Arguably, accepting incomplete accounting rules for LULUCF was a political compromise without which international agreement at Kyoto might not have been reached. The challenge policy makers now face is ensuring that bioenergy policies do not inadvertently incentivise reductions in carbon stocks. In the case of the EU RED and US RFS2, this is addressed, at least in part, by the introduction of supplementary sustainability criteria that seek to exclude the worst performing value chains.

The 'accounting error' rhetoric gives the impression that an ideal alternative framework exists, and that policy makers have in some way been negligent or myopic. The reality, however, is a more prosaic story of mixed motivations, compromise, and incremental improvement in the face of a daunting and complex reporting endeavour.

2.5 DIRECT AND INDIRECT LAND-USE CHANGE—THE PROBLEM WITH TRANSPORT BIOFUELS

Almost all the biomass used to produce liquid transport fuels (sugar, starch, and vegetable oil) is grown on agricultural land. An increase in biofuel consumption could therefore lead to cropland expansion in one of two ways:

- *directly*, for example, converting forested or natural grassland areas to energy crops—known as direct land use change (DLUC); or,
- *indirectly*, when the production of biomass on existing cropland displaces food crop production to other areas—known as indirect land use change (ILUC).

The GHG impact of ILUC and DLUC is potentially very significant because conversion of forest or grassland to cropland can lead to large releases of GHG to the atmosphere, for example, through increasing the rate of soil carbon oxidation or permanent reduction in aboveground carbon stocks.

The potential for increased bioenergy demand to drive land-use change was first discussed in the 1990s (see, e.g. [37,38]), but received little attention. At this time, bioenergy policy was in its infancy, food prices were low, and large areas of cropland had been taken out of production in Europe and the United States to reduce overcapacity. Ten years later, in 2008, when two studies [22,39] argued that biofuel policies could result in substantial land-use change, the reaction was dramatically different. Policy makers were rapidly faced with a vocal lobby that asserted that bioenergy policies were ill conceived, could result in welfare losses, and would not contribute to the GHG mitigation goals on which they had been predicated. A political response including policy reform was considered essential.

Unlike DLUC which can at least in theory be quantified using surveying and earth observation techniques, one of the major problems with including ILUC in policy is that it cannot be observed, or easily quantified [40,41]. For example, farmers in Europe deciding to grow wheat for ethanol instead of food will not see any effect on their direct GHG emissions. Impacts on other farmers in the United Kingdom or abroad will be mediated indirectly through changing market prices. It is also impossible to attribute cause and effect to a particular project or policy. For instance, it can never be proven that a certain change in land-use, e.g. in Brazil, is the result of the decision to proceed with a wheat-to-ethanol project in Europe. The challenge which policy makers have had to confront is how to put in place appropriate safeguards to protect against something for which there is only limited scientific evidence concerning the scale and severity, and for which there is no way to monitor the effectiveness of the policy once introduced [40].

Since 2008, there has been a surge in the number of studies attempting to estimate the ILUC impacts of bioenergy and especially biofuel expansion. The most common approach to estimating ILUC is using economic equilibrium models. These complex models may include the entire global economy or a specific sector, e.g. agriculture,[9] and calculate ILUC as the difference between scenarios with and without bioenergy policies. They also assume that perfect markets exist and that an equilibrium is reached when supply equals demand, assumptions that are often criticised as oversimplistic [42]. Several alternatives to economic optimisation models have been developed relying, primarily, on descriptive methods informed by expert opinion [43] or extrapolating statistics on past land-use change to predict future trends.

As additional studies have been undertaken, estimates of ILUC associated with biofuel policies in the EU and United States have to some extent converged [44]; however, there remains a conflict between policymakers' demands for exact values and the capacity of current models to supply results with the desired level of precision.

After around 7 years of intense debate, EU policy makers amended the RED by adopting the ILUC Directive in 2015 [45]. This Directive limited the way member states could meet the RED 2020 target of 10% renewables in transport fuels by introducing a cap (7% of target) on the contribution of 'food crop'-based biofuels and requiring minimum contribution (3% of target) from alternatives including used cooking oil, electricity, and advanced biofuels. In this way, policy makers sought to promote biofuels presenting a 'low ILUC risk' and limit those

[9]Well-known examples are the GTAP-model developed by Purdue University, the FAPRI-CARD model developed by Food and Agricultural Policy Research Institute together with Iowa State University, and the MIRAGE-model developed by the European Commission, INRA, the UN and the World Trade Organization.

considered to present a 'high ILUC risk.'[10] The agreement also included the requirement to report data on ILUC-related emissions of GHG on both national and European levels. For reporting purposes, member states were required to add an ILUC factor to the inventory of GHG emissions for each value chain. The agreed factors placed ILUC emissions on the same scale as the potential GHG savings and penalised oil crops most severely as they could be associated with tropical deforestation for palm oil.[11] In the United States, the US Environmental Protection Agency (EPA) issued rules for incorporating ILUC emissions into the Renewable Fuel Standard (RFS) accounting rules in 2010 [47]. Using economic models developed for the task, the EPA assigned each type of biofuel to one of the three categories based on the overall GHG savings compared to fossil fuels: (i) renewable fuels (20% savings), (ii) advanced biofuels (50% savings), (iii) cellulosic biofuels (60% savings). The decision process led by the EPA lasted only one year and the final decision, although contested by several stakeholder groups, did not attract the level of criticisms experienced by EU policy makers [48]. Thus far, however, biofuel policies in China, India, Canada, Mexico, Japan, Australia, and Brazil have not incorporated accounting or reporting of indirect land-use change emissions [41].

The debate around indirect land-use change has arguably increased political attention on the importance of land as both a source and sink of GHG emissions. The role of land in climate mitigation is also the focus of a special report by the Intergovernmental Panel on Climate Change (IPCC) expected to be published in 2019. Modelling ILUC may not provide straightforward answers to policy makers' questions, but despite the limitations it provides an opportunity to identify biofuel (and other agricultural production) pathways that lead to the greatest overall GHG emission reductions.

2.6 DYNAMIC CHANGES TO FOREST CARBON STOCKS AND THE PROBLEM OF CARBON DEBT

The majority of biomass used to produce heat and power is derived from forests. Unlike agricultural crops where growth and harvest typically occur within a single year (or ~3 years for coppice systems), forest management cycles are usually measured in decades. The timing of forest carbon stock changes and how they relate to management practises has been discussed in the academic and silviculture literature for at least 20 years, but the issue received comparatively little attention from the energy policy community [49]. This changed in 2012 when a group of non-governmental organisations including Greenpeace, Friends of the Earth, and the RSPB published a report claiming that electricity produced in Europe from wood pellets imported from the United States could be 'dirtier than coal' [50]. Once again, a political and policy response was demanded and the legitimacy of bioenergy policies challenged.

[10] It should also be noted that ILUC is not exclusively related to biofuel production since all other land using sectors may also cause ILUC.

[11] The EU ILUC factors agreed were 12 g CO_2equiv./MJ for cereals and other starch-rich crops, 13 g CO_2equiv./MJ for sugars and 55 g CO_2equiv./MJ for oil crops based on the results of the general equilibrium economic model IFPRI-MIRAGE-BioF [46]. This compares to the full lifecycle emissions intensity of fossil gasoline and diesel of around 90 g CO_2equiv./MJ.

The core argument outlined in the 'dirtier that coal' report was that if you cut down and burn a 60-year-old tree, and it takes 60 years for another one to grow, you have increased the amount of carbon in the atmosphere in the short term. In other words, burning trees now creates a 'carbon debt' that will take many decades to repay. The assertion is that this limits the role that forest biomass can play in helping meet short-term political targets, especially given the shortcomings in GHG accounting frameworks.

The term 'carbon debt' is not precisely defined, but rather it has become convenient shorthand to refer to three different, though related, phenomena. The first type of debt is the single tree example given above. In this case, the debt only appears because of the scale at which the calculation is done. If, instead of a single tree, you consider a population of different age trees in a landscape, and each year you only harvest the annual growth increment, no debt arises. The second type of debt occurs where you increase the intensity of harvest at landscape level and this reduces the total mass of trees across the entire forest; for instance, if you stepped up harvesting from every 100 years to every 60 years. In this case, the reduction in the average mass of the trees would result in a one-off carbon emission, which would need to be set against the benefits of energy production. This type of debt is most likely to occur when unmanaged forest is brought into management, but it also assumes that nothing is done to increase the productivity of the forest, such as increased replanting. The third type of debt is where you have a choice between continuing to manage a forest or stopping and allowing it to carry on growing and sequestering carbon. If you decide to keep managing the forest, then the opportunity cost of your decision is the carbon sequestration foregone. The problem with this argument is that, as a forest matures, the rate of carbon sequestration declines and it becomes more vulnerable to natural disturbance that would result in some of the carbon being released anyhow. In the absence of a real opportunity to stop existing forest management, this scenario is essentially hypothetical [51].

The 'Dirtier than Coal' report [50] compares a scenario in which forest biomass is harvested with one in which it is assumed that no harvesting occurs and the forest matures to maximise the carbon content of the landscape. This comparison is arguably an oversimplification because it ignores the impact of episodic natural disturbances, which means that the theoretical maximum carbon content of the landscape is never attained in practise. Making such a comparison also exaggerates the apparent size of any carbon debt that might occur. Nevertheless, it is clear that the impact of increased harvesting of bioenergy on long-term carbon stocks in a forest is something that should be taken into account in the overall GHG balance. This has been assessed for some forestry systems [59–63], but more analysis and evidence is required to be confident that forest harvesting is not having a negative impact on forest carbon stocks.

When considering 'foregone sequestration'. It is also essential to consider the appropriate counterfactual and the alternative land-use that would prevail if forests were not supplying wood for bioenergy. In regions such as the SE United States, this could be an alternative crop or residential or leisure developments. The aphorism often quoted by the United States forest industry is 'the forest that pays is the forest that stays'. Following this logic, it could be argued that forest growth and continued sequestration is only likely if harvesting continues and there is a financial income stream. The choice of counterfactual ultimately reflects societal preferences for how land and forests should be used. Whether increasing demand for forest products—of which bioenergy is just one—ultimately results in an increase or decrease in the global forest area can be modelled economically in similar way to indirect land-use change, but cannot easily be assessed on a project by project basis.

It is also important to recognise that carbon sequestration is only one of the many ecosystem services provided by forestry and forest management. There are many reasons why forest management decisions, for example, to increase biodiversity or maximise amenity value, may impact forest carbon stocks. Carbon storage is an important function of forestry systems, but not the only one.

Carbon debt has only recently entered the public consciousness. It has gained salience because of the rapid growth of the wood pellet market and because the timeframes for policy and forest management decisions are so dramatically different. It has yet to be seen how this heightened concern will affect political support for co-firing projects in the United Kingdom and Europe and is currently the subject of ongoing policy analysis and review in Europe and the United States.

2.7 MEETING THE GOVERNANCE CHALLENGE

Although many motivations have been cited for supporting bioenergy deployment over the last 40 years, the perceived risks of undesirable side effects on ecosystems, food production, and human well-being have forced the issue of sustainability high up the political agenda. Many countries have responded by introducing changes in their national bioenergy policies and programmes. These sustainability initiatives can be grouped into three main categories:

(i) *Setting minimum sustainability requirements for bioenergy mandates.* Mandates have been particularly successful in promoting transport biofuels and work by setting a minimum share of liquid biofuels to be blended with traditional fossil-based fuels. Compliance can be restricted to producers who will be able to demonstrate that their products meet minimum criteria such as a minimum GHG reduction threshold, or excluding the use of feedstock from land with high carbon stock or high biodiversity value. The US RFS2 and the EU RED both adopt this approach.

(ii) *Providing financial incentives for low impact sources of bioenergy.* Incentives schemes include grants, direct payments, tax credits, and payments for environmental services. Payments can be made conditional on compliance with specific environmental and socio-economic good practises. To be effective, a substantial and long-term government commitment is often required.

(iii) *Supporting capacity-building initiatives.* These include policies and programs intended to create an enabling environment for the development of sustainable bioenergy systems. Initiatives include training and education, information sharing and dissemination, best practise guidelines, and research. Success often depends on effective engagement with stakeholders, and again a long-term financial commitment by governments is usually required [52].

All three of these approaches are underpinned by voluntary and government-led certification and standardisation initiatives. Certification schemes aim to translate broad sustainability principles into decision-making criteria that can be evaluated on the basis of detailed and specific indicators. The EU, for example, has championed the use of certification for biofuels with 19 voluntary certification schemes recognised for compliance with the sustainability requirements set by the RED [53].

Although the concept of certification is not new—familiar examples include fair-trade coffee and the Forestry Stewardship Council (FSC) standard for wood products—the introduction and design of certification standards for bioenergy as a means to ensure sustainability is not straightforward. Identifying appropriate criteria presents a trade-off between efficacy and ease of adoption. If criteria are overly detailed and too stringent, compliance may be difficult to demonstrate or they may act as a barrier to trade as reporting costs may become excessive. Conversely, if criteria are too general, they may become meaningless. There is also a risk of leakage if measures are applied to bioenergy production in isolation from the rest of the agricultural and forestry system.

One of the most comprehensive sets of sustainability criteria and indicators has been developed by the Global Bioenergy Partnership—summarised in Fig. 2.3. These indicators aspire to be value-neutral and do not provide thresholds or limits. Nonetheless, they show the breadth of impacts that policy makers might wish to incorporate in bioenergy policy.

	Environmental	Social	Economic
Themes	Greenhouse gas emissions. Productive capacity of the land and ecosystems. Air quality. Water availability, use efficiency and quality. Biological diversity. Land use change, including indirect effects	Price and supply of a national food basket. Access to land, water, and other natural resources. Labour conditions. Rural and social development. Access to energy. Human health and safety	Resource availability and use efficiencies in bioenergy production, conversion, distribution and end use. Economic development. Economic viability and competitiveness of bioenergy. Access to technology and technological capabilities. Energy security/Diversification of sources and supply. Energy security/ infrastructure and logistics for production and use
Indicators	• Lifecycle GHG emissions • Soil quality • Harvest levels of wood resources • Emission of non-GHG air pollutants, including air toxics • Water use and efficiency • Water quality • Biological diversity in the landscape • Land use and land-use-change related to bioenergy feedstock production	• Allocation and land tenure for new bioenergy production • Price and supply of a national food basket • Change in income • Jobs in the bioenergy sector • Change in unpaid time spent by women and children collecting biomass • Bioenergy used to expand access to modern energy services • Change in mortality attributable to indoor smoke • Incidence of occupational injury, illness and fatalities	• Productivity • Net energy balance • Gross added value • Change in consumption of fossil fuels and traditional use of biomass • Training and requalification of the workforce • Energy diversity • Infrastructure and logistics for distribution of bioenergy • Capacity and flexibility of use of bioenergy

FIG. 2.3 The Global Bioenergy Partnership's (GBEPs) sustainability indicators for bioenergy. *Modified from GBEP. The global bioenergy partnership sustainability indicators for bioenergy. 1st ed. Rome: Food and Agricultural Organization of the United Nations (FAO); 2001.*

Despite considerable progress in modifying bioenergy policies to promote sustainability, government and industry initiatives have not been able to address the criticisms towards expanding the bioenergy sector or avoid the backlash illustrated earlier in this chapter. Many stakeholders, and the public at large, continue to question the capacity of these policies to effectively steer bioenergy deployment in a positive direction and still perceive bioenergy as too risky. Governance has undoubtedly become more challenging in the past decades as the scale, complexity, and diversity [54][12] of the system have increased—driven by international trade, the interlinkage of energy, forestry and agricultural markets, the involvement of ever more stakeholders, and the emergence of new scientific knowledge [57]. Clear limitations in governance capacity have also emerged. Not least the fact that state and social actors still work primarily at national–regional scale due to strong political and jurisdictional boundaries.

Governing bioenergy deployment shares many characteristics of a wicked problem, in that the problem definition is (at least in part) socially constructed and there is no well-described set of potential solutions [58]. For such problems, there is no simple policy prescription and governance must instead rely on the collective judgment of stakeholders involved in a process that is experiential, interactive, and deliberative.

2.8 CONCLUSIONS

Over the last 40 years, bioenergy has gone from being viewed as a simple substitute for fossil fuels to an essential tool in the fight against climate change. In the 1990s, bioenergy was generally regarded as an uncontroversial and environmentally benign alternative to more polluting forms of energy. It was seen by policy makers and politicians as an attractive and low-risk option and one that could help meet a wide range of policy goals. This favourable view, however, was largely un-tempered by experience. Outside of a small number of industrial sectors such as pulp and paper, and countries such as Brazil, which had been an early adopter of ethanol for transport, there was very limited practical understanding of deploying bioenergy technologies at scale.

As we approach 2020, one of the key dates by which countries have committed to reduce carbon emissions, this situation has to a large extent reversed. Policy initiatives have successfully incentivised international trade in biomass and large-scale production of heat, power, and liquid biofuels. Companies and policy makers have accumulated a wealth of knowledge from real projects. Accounting frameworks and reporting obligations have been implemented and continue to be improved as performance data is collected. Yet as an increasingly complex and contested picture of the potential impacts and benefits has emerged, public controversy around the sustainability of biomass supply shows no sign of abating. Against this back-

[12]*Scale* pertains to the spatial dimension of the system, the size, range and boundaries of its natural and socio-technical components. Large-scale systems are believed to be less governable than systems with similar properties, but of smaller scale [55]. *Diversity* points to the nature and degree to which the entities that form the system differ. Systems with high diversity require large amounts of data of high resolution and are therefore expected to be less governable [56]. Complex systems are characterised by multiple relations (and feedback) between system components. More complex systems require more in-depth analysis and understanding, and can be expected to be less governable [56].

ground of mistrust and opposition, it remains to be seen whether the policy community will maintain the drive for increased bioenergy production.

The greater the role that bioenergy has in meeting future energy demand, the more important it is to ensure that sustainability benefits can be achieved. Perhaps the greatest opportunity lies in the continued co-evolution of bioenergy production and governance systems. To this end, the way in which issues such as indirect land-use change, carbon accounting, and carbon debt are tackled by policy makers will be critical to ensuring the political and social legitimacy of continued bioenergy deployment. Conversely, the greatest risk lies in political loss of confidence and institutional paralysis. The benefits of adopting bioenergy production systems cannot be taken for granted and need to be demonstrated and proven, but as the IPCC global climate change mitigation scenarios demonstrate, there are very few alternatives to achieving the deep carbon emissions reductions required.

KEY POINTS

- Policy instruments have been a key driver of the implementation of bioenergy systems
- The motivations for bioenergy policy development are multifaceted with GHG reduction, now arguably the primary driver; however, other factors are often equally important such as energy security, rural development, renewable energy, and economic advancement.
- Carbon accounting methodologies and frameworks are a crucial mechanism to ensure that genuine GHG savings are made. Existing frameworks are imperfect but they continue to be extended and improved.
- Direct and Indirect land-use change is one key area that has proven difficult to resolve, leading to a wide variability in estimates and uncertainty in GHG savings from bioenergy.
- Carbon debt is an issue for forest biomass which demonstrates the importance of forest management decisions in achieving genuine GHG savings.
- Regions such as the EU now have fairly robust carbon accounting methodologies which endeavor to ensure that only appropriate bioenergy systems are being implemented. Sustainability criteria introduced have ensured bioenergy operators have to comply with land criteria and achieve minimum GHG savings.
- The greatest opportunity lies in the continued co-evolution of bioenergy production and governance systems.

References

[1] Scott R. The history of the international energy agency 1974–1994: major policies and actions. Paris: OECD Publications; 1994.

[2] UNFCCC. First steps to a safer future: introducing The United Nations Framework Convention on Climate Change. United Nations Framework Convention on Climate Change (UNFCC): UNFCC; Available from: http://unfccc.int/essential_background/convention/items/6036.php.

[3] WEC. World energy trilemma: priority actions on climate change and how to balance the trilemma. London: World Energy Council (WEC); 2015. cartographer.

[4] IEA. Energy technology perspectives 2010: scenarios and strategies to 2050. Paris: International Energy Agency; 2010.

[5] IPCC. Contribution of working group III to the fourth assessment report of the intergovernmental panel on climate change. In: Metz B, Davidson OR, Bosch PR, Dave R, Meyer LA, editors. Cambridge, New York: Cambridge University Press; 2007.
[6] IEA. World energy outlook. Paris: International Energy Agency; 2012.
[7] Rose SK, Kriegler E, Bibas R, Calvin K, Popp A, Vuuren DPV, et al. Bioenergy in energy transformation and climate management. Clim Chang 2014;123:477–93.
[8] Edenhofer O, Pichs-Madruga R, Sokona Y, Kadner S, Minx JC, Brunner S, et al. Contribution of working group III to the fifth assessment report of the intergovernmental panel on climate change. In: Edenhofer O, Pichs-Madruga R, Sokona Y, Farahani E, Kadner S, Seyboth K, et al., editors. Technical summary—climate change 2014: mitigation of climate change. Cambridge, New York: Cambridge University Press; 2014.
[9] GBEP. A review of the current state of bioenergy development in G8 + 5 countries. Rome: Global Bioenergy Partnership (GBEP)/Food and Agriculture Organisation of the United Nations (FAO); 2007.
[10] Faaij APC. Bio-energy in Europe: changing technology choices. Energ Policy 2006;34(3):322–42.
[11] REN21. Renewables 2016: Global Status Report. France: Renewable Energy Policy Network for the 21st Century (REN21); 2016.
[12] COM(97)599 final, White Paper for a Community Strategy and Action Plan, Available from: http://europa.eu/documents/comm/white_papers/pdf/com97_599_en.pdf.
[13] Directive 2001/77/EC of the European Parliament and of the Council of 27 September 2001 on the promotion of electricity produced from renewable energy sources in the internal electricity market.
[14] Directive 2003/30/EC, Directive 2003/30/EC of the European Parliament and of the Council of 8 May 2003 on the promotion of the use of biofuels or other renewable fuels for transport.
[15] COM(2004) 366 final, The share of renewable energy in the EU.
[16] Slade R, Panoutsou C, Bauen A. Reconciling bio-energy policy and delivery in the UK: will UK policy initiatives lead to increased deployment? Biomass Bioenerg 2009;33(4):679–88.
[17] Beurskens LWM, Hekkenberg M. Renewable Energy Projections as Published in the National Renewable Energy Action Plans of the European Member States—Covering all 27 EU Member States. Energy Research Centre of the Netherlands (ECN); 2011. Contract No.: ECN-E--10-069.
[18] Directive 2009/28/EC. On the promotion of the use of energy from renewable sources and amending and subsequently repealing Directives 2001/77/EC and 2003/30/EC.2009.04.23. Off J EU 2009.L 140: 16-62.
[19] Müller S, Brown A, Ölz S. Renewable energy policy considerations for deploying renewables. France: OECD/IEA; 2011.
[20] Morgera E, Kulovesi K, Gobena A. Case studies on bioenergy policy and law: options for sustainability. Rome: Food and Agriculture Organisation of the United Nations (FAO); 2009.
[21] WBA. WBA global bioenergy statistics 2016. Stockholm, Sweden: World Bioenergy Association (WBA); 2016.
[22] Searchinger T, Heimlich R, Houghton RA, Dong F, Elobeid A, Fabiosa J, et al. Use of U.S. croplands for biofuels increases greenhouse gases through emissions from land-use change. Science 2008;319(5867):1238–40.
[23] Eide A. The right to food and the impact of liquid biofuels (agrofuels). Rome: Food and Agriculture Organization of the United Nations; 2008.
[24] Agostini A, Giuntoli J, Boulamanti A. Carbon accounting of forest bioenergy—conclusions and recommendations from a critical literature review. In: Marelli L, editor. Luxembourg: European Commission Joint Research Centre (JRC); 2013.
[25] Creutzig F, Popp A, Plevin R, Luderer G, Minx J, Edenhofer O. Reconciling top down and bottom-up modellling on future bioenergy deployment. Nat Clim Chang 2012;2.
[26] Slade R, Bauen A, Gross R. Global bioenergy resources. Nat Clim Chang 2014;4:99–105.
[27] Haberl H, Erb K-H, Krausmann F, Running S, Searchinger T, Smith K. Bioenergy: how much can we expect for 2050? Environ Res Lett 2013;8. Article No.: 031004.
[28] Bright R, Bogren W, Bernier P, Astrup R. Carbon-equivalent metrics for albedo changes in land management contexts: relevance of the time dimension. Ecol Appl 2016;26(6):1868–80.
[29] Giuntoli J, Agostini A, Caserini S, Lugato E, Baxter D, Marelli L. Climate change impacts of power generation from residual biomass. Biomass Bioenergy 2016;89:146–58.
[30] Bowyer C, Baldock D, Kretschmer B, Polakova J. The GHG emissions intensity of bioenergy: does bioenergy have a role to play in reducing Europe's GHG emissions? London: Institute for European Environmental Policy (IEEP); 2012.
[31] Bird DN, Pena N, Zanchi G. Zero, one, or in between: evaluation of alternative national and entity-level accounting for bioenergy. GCB Bioenergy 2011;4:576–87.

[32] Haberl H, Sprinz D, Bonazountas M, Cocco P, Desaubies Y, Henze M, et al. Correcting a fundamental error in greenhouse gas accounting related to bioenergy. Energ Policy 2012;45:18–23.
[33] EEA. Opinion of the EEA scientific committee on greenhouse gas accounting in relation to bioenergy. Copenhagen: European Environment Agency (EEA); 2011.
[34] Mackey B, Prentice C, Steffen W, House JI, Lindenmayer D, Keith H, et al. Untangling the confusion around land carbon science and climate change mitigation policy. Nat Clim Chang 2013;3:552–7.
[35] Nesbit M, Paquel K, Illés A, Maréchal A, Allen B. Designing a LULUCF pillar that works for forests and climate: report and reccomendations. London: Institute for European Environmental Policy (IEEP); 2015.
[36] Decision No 529/2013/EU of the European Parliament and of the Council of 21 May 2013 on accounting rules on greenhouse gas emissions and removals resulting from activities relating to land use, land-use change and forestry and on information concerning actions relating to those activities, Available from: http://eur-lex.europa.eu/legal-content/EN/TXT/?uri=CELEX%3A32013D0529.
[37] Marland G, Schlamadinger B. Forests for carbon sequestration or fossil fuel substitution? A sensitivity analysis. Biomass Bioenergy 1997;13:389–97.
[38] Leemans R, van Amstel A, Battjes C, Kreileman E, Toet S. The land cover and carbon cycle consequences of large-scale utilizations of biomass as an energy source. Glob Environ Chang 1996;6:335–57.
[39] Fargione J, Hill J, Tilman D, Polasky S, Hawthorne P. Land clearing and the biofuel carbon debt. Science 2008;319(5867):1235–8.
[40] Di Lucia L, Ahlgren S, Ericsson K. The dilemma of indirect land-use changes in EU biofuel policy—decision-making in the context of scientific uncertainty. Environ Sci Policy 2012;16:9–19.
[41] Malins C, Searle S, Baral A. A guide for the perplexed to the indirect effects of biofuels production. Washington, DC: International Council on Clean Transportation (ICCT); 2014.
[42] Böhringer C, Löschel A. Computable general equilibrium models for sustainability impact assessment: status quo and prospects. Ecol Econ 2006;60:49–64.
[43] Bauen A, Chudziak C, Vad K, Watson P. A causal descriptive approach to modelling the GHG emissions associated with the indirect land use impacts of biofuels—Final report—A study for the UK Department for Transport. London: E4tech; 2010.
[44] Ahlgren S, Di Lucia L. Indirect land use changes of biofuel production—a review of modelling efforts and policy developments in the European Union. Biotechnol Biofuels 2014;7(35). https://doi.org/10.1186/1754-6834-7-35.
[45] OJ L239/2, 1592015, Directive (EU) 2015/1513 of the European Parliament and of the Council of 9 September 2015 amending Directive 98/70/EC relating to the quality of petrol and diesel fuels and amending Directive 2009/28/EC on the promotion of the use of energy from renewable sources.
[46] SWD(2012) 343, Impact assessment accompanying the proposal for a Directive amending Directive 98/70/EC and Directive 2009/28/EC (Commission staff working document).
[47] 40 CFR Part 80, Regulation of Fuels and Fuel Additives: Changes to Renewable Fuel Standard Program; Final Rule, Stat. EPA–HQ–OAR–2005–0161.
[48] Farber DA. Indirect land use change, uncertainty, and biofuels. Policy, U. Ill. L. Rev. 381; 2011.
[49] Marland G, Marland S. Should we store carbon in trees? Water Air Soil Pollut 1992;64:181–95.
[50] RSPB. Dirtier than coal? Why Government plans to subsidise burning trees are bad news for the planet. Sandy, Bedfordshire: The Royal Society for the Protection of Birds (RSPB); 2012.
[51] Matthews R, Mortimer N, Lesschen JP, Lindroos TJ, Sokka L, Morris A, et al. Carbon impacts of biomass consumed in the EU: quantitative assessment—part A: main report—Project: DG ENER/C1/427. Farnham: Forest Research; 2015.
[52] FAO. Policy instruments to promote good practices in bioenergy feedstock production. Rome: Food and Agriculture Organization of the United Nations (FAO); 2012.
[53] European Commission. Voluntary-schemes. Available from: http://ec.europa.eu/energy/en/topics/renewable-energy/biofuels/voluntary-schemes.
[54] Di Lucia L. Too difficult to govern? An assessment of the governability of transport biofuels in the EU. Energ Policy 2013;63:81–8.
[55] Jentoft S, Chuenpagdee R. Fisheries and coastal governance as a wicked problem. Mar Policy 2009;33:553–60.
[56] Chuenpagdee R. Interactive governance for marine conservation: an illustration. Bull Mar Sci 2011;87(2):197–211.
[57] Junginger M, Goh CS, Faaij A. International bioenergy trade—history, status & outlook on securing sustainable bioenergy supply, demand and markets. Dordrecht, Heidelberg, New York, London: Springer; 2014.
[58] Fast S, McCormick K. Biofuels: from a win–win solution to a wicked problem? Biofuels 2012;3(6):737–48.

[59] Burger J. Management effects on growth, production and sustainability of managed forest ecosystems: past trends and future directions. For Ecol Manage 2009;258:2335–46.

[60] Lattimore B, Smith CT, Titus BD, Stupak I, Egnell G. Environmental factors in woodfuel production: opportunities, risks, and criteria and indicators for sustainable practices. Biomass Bioenergy 2009;33:1321–42.

[61] Nave LE, Vance ED, Swanston CW, Curtis P. Harvest impacts on soil carbon storage in temperate forests. For Ecol Manage 2010;259:857–66.

[62] Schlamadinger B, Spitzer J, Kohlmaier GH, Deke ML. Carbon balance of bioenergy from logging residues. Biomass Bioenergy 1995;8:221–34.

[63] EEA. European forests—ecosystem conditions and sustainable use. Copenhagen: European Environment Agency (EEA); 2008.

Further Reading

[1] GBEP. The global bioenergy partnership sustainability indicators for bioenergy. 1st ed. Rome: Food and Agricultural Organization of the United Nations (FAO); 2001.

CHAPTER

3

Greenhouse Gas Balances of Bioenergy Systems: The Role of Life Cycle Assessment

Marcelle C. McManus, Caroline M. Taylor†*

*University of Bath, Bath, United Kingdom †EarthShift Global, Kittery, ME, United States

3.1 INTRODUCTION

Life cycle assessment (LCA) is an environmental management tool that considers the impact towards certain pre-determined environmental issues over the life time or a product or a system. The broader term and concept, life cycle thinking, is increasingly popular for policy uses, especially in the field of bioenergy.

Biomass energy sources such as energy crops have some potential to be low carbon energy systems, and some hope with bio-CCS to be 'carbon negative' over their life cycle. They, therefore, have the potential to make a significant contribution to renewable energy and climate change targets. The growth of biofuel crops, such as oilseed rape, sugar beet, and cereals, is considered by some to be an opportunity for farmers to diversify and so bioenergy also has the potential to create and sustain jobs in rural areas. In substitution for fossil-derived transport fuels, liquid biofuels can contribute to the reduction of greenhouse gas (GHG) emissions. Bioenergy production's inter-section with other sectors, especially the critical one of agriculture upon which society places a particular premium, however, heightens concerns about balancing its risks and benefits. For these reasons, the use and production of bioenergy has been widely studied in recent years. With that has come the expansion of modelling techniques and frameworks such as LCA to determine their potential impact.

Many countries and regions have targets to increase the amount of bioenergy and biofuels in order to help minimise GHG emissions and meet climate change targets. In order to ensure that their use helps meet these targets, it is important that their impact can be accurately, transparently, and consistently measured. As such, LCA has come to be considered a very useful tool to help measure the GHG emissions from various bioenergy pathways and systems. Its ability to calculate GHG savings and emissions, and/or to improve processes, has

Greenhouse Gas Balances of Bioenergy Systems
https://doi.org/10.1016/B978-0-08-101036-5.00003-3

encouraged its use in helping to set and meet targets, select energy options and, most recently, determine policy by helping to assess possible outcomes of policy decisions. It is a key tool in the development and uptake of bioenergy systems.

This chapter outlines the evolution and use of LCA within the bioenergy sector, highlighting some of the specific advantages of its use as well as limitations. It then discusses some of the developments in the methodology (including attributional LCA (aLCA) and consequential LCA (cLCA)) as well as looking into the future to outline how it can be used as part of and alongside wider sustainability criteria to give a wider perspective on the suitability of the implications of particular bioenergy pathways, technologies, and policies.

Delivering energy from biomass requires a conversion facility to be established which converts the biomass feedstock into an appropriate energy vector. However, it also requires a feedstock to be delivered to the conversion plant. That feedstock may require processing before use and will normally be produced at a location remote from the conversion facility. Doing this may interrupt other systems: farming, land-use, and forestry and the consequences of this may need to be taken into account in considering the real impact of a bioenergy system. The energy produced may be used at the conversion facility or may be exported elsewhere. When we consider renewable energy systems, it is normally as part of a shift from incumbent fossil fuel-based systems and so the utility or value of the bioenergy system should be considered against an appropriate comparator.

So, it is immediately apparent that a bioenergy system extends significantly beyond the conversion of biomass to energy which occurs at the facility. It should take into account all of these impacts, even though some are remote from the energy conversion system, some are separated in time from the point of conversion, and some many not, at first inspection, appear very obviously 'energy-related'.

3.2 LCA METHODOLOGY

In the environmental management toolkit, LCA determines the potential impact of a product or system over its production, use, and disposal, see Fig. 3.1, for a variety of pre-selected environmental issues.

There are four main methodological steps within an LCA, as shown in Fig. 3.2. Further detail can be found in many text books, such as Baumann and Tillman [1], but a brief description is included here. *Goal definition* is the stage in which the scope of the project is outlined. Here the study boundaries are established and the environmental issues that will be considered are identified. The *inventory* stage is where the bulk of the data collection is performed. This can be done via literature searches, practical data gathering, or through the use of databases (sometimes) contained within LCA software. Most commonly, a combination of the three sources is required. *Impact assessment* is where the actual effects on the chosen environmental issues are assessed. This stage is further sub-divided into two to three elements: classification, characterisation, and valuation. Classification is where the data in the inventory is assigned to the environmental impact categories. Within each classification group, there will be several different emission types, all of which will have differing effects in terms of the impact category in question. A characterisation step is therefore undertaken to enable these emissions to be directly compared and added together, for example, methane and CO_2 are both GHGs;

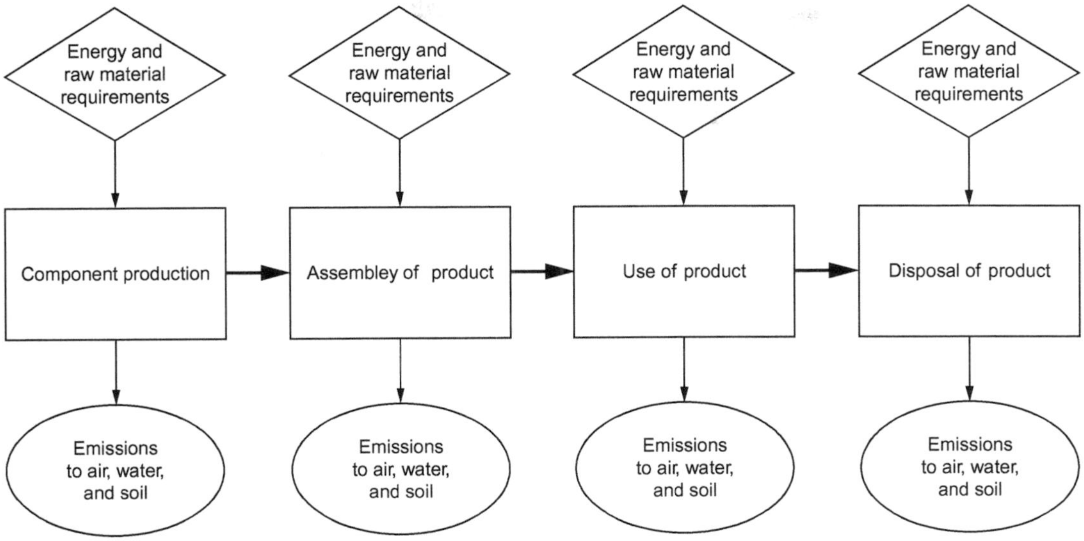

FIG. 3.1 Data and requirements within LCA.

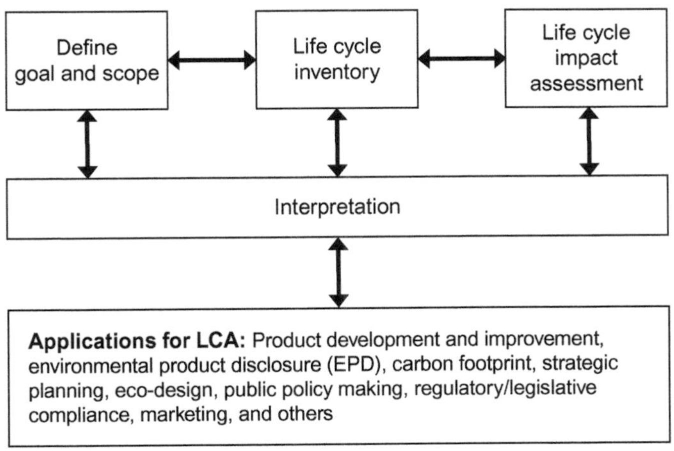

FIG. 3.2 Stages within LCA. *Modified from ISO standard 14040.*

in the characterisation stage, these will both be allocated their Global Warming Potential factor (1 for CO_2 and 25 for methane). The characterisation stage yields a list of environmental impact categories, for example, GHG, acidification, eutrophication, etc., and to each one a single number can be allocated. These impact categories are very difficult to compare directly and so the valuation stage is sometimes employed so that their relative contributions can be weighted and combined to produce a single number. This is subjective and difficult to undertake and many studies omit this stage from their assessment. Instead, they employ normalisation as an intermediate step. *Improvement assessment* is the final phase of an LCA in which areas for potential improvement are identified and implemented. Although it is the final step, it is common, and best practise, for this step to be iterative throughout the study.

One aspect of LCA methodology that is of particular interest to complex supply chains such as bioenergy is allocation. Allocation involves partitioning the input or output flows of a process or product system under study if there are more than one output. For example, if examining the impact of wheat straw to produce, for example bioethanol, one would have to determine the impact of growing a field of wheat and allocate a proportion of the impact to the straw and to the wheat. How this is done varies, with the most common methods of allocation being by economic value, by mass of the products, or by energetic content of the products. Results can differ significantly as a result of allocation choices and so it is important that any choice is transparent, and where possible, a range of results are given.

It is not possible, in practise, to analyse all known environmental problems and so it is important that the most appropriate ones are adopted for each given study. During the goal definition, or scoping, stage, the relevant environmental impact categories should be outlined together with the reasons for their selection over other possible categories; often this is policy- or legislatively driven, e.g. the main driver for analysing GHG balances (the focus of this book) is to support policy commitments on GHG reductions.

3.3 THE HISTORICAL EVOLUTION, UTILITY, AND LIMITATIONS OF LCA

The use of LCA has increased rapidly since its conception, so that it is now a well-known and used tool across industry, academia, and policy [2]. Fig. 3.3A shows the steep increase of academic publications in the area in the last 25 years. The timeline stages for LCA are summarised in Table 3.1. When LCA emerged in the late 1960s, it was as a tool used and developed by companies for resource management [3,4]. It was predominately single issue, such as waste, or single product-based. In the United States, this was largely linked to Resource and Environmental Profile Analyses (REPAs), and in the early 1970s, solid waste management was a primary driver. Later in that decade, the energy crisis drove companies to adopt an approach of energy management based on life cycle thinking. LCA has evolved from its origins in energy analysis in the 1960s and 1970s into a wide ranging tool that is used to determine impacts of products or systems over several environmental and resource issues [2].

Many of these early LCAs were not published, but one of the first of these that encompassed a wider range of environmental impact analysis was produced for Coca Cola in the 1960s [5]. Although the lack of publications limited their availability, these initial early studies set the scene for the wider ranging assessments that were to come after and helped to develop the methodology [2].

To date, there have been three main drivers for LCA development (see Table 3.1); company-driven in the 1960s and 1970s, regulatory and compliance-driven from the end of the 1970s until the 2000s, and most recently policy drivers.

In the early 1990s, the Society of Environmental Toxicology and Chemistry (SETAC) standards was developed. These were based on the establishment of a retrospective tool to quantify the impacts of a particular product. At this stage, the tool began to be used for regulatory purposes, but was still based on a retrospective approach. It was also associated with marketing, eco-labelling, packaging legislation, and the suite of integrated product policy-based (IPP) regulations in the EU [2]. The SETAC standards were adopted and amended to ISO standards in the late 1990s, forming the ISO 14040 series (ISO 14040–49 in 1997–2000). These were later revised in 2006.

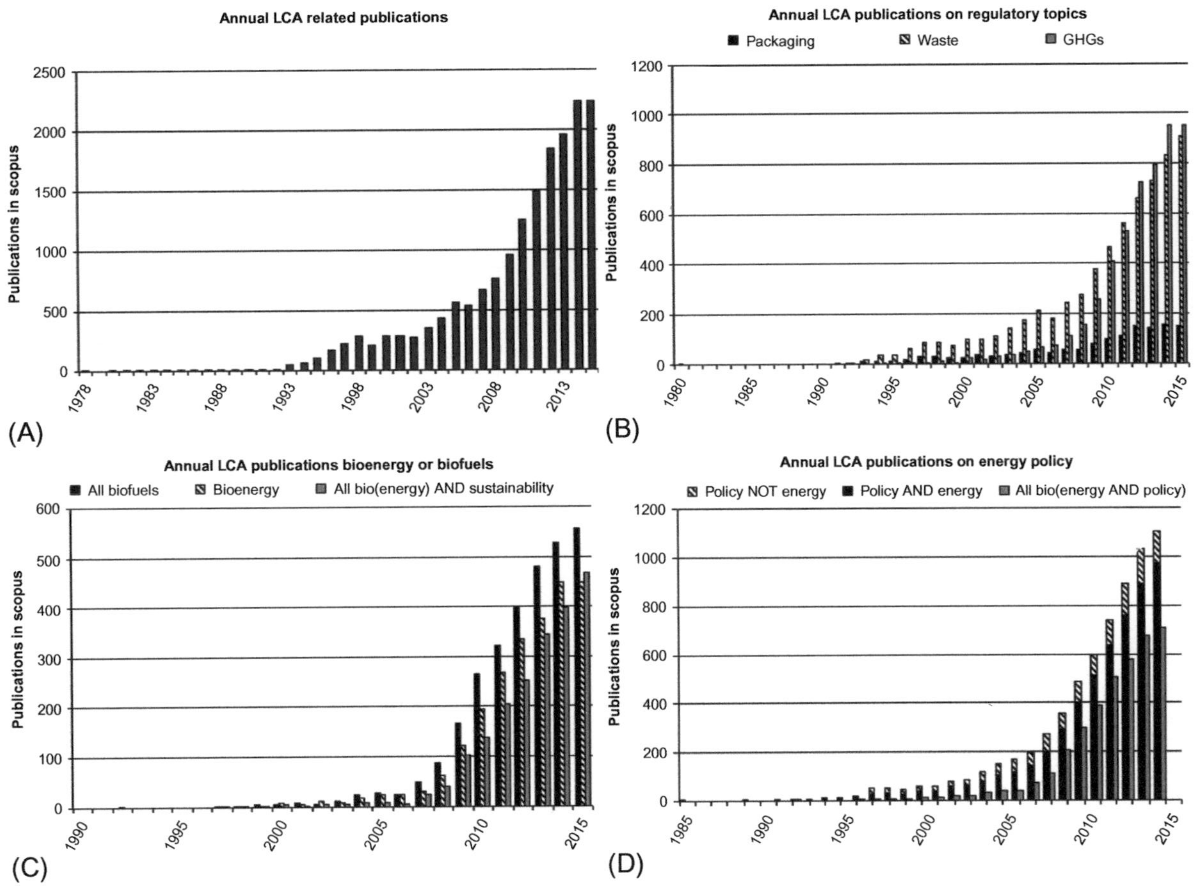

FIG. 3.3 LCA-related publications. (A) Annual total publications in the Scopus database; (B) publications on regulatory topics by year; (C) publications on bioenergy or biofuels by year; (D) publications on energy policy. *Modified from McManus MC, Taylor CM. The changing nature of life cycle assessment. Biomass Bioenergy 2015;82:13–26.*

TABLE 3.1 Trajectories and Drivers Along LCA's Development Timeline

	1960s	1970s	1980s	1990s	2000s	2010s
	Single Issues and Products		**Product Policy**	**Pollution Prevention**	**(Energy) Policy Development**	
Early		Solid waste driver in product development. Methodologies developed (private clients)	Slowdown in interest	First SETAC workshop. SETAC LCA framework developed. First peer-reviewed papers produced	Begins to be used more widely. Green Public Procurement IPP	LCA in energy policy, especially biomass, biofuel. United States: LCA for market access across state lines. EU:RED include iluc calculations
Mid		Continued, but limited company interest	Concern shifted to waste management	SETAC methodology	Revised ISO standards	
Late	Coca Cola	More interest during energy crisis	Waste becomes global issue and life cycle thinking expands again	First ISO standards	Energy Policy and Regulations	
Policy instruments (internal)		EPA		SETAC ISO EPDs	IPP EISA LCFS RFS2 RED	RTFO LCFS LCA upheld RED LUC decided
	Company-driven					Policy-driven
			Regulatory/compliance-driven			

Notes: EISA/US Energy Independence and Security Act 2007. EPA/US Environmental Protection Agency. EPDs/Environmental Product Declarations (ISO 14025). IPP/Integrated Product Policy concept initiated 2003. ISO/International Standards Organization. LCFS/California's Low Carbon Fuel Standard, 2007. RED/EU Renewable Energy Directive, 2009. RFS2/Renewable Fuel Standard, overseen by the US EPA 2007/2008. RTFO/UK Renewable Transport Fuel Obligation. 2012. SETAC/Society of Environmental Toxicology and guidelines.

Modified from McManus MC, Taylor CM. The changing nature of life cycle assessment. Biomass Bioenergy 2015;82:13–26.

At the turn of the millennium, the biggest increase and changes in LCA start to emerge (see Fig. 3.3A and B). The increase can roughly be split into two, between publications relating to regulatory topics, such as packaging, waste, and GHGs. This shows the increase of LCA's use and significance in research and industry. Although it is a tool that can explore a wide range of impacts, its use is often limited to GHG evaluation.

The predominant driver for much energy policy, and hence many energy-related LCAs, is GHG accounting (e.g. Renewable Energy Directive (RED) [6], etc.). The change in drivers is fairly recent; GHGs overtook waste in the LCA publication record only in 2012 (Fig. 3.3B). While recent, the speed of the shift is significant, clear in the volume of papers and studies produced in this period (from 24 in 2000 to 954 in 2013). The rate of LCA studies also shifts, from a slower, steadier accumulation relying mostly on additional waste and packaging analyses up to the early-mid 2000s, into a rapid explosion of papers dominated by GHG comparisons. This suggests that not only has the driver for using LCA changed, so has the force of that driver [2].

Across industry, policy, legislation, and academia, LCA's use continues to expand in both context and frequency as we seek to encompass impacts as diverse as resource accounting and social well-being. Fig. 3.3C shows how strongly energy, particularly bioenergy and an emphasis on sustainability, has influenced these expansive changes in use and methodology [2].

The beginning of more policy development and use associated with the use of LCA from the mid-1990s is clear in Fig. 3.3D. This use is particularly prevalent within the energy area, suggesting energy as a major driver for the expansion. Approximately half of the energy-related policy LCAs relate to bioenergy (Fig. 3.3C and D), showing the special influence of bioenergy in this field.

Biofuels and bioenergy alone have had a notable influence on the rise of LCA. Much of this is driven by the aforementioned desire for a low carbon, dispatchable[1] fuel. With the recognition that complex supply chains and technologies have to be monitored and measured in terms of sustainability, LCA use has significantly increased. In addition, a number of bio-based LCA tools, such as GREET [7], GHGenius [8], and BEAT [9], have all become available, enabling a wider audience to use the technique.

3.4 THE CHANGING NATURE OF LCA

LCA use has changed dramatically over recent years. Traditionally, LCA was used to answer specific questions [10] that are directly attributable to the life cycle of a product in the existing technological and economic climate [11]. This is now known as aLCA and is the process described at the beginning of this chapter. aLCA is generally (and traditionally) retrospective, but can also be used on novel pathways to help predict potential impacts. This use has become more common within the bioenergy area to help determine the potential impacts of novel pathways and technologies. It is often presented in literature that 80% of all environmental effects associated with a product are determined during the design stage [12], so the trend to increasingly use LCA at the early stages of research and design is relatively un-surprising. Use at this stage enables the practitioner to explore options for minimising impact from the earliest stage of a product or system's life. LCA practitioners and researchers can work together to select the most environmentally benign materials and processes; hence reducing impact from the outset (e.g. [13]). However, there are some problems associated with this approach, particularly regarding data availability [14]. Fig. 3.4 illustrates the shift in data needs and target use between LCA for the two arenas.

LCA is also increasingly being asked to help to anticipate future impacts and illustrate complex intersections, impacting system boundaries and the overall complexity of LCA. Historically, LCA has been conceptually simple, but challenging in execution. However, its expansion in this way and use to influence and evaluate policy options as part of governance for sustainability in energy, emerging technology, and resources is changing that [15]. This is cLCA and allows impacts to be considered in a wider, even global, context of producers and consumers [16]. This more consequential approach is considered to be the appropriate method for policy makers [17]. However, this expansion has proven challenging, and in some cases, controversial (see, e.g. discussion in Breetz [18]).

[1]A dispatchable fuel is one that can be used to generate power as and when it is required within a relatively short timescale (seconds). It is not reliant on weather, as solar and wind are, and is therefore not an intermittent source of power. Bioenergy, as it can be stored for use when required in a similar manner to coal and gas, is a dispatchable power source.

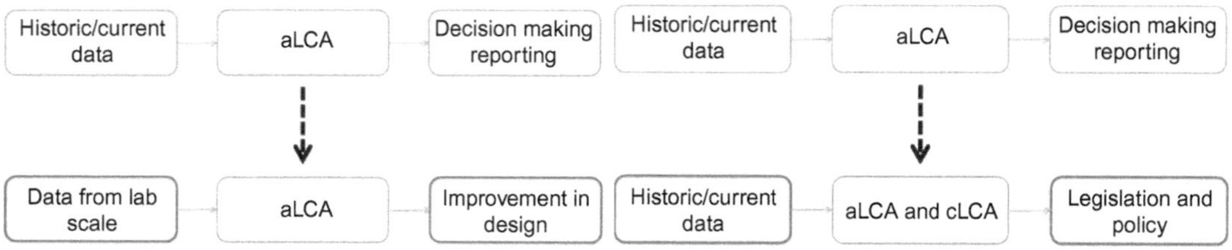

FIG. 3.4 Evolution of scope and use trends in LCA. *Modified from McManus MC. The use of LCA for the development of bioenergy pathways. In: Pinto A, Zilberman D, editors. Modeling, dynamics, optimization and bioeconomics II. Springer; 2016. In Press.*

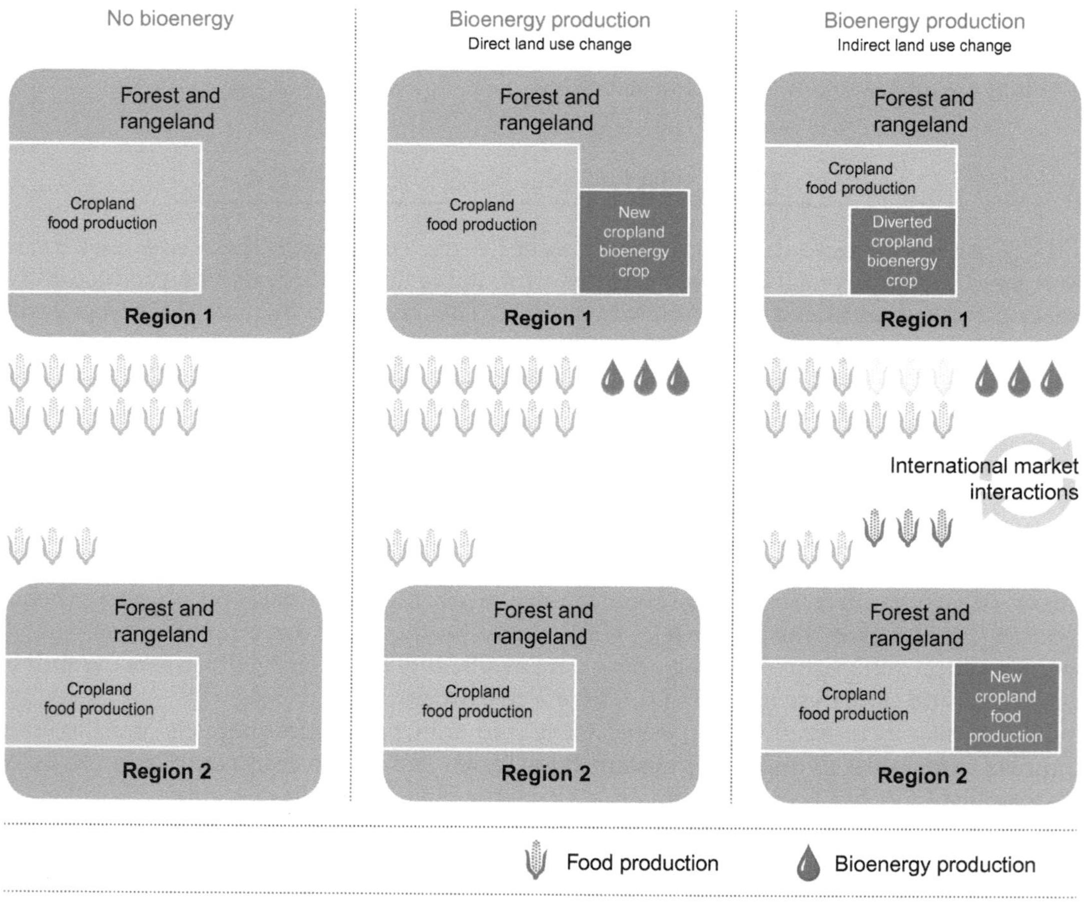

FIG. 3.5 Land inter-section pathways relevant for bioenergy. *From BioEnergy Connection, v3.2, used by permission.*

The three panels in Fig. 3.5 illustrate why. Bioenergy's position directly linking agricultural and energy sectors brought this to the fore with concerns that expanding bioenergy production could correlate to market-mediated land-use change adversely impacting food/feed production and prices. The left panel in Fig. 3.5 shows the case without bioenergy, with agricultural production in two regions, one high productivity, one lower; with bioenergy, the system response would depend on management strategy, maturation, and policy instruments and enforcement. Assessing these linkages is highly sensitive to speculation about many aspects of the system's development with time and other actors. The inter-section has induced one of the most significant shifts in LCA's development, leading LCA to strongly influence selection and survival of bioenergy deployment projects at planning stages and policy barriers or support.

3.5 DATA AND THE USE OF LCA RESULTS AS GENERIC INDICATORS

As discussed in this chapter, Chapter 4, and others, the bioenergy system is complex. There are a myriad of potential feedstocks, several conversion technologies, and it can be used to produce heat, fuel, and electricity. The combinations of these result in many different potential pathways for bioenergy production, even for the same end product. This can lead to some difficulty in determining the impact of any particular bioenergy type. This has historically been the case: in one of the earlier LCAs of biodiesel, for example, Cueterick and Spirinkx [19] found that the biodiesel had a higher environmental impact than did the fossil-based diesel. Although biodiesel only had smaller impacts in the categories designated 'fossil fuels' and 'greenhouse effect', it exhibited higher impacts in all the other fields considered; inorganic raw materials, water, acidification, eutrophication, photochemical oxidants formation, and both radioactive and non-radioactive waste. The cause of this was primarily due to the use of fertilisers. However, a similar study undertaken by Stelzer [20] indicated that there were far more benefits than drawbacks to biodiesel. The primary reason for this is the more favourable rapeseed growing conditions found in Germany compared with Belgium; requiring less fertiliser to be used. This shows that the environmental impact of these products is highly dependent on where they are grown and the methods used for their production. In terms of emissions, LCA was originally unable to determine differing impacts (leaving such issues to methods such as Environmental Impact Assessment, etc.). However, as the tool develops, it is becoming more geographically and temporally sophisticated [21].

A lack of common boundaries, allocation issues, and transparency makes it difficult to compare different LCA studies [15]. A number of regulatory agencies have attempted to overcome this by releasing detailed guidelines for products under their purview (for example, DECC: approval under RTFO [22]; EPA: pathway for RINs under RFS2 [23]; and CARB: pathway under LCFS [24]). However, this has resulted in a proliferation of non-transferable guidelines, as each was originally designed for a specific product or purpose. For example, the European Renewable Energy Directive calculation methodology was designed to account for the GHG emissions from biofuels. The allocation method specified in the RED [6] (energy content) cannot easily be employed by those performing similar GHG assessments in other bioenergy systems, for example anaerobic digestion, as the technology generates large quantities of a valuable co-product, digestate, but has negligible energy content [24]. As a result, the methods of assessment between each set of guidelines differ considerably in their system boundaries, co-product and waste definitions, and methods of allocation of environmental impacts [25].

These methods all comply with the ISO standards 14040:2006 and ISO 14044:2006 [26,27] that govern the LCA practise, but the ISO standards leave a great deal of scope for interpretation and flexibility to the LCA practitioner [28]. As a result, it is virtually impossible to make confident comparisons between studies as they can differ not only due to inherent variability between systems, but also due to the methods in which they are examined [29]. This lack of consistency, and hence credibility in places, makes it difficult for government departments and policy makers to use LCA studies [15].

Failure to transparently report or manage the complexities in bioenergy GHG assessments can lead to confusion and misleading results [30] and facilitate misrepresentation that has the potential to focus policies on the wrong areas. The 'Carbon Sink or Sinner' report [31] reviewed the sensitivity of the results in the Biomass Environmental Assessment Tool (BEATv2 [9]) to biomass supply chain parameter assumptions, concluding, among other things, that the lifecycle GHG emissions were strongly influenced by how the fuel was produced, transported, and processed. They stated that 'bad practises' involving transporting fuels very long distances and excessive use of nitrogen fertilisers could reduce the emissions savings for the same fuel by 15%–50% [32]. Collectively assessing these two stages of the supply chain created a misinterpretation that transport emissions were a major cause for concern in net GHG emissions of bioenergy supply chains. Media reports and local anti-biomass websites then reproduced this, emphasising the less significant transportation impacts rather than nitrogen fertiliser use [15].

3.6 CURRENT AND FUTURE USE OF LCA

LCA is now becoming more heavily relied upon to reveal potential unintended consequences of bioenergy. This use is almost unique to the bioenergy field, and in some cases, masks a high degree of uncertainty [15]. Since biofuel policies were established in the United States and European Union (EU), there has been concern that unintended consequences could arise and threaten climate and other sustainability goals (e.g. [33,34]). Among these have been concerns that a major switch to biofuels produced from food crops will lead to competition between the use of crops for food/feed or for fuel (illustrated in Fig. 3.5); or that using land for bioenergy could potentially contribute to other effects through global markets. There have also been concerns that the production of bioenergy crops could be carried out on environmentally sensitive lands; or that where land has been acquired in the Global South, leads to the loss of livelihoods and local food and energy production [35]. Interest in these factors is clear in the growth of publications shown in Fig. 3.6. Significantly, many of the potential sustainability impact categories for biofuels occur at the systems level, for example land-use change and deforestation, and so cannot be easily separated into distinct categories among environmental, economic, or social impacts [36], thus making LCA's task of anticipating the unanticipated all the more challenging.

As engagement with broader sustainability standards has grown, the metrics space has begun to broaden beyond the initial emphasis on GHG balances [2] to include social and ecosystem impacts, among others (see Fig. 3.6), suggesting clearly where LCA's future lies. Although these wider issues are within the purview of LCA, the particular challenge here is that approaches to do so are still in early stages and have correspondingly higher uncertainty. LCA is balanced between its past and continuing future as a data-based, rigorous retrospective tool for product and process optimization and its parallel future as a policy or strategic planning tool.

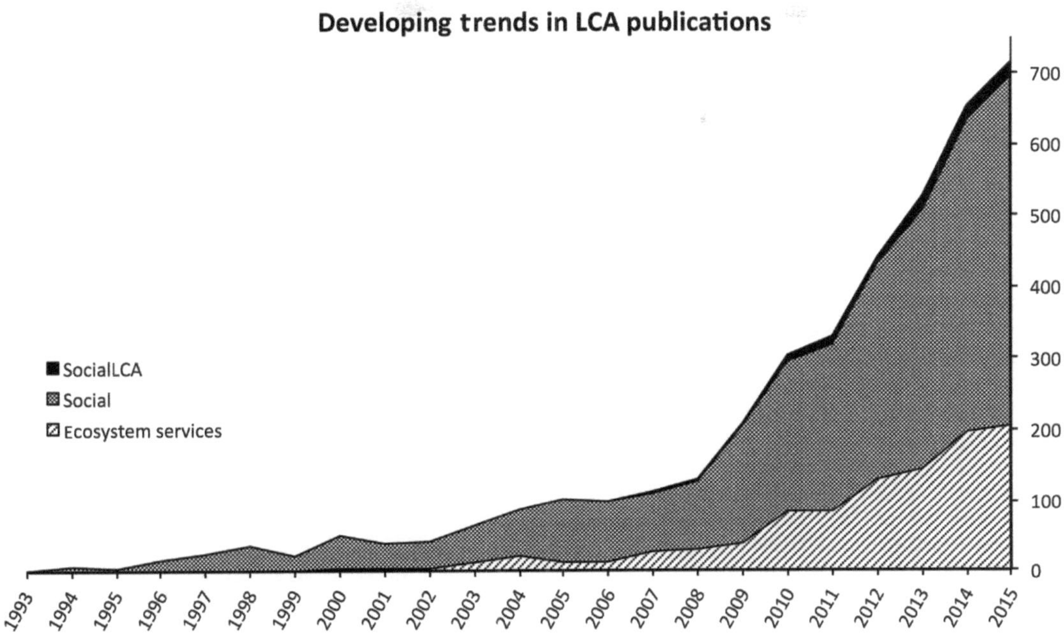

FIG. 3.6 Increased emphasis on wider impacts: LCA publications on ecosystem services, social factors, and socialLCA by year in the Scopus database.

3.7 CONCLUSIONS

Bioenergy has contributed to the fastest era of change within LCA's history. LCA's use, demands, and methodologies are all evolving in response to the new needs and challenges associated with bioenergy expansion, particularly at scale. Nevertheless, it remains a valuable tool to help determine impacts associated with existing products and systems, as well. Its use in both early stage research and to help determine potential impacts of the use of a fuel into the future require the tool to develop in different directions at the same time, which can result in uncertainty and rapid evolution. These intersections, driven and informed by bioenergy sustainability needs, are ushering in a golden era for life cycle thinking and assessment. Both the LCA and the bioenergy communities, along with a range of other stakeholders upon whom results could impact, are vital in the effective development of LCA as a robust tool into the future.

KEY POINTS

- Life cycle thinking can aid reduction in emissions and impact
- LCA and its application are changing rapidly. Bioenergy and energy are drivers in this progression and evolution
- Allocation can have significant impact on results and such methods need to be transparent
- A lack of common boundaries, allocation issues, and transparency makes it difficult to compare different LCA studies
- Consequential analysis is growing in the bioenergy area to aid policy making

Acknowledgements

This chapter is produced primarily from the work presented in previous publications: McManus &Taylor, 2015 and McManus, Taylor et al, 2015 and McManus, 2016. The authors would like to thank all the other co-authors on McManus, Taylor et al. and their funders, the BBSRC UK/US partnership grand BB/JO20184/1, the EPSRC Supergen Bioenergy Hub (EP/JO17302/1), and the EU Marie Curie Grant EnvBio Grant number: 318927 FP7-PEOPLE-2012-IRSES.

References

[1] Baumann H, Tillman AM. The hitchhiker's guide to LCA, an orientation in life cycle assessment methodology and application. Sweden; 2001.
[2] McManus MC, Taylor CM. The changing nature of life cycle assessment. Biomass Bioenergy 2015;82:13–26.
[3] Hunt RG, Franklin W. LCA—how it came about. Int J Life Cycle Assess 1996;1(1):4–7.
[4] Curran MA. Broad-based environmental life cycle assessment. Environ Sci Technol 1993;27(3):430–6.
[5] Guinée JB, Heijungs R, Huppes G, Zamagni A, Masoni P, Buonamici R, et al. Life cycle assessment: past, present, and future. Environ Sci Technol 2011;45(1):90–6.
[6] Renewable Energy Directive: EU (2009) Directive 2009/28/EC of the European Parliament and of the Council of 23 April 2009 on the Promotion of the Use of Energy from Renewable Sources and Amending and Subsequently Repealing Directives 2001/77/EC and 2003/30/EC.
[7] GREET CA-GREET. California Air Resources Board. Available from: http://www.energy.ca.gov/ab1007/ca-greet_model/ [Accessed 22 July 2014].
[8] GHGenius. Available from: http://www.ghgenius.ca/downloads.php [Accessed 21 July 2014] Hunt & Franklin; 1996.
[9] BEATv2 AEA Technology & Associates, N. E. Biomass Environmental Assessment Tool. 2.1 ed. Oxford and Sheffield; 2010.
[10] Sandén BA, Karlström M. Positive and negative feedback in consequential life-cycle assessment. J Clean Prod 2007;15:1469–81.
[11] Sanchez ST, Woods J, Akhurst M, et al. Accounting for indirect land-use change in the life cycle assessment of biofuel supply chains. J R Soc Interface 2012. https://doi.org/10.1098/rsif.2011.0769 rsif20110769.
[12] Tischner U, Schmincke E, Rubik F, Prosler M. How to do ecodesign?: A guide for environmentally and economically sound design. German Federal Environmental Agency, editor. Verlag form (Praxis); 2000. ISBN: 3898020258/9783898020251.
[13] Griffiths OG, Owen RE, O'Byrne JP, Mattia D, Jones MD, McManus MC. LCA using life cycle assessment to measure the environmental performance of catalysts and directing research in the conversion of CO_2 into commodity chemicals: a look at the potential for fuels from 'thin-air'. RSC Adv 2013;30:12244–54.
[14] Hetherington A, Borrion AL, Griffiths OG, McManus MC. The use and implications of LCA in early stage research. Int J Life Cycle Assess 2014;19(1):130–43.
[15] McManus MC, Taylor CM, Mohr A, Whittaker C, Scown CD, Borrion AL, et al. Challenge clusters facing LCA in environmental decision-making—what we can learn from biofuels. Int J Life Cycle Assess 2015;20:1399–414.
[16] Nuffield Council on Bioethics. Biofuels: ethical issues. London: Nuffield Council on Bioethics; 2011. Available from: http://www.nuffieldbioethics.org/sites/default/files/Biofuels_ethical_issues_FULL%20REPORT_0.pdf; [Accessed 22 July 2014].
[17] Brander M, Tipper R, Hutchison C, Davis G. Consequential and attributional approaches to LCA: a guide to policy makers with specific reference to greenhouse gas LCA of biofuels; 2009.
[18] Breetz HL. Science, values, and the political framing of indirect land use change (ILUC). In Science and the law: how the communication of science affects policy development in the environment, food, health, and transport sectors, 1207:95–122. ACS Symposium Series 1207. American Chemical Society; 2015. https://doi.org/10.1021/bk-2015-1207.ch007.
[19] Ceuterick D, Spirinkx C. Comparative LCA of biodiesel and fossil fuel. Brussels: Flemmish Institute for Technological Research; 1997.

[20] Stelzer T. Biokraftstoffe im Verleich zu konventionellen Kraftstoffen—Lebensweganalysen von Umweltwirkungen, Research Report, University of Stuttgart, Institut fu¨ r Engergiewirtschaft und Rationelle Energieanwendung, Stuttgart; 1999. Cited in: Puppa´n D. Environmental evaluation of biofuels. Period Polytech Ser Soc Man Sci 2002;10(1):95–116.
[21] Morais TG, Teixeira RFM, Domingos T. Regionalization of agri-food life cycle assessment: a review of studies in Portugal and recommendations for the future. Int J Life Cycle Assess 2016;21:875–84. https://doi.org/10.1007/s11367-016-1055-3.
[22] DECC: approval under RTFO; UK DTI. RTFO. Renewable Transport Fuel Obligation (2012) Available from: https://www.gov.uk/renewable-trans-port-fuels-obligation [Accessed 23 July 2014].
[23] EPA: pathway for RINs under RFS2; US EPA. RFS2. US Environment Protection Agency. Renewable Fuel Standard, overseen by the US EPA. Available from: http://www.epa.gov/OTAQ/fuels/renewablefuels/ [Accessed 21 July 2014].
[24] Manninen K, Koskela S, Nuppunen A, et al. The applicability of the renewable energy directive calculation to assess the sustainability of biogas production. Energy Policy 2013;56:549–57.
[25] Whitaker J, Ludley KE, Rowe R, et al. Sources of variability in greenhouse gas and energy balances for biofuel production: a systematic review. GCB Bioenergy 2010;2:99–112.
[26] CEN. BS EN ISO 14040:2006. Environmental management—life cycle assessment—principles and framework. Brussels: European Committee for Standardisation; 2006.
[27] CEN. BS EN ISO 14044:2006. Environmental management—life cycle assessment—requirements and guidelines. Brussels: European Committee for Standardisation; 2006.
[28] Aylott M, Higson A, Evans G, et al. What is the most appropriate LCA method for measuring greenhouse gas emissions from bioenergy? Biofuels Bioprod Biorefin 2011;5:122–4. https://doi.org/10.1002/bbb.282.
[29] Menichetti E, Otto M. Energy balance and greenhouse gas emissions of biofuels from a life-cycle perspective. Biofuels: environmental consequences and interactions with changing land use. In: Proceedings of the Scientific Committee on Problems of the Environment (SCOPE) International Biofuels Project Rapid Assessment. Gummersbach, Germany. Cornell University, Ithaca, NY USA; 2008. p. 89–109.
[30] Adams PWR, Bows A, Gilbert P, Hammond J, Howard D, Lee R, et al. Understanding greenhouse gas balances of bioenergy systems. Supergen Bioenergy Hub Report; 2013.
[31] 'Carbon Sink or Sinner' report (AEA Technology/Ricardo-AEA; 2009).
[32] Bates J, Edberg O, Nuttall C. Minimising greenhouse gas emissions from biomass energy generation. Didcot: AEA Technology; 2009.
[33] Mol AP. Boundless biofuels? Between environmental sustainability and vulnerability. Sociol Rural 2007;47:297–315.
[34] Jaeger WK, Egelkraut TM. Biofuel economics in a setting of multiple objectives and unintended consequences. Renew Sust Energ Rev 2011;15:4320–33.
[35] Van Eijck J, Romijn H. Prospects for Jatropha biofuels in Tanzania: an analysis with strategic niche management. Energy Policy 2008;36:311–25.
[36] Mohr A, Raman S. Lessons from first generation biofuels and implications for the sustainability appraisal of second generation biofuels. Energy Policy 2013;63:114–22.

Further Reading

[1] CARB: pathway under LCFS California Air Resources Board. CARB, within the California Environmental Protection Agency (Cal/EPA) Title 17, California Code of Regulations (CCR), sections 95480 through 95490 CA-GREET California Low Carbon Fuel Standard (CA-LCFS) Available from: http://www.arb.ca.gov/fuels/lcfs/2a2b/2a-2b-apps.htm [Accessed 22 July 2014].
[2] McManus MC. The use of LCA for the development of bioenergy pathways. In: Pinto A, Zilberman D, editors. Modeling, dynamics, optimization and bioeconomics II. Springer; 2016. In Press.

CHAPTER

4

Scope of System for Analysis

Patricia Thornley

University of Manchester, Manchester, United Kingdom

4.1 INTRODUCTION

Delivering energy from biomass requires a conversion facility to be established which converts the biomass feedstock into an appropriate energy vector. However, it also requires a feedstock to be delivered to the conversion plant. That feedstock may require processing before use and will normally be produced at a location remote from the conversion facility. Doing this may interrupt other systems: farming, land-use, forestry, and the consequences of this may need to be taken into account in considering the real impact of a bioenergy system. The energy produced may be used at the conversion facility or may be exported elsewhere. When we consider a renewable energy systems, it is normally as part of a shift from incumbent fossil fuel-based systems and so the utility or value of the bioenergy system should be considered against an appropriate comparator.

So, it is immediately apparent that a bioenergy system extends significantly beyond the conversion of biomass to energy which occurs at the facility. It should take into account all of these impacts, even though some are remote from the energy conversion system, some are separated in time from the point of conversion, and some many not, at first inspection, appear very obviously 'energy-related'.

Chapter 3 has described the development, evolution, and use of life cycle assessment as an analysis tool for bioenergy systems. It explored the challenges associated with applying life cycle assessment techniques to bioenergy, including the issues associated with a lack of common boundaries, making it difficult to compare LCA studies and the importance of consequential LCA for policy evaluation. So, it is important that the scope of analysis for life cycle evaluation of bioenergy systems is consistent when making cross comparison, but ultimately the scope should be informed by the 'LCA question' being asked, i.e. the scope of system being analysed should be carefully matched to the goal of the study and this requires careful consideration of whether different system components should be included in the LCA assessment.

Greenhouse Gas Balances of Bioenergy Systems
https://doi.org/10.1016/B978-0-08-101036-5.00004-5

4.2 THE SCOPE OF BIOENERGY SYSTEMS

Bioenergy systems can often usefully be divided into three stages: biomass production, biomass conversion, and bioenergy utilisation.

4.2.1 Biomass Production

It is essential that greenhouse gas (GHG) balances for bioenergy systems take into account the greenhouse gases associated with production of the biomass. How this should be done may vary depending on the type of biomass being considered. For the purposes of this chapter, we will consider the following types of biomass:

Annual crops—these are crops which grow to harvest readiness within a 1 year period.
Perennial crops—these are crops which are planted 1 year and remain productive for many years. Some will be harvested annually; others less frequently.

For a purpose grown annual crop, in which case the system scope would normally include ground preparation, crop planting, agrochemical application, and harvesting, it will often require post-harvest processing (e.g. drying or silaging, which may require energy and transform the feedstock properties) and transportation.

When dealing with perennial crops we need to think about all of the same issues, but need to be particularly careful about the boundaries being set, e.g. how does extraction of the material affect growth and longer term carbon sequestration, including soil carbon levels. There is potential for perennial crops to increase soil carbon levels in the long term, as discussed in Chapter 5. Perennial crops may also be more likely to have interfaces with the food and land-use system that have wider, long-term impacts, e.g. removal of land from a previous crop rotation system.

Forestry systems present an extreme example of a perennial crop since the timescale involved is very long in comparison to even the lifetime of a bioenergy conversion plant/system. Therefore, we are unlikely to be considering establishment, but will be much more focused on the impact of biomass removal on the remaining forest system, including the carbon stock of the forest.

4.2.2 Biomass Conversion

Some bioenergy systems convert biomass directly for energy use, e.g. combustion of wood chips in a boiler directly delivers heat in the form of hot water. However, many systems convert biomass to intermediate products, e.g. gasification produces a syngas which may be used for a variety of energy or nonenergy purposes. Therefore, we need to consider the biomass conversion step as well as any other subsequent conversion steps to the final energy product, e.g. in production of substitute natural gas or aviation biofuels, there may be an initial anaerobic digestion or gasification step, then more than one upgrading and purification step to include in the greenhouse gas balance calculation. Information relating to different biomass conversion technologies which will be useful for LCA is presented in Chapter 8.

4.2.3 Bioenergy Use

When considering a bioenergy system, we must follow the biomass path through to actual end-use of the energy and consider what utility or function the end energy product is providing, including whether it is displacing an alternative energy source. This may require consideration of whether a different appliance is required to make use of the bioenergy and the nature of the existing dominant technology. The timing of energy provision can also be important, e.g. thinking about whether the bioenergy is displacing energy use at a time of peak demand and whether the new bioenergy capacity is avoiding construction of new energy plant and, if so, of what type, e.g. is it displacing a natural gas boiler than would be used on demand or a district heating scheme?

4.3 LIFE CYCLE ASSESSMENT: GOAL AND SCOPE DEFINITION

Life cycle assessment is a tool that allows the user to evaluate the impact of a product or service from the 'cradle to the grave'. As described in Chapter 3, life cycle assessment was originally designed to support companies in resource management and provide retrospective information on the impact of a particular product, taking into account both upstream and downstream impacts. For example, LCA might be used to quantify the energy use in production of aluminium foil (all focused on the upstream elements before the product gets to the consumer) or the ozone depletion caused by antiperspirant use (an impact initiated at the point of use) or the aquatic damage caused by detergent use (mostly focused on impacts post-consumer use after discharge). So, life cycle assessment is based around saying that all of these things are equally important and need to be considered in any holistic appraisal of the product or service being considered.

It is, however, challenging to cross-compare the impacts of what are sometimes very different approaches to delivering the same consumer utility. For this reason, a set of standards has been developed to try to normalise the use of the tool and set guidelines for its application. The ISO standards provide an overarching framework for execution of life cycle assessment for practitioners [1]. However, even these recognise that there is no one scope of system that could be applied to all areas of interest, e.g. if I am considering bioenergy I may be very interested in land-use change, but that might be quite peripheral to other products whose land use is minor. Therefore, the guidelines actually state that the first step in any life cycle assessment is to define the goal and scope of the study and the system boundary should be chosen to suit the goal, scope, and study objectives.

This is one of the most important and yet one of the most commonly overlooked elements of LCA. It simply requires that the practitioner stop to think about what he is calculating and why. That understanding will then guide the scoping of an appropriate system for the LCA study and will also guide the appropriate calculation methodology. It is important to realise that different questions will give different results and both results could be equally valid, but the context must be clearly understood [2].

LCA is a tool that has many choices and variations. In the same way that there is no one single recipe for baking bread, but many variations that can all achieve the same nominal end product; so is there no one unique way to carry out an LCA. In some cases, it may be appropriate to consider land-use change; in some, it may be appropriate to consider displaced energy sources. The key thing is that the practitioner thinks about the question being asked and its context and makes appropriate decisions about the LCA scope based on that information.

There are a few basic principles that should be followed: it is always helpful to compare like with like: so compare one energy vector with another similar vector (comparison of natural gas with substitute natural gas is fine, but comparing 1 MJ of high pressure steam with 1 MJ of low-grade heat is less instructive). Similarly, it is always a good idea in principle to close the system being considered by returning it to the initial state, e.g. if starting with an arable field and planting a perennial crop, the user should include removal of the crop and returning the field to its original state at the end. In theory, both construction and decommissioning of a bioenergy facility could/should be included in life cycle assessment scope. However, work in related sectors [3] has shown that this is a very small component of overall life cycle impact; sensitive to specific information about the amount of concrete and steel used in construction, which is extremely difficult to extrapolate for prospective plants that have not yet been built. Therefore, this element of the life cycle assessment is often put to one side since the impact on overall results is not expected to justify the data collection effort.

In a sense LCA can be adapted to answer any question that the user wishes, but it is essential that the user knows what question they are posing. Therefore, researchers need to stop and carefully think about what research question they are actually trying to address in their LCA study. There is often a tendency to base LCA studies around the methods developed for legislative or certification purposes, e.g. to use the Renewable Transport Fuel Obligation (RTFO) methodology or the Renewable Energy Directive (RED) methodology [4–6]. This can sometimes provide the advantage of having a completely consistent framework within which to compare systems. However, these frameworks are generally based on policy objectives, which may or may not be firmly rooted in scientific understanding and will often adopt simplifications, such as ignoring supply chain emissions for materials deemed to be wastes. The research challenge is to properly frame the question being asked, adjust basic methodologies to respond to that question, and interpret results in a way that presents impartial scientific evidence related to actual environmental impact.

A bioenergy system boundary for an LCA study of forest residues to electricity is given as an example in Fig. 4.1 [7] and this case is discussed in more detail in Chapter 14. In this case, we are interested in the impact of taking residues that are generated during normal forest harvesting, processing them, transporting them to the United Kingdom, and generating electricity. Therefore, a comprehensive scope that included all steps from forest establishment through to the generation of electricity was used. If the objective had instead been to consider the alternative disposal/management options of unwanted forest residues, a reduced system scope that did not include the forest establishment might be more appropriate.

Differences in system boundary can make very significant differences in system results and a key issue for many energy crop systems is whether and which land-use changes are taken into account. For example, Upham and Thornley [8] showed that inclusion of different land-use changes could increase the carbon payback time on a biodiesel system by over 5000 years. If considering the long-term impact of a dedicated biofuel plantation, it would be appropriate to take this land-use change into account, but if the crop establishment is

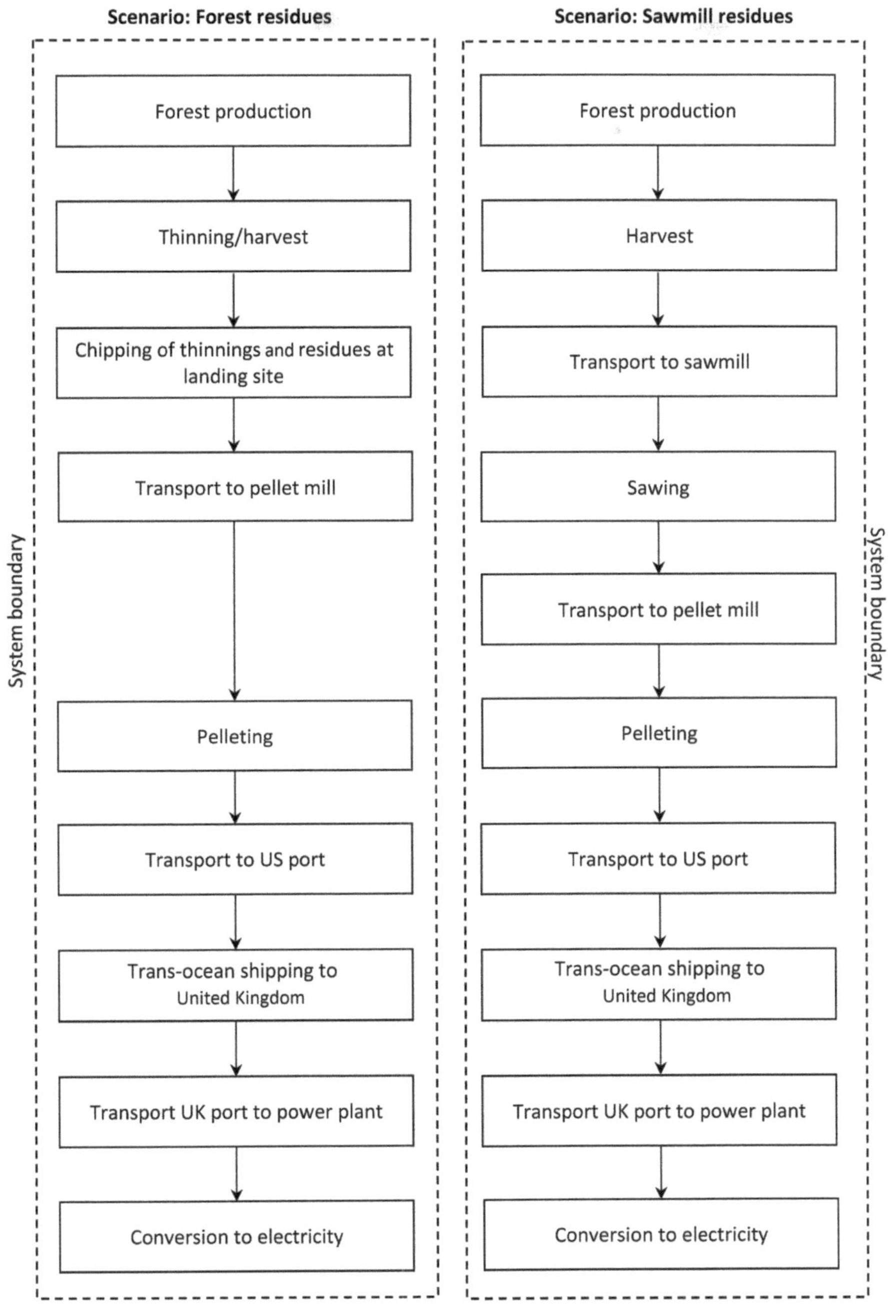

FIG. 4.1 Scope of system for LCA of bioenergy from forest residues. *From Röder M, Whittaker C, Thornley P. How certain are greenhouse gas reductions from bioenergy? Life cycle assessment and uncertainty analysis of wood pellet-to-electricity supply chains from forest residues. Biomass Bioenergy 2015;79:50–63.*

primarily being driven by another market or another product, it may make more sense to exclude or only partly account for the impacts of land-use change, since the bioenergy system may not be considered causally responsible for the land-use change.

It is key, therefore, to ensure that the goal of a study is clearly understood at the outset. For example, the objectives of achieving a percentage reduction in greenhouse gases (a common target of many policy objectives, including commitments under the United Nations Framework Convention on Climate Change (UNFCC)) would require a slightly different calculation approach from the objective of achieving an absolute reduction of greenhouse gas emissions. Life cycle assessment can be used to address both these questions, but the rank ordering of the preferred systems will be different for the subtly different questions and associated methods [9].

It is, therefore, extremely important to consider the research objective and appropriate metric when using LCA. The system boundary should then be chosen consistent with this objective. For example, if considering the potential impact of bioenergy expansion on a country's GHG emissions, it would be appropriate to take into account direct land-use change and indirect land-use change or some other means of evaluating the impact of any displaced land function. However, if considering the benefit for a company of switching to self-supply of bioenergy using their own previous recreational land, then direct land-use change is relevant but indirect is unlikely to be important. It should be noted that it is also important to consider whether attributional or consequential LCA is most appropriate for the research question being studied, as already discussed in Chapter 3.

4.4 SOME KEY SYSTEM PARAMETERS

Adequately addressing the research challenge outlined at the end of Section 4.3 is extremely time-consuming. The framework and method require appropriate consideration and a vast number of data inputs are required. Many LCA practitioners resort to specialist software linked to databases which have a huge advantage in providing ready access to very detailed information on the environmental impacts of different systems. For example, it is reasonable to expect bioenergy researchers to have knowledge in and take an interest in the yields and conversion efficiencies for the plants systems they are evaluating and to check that these are consistent and relevant. However, it is unlikely that bioenergy researchers will be well-informed about the heavy metal releases to soil associated with slow erosion of the tyres on trucks as they make land journeys transporting biomass. On the one extreme, each step of a bioenergy system can be simply viewed as a 'black box', which takes specified inputs and transforms them to specified outputs and greenhouse gas balances of systems can be accurately computed drawing upon a comprehensive set of such information. However, there is a risk that if the LCA practitioner does not fully understand the process step, they may make inappropriate assumptions or incorrectly transpose data, e.g. not realising how conversion efficiency might vary with scale or process conditions could result in incorrect conclusions being drawn.

In such situations, specialist software and databases are invaluable in allowing a comprehensive and consistent approach. However, there is always a danger in using data sets with which the user has limited familiarity. It, therefore, behoves researchers to ensure that they consider which aspects of a bioenergy system are likely to have a significant impact

TABLE 4.1 Summary of Key Issues Most Likely to Influence the GHG Balance of Different Feedstock Types (+ = Relevant Factor, ++ = Can Be a Key Determining Factor, – = Usually Not a Dominant Factor)

	Embodied Emissions Associated With Agrochemical Inputs	Land Emissions	Role of Co-products	Carbon Stocks	Land-Use Change Emissions	Indirect Land-Use Change Emissions	Accessible Yield of Crop
Annual crops	++	++	++	–	+	+	++
Perennial crops	–	–	–	+	++	++	–
Forestry systems	–	–	+	++	–	–	–
Waste and residue systems	–	++	++	++	–	–	–
Algal systems	++		+				++

on end-results and to ensure they are using appropriate, up-to-date figures for those inputs, which should also be considered for sensitivity analysis once results are computed.

With an area as diverse as bioenergy, it is therefore useful to start with a high level indication of which factors are likely to be significant for different types of bioenergy system. Table 4.1 (taken from Adams et al. [10]) shows the factors that are likely to dominate the greenhouse gas balance calculations for different types of bioenergy system, disaggregated by feedstock. This can be used as a guide to focus attention on ensuring appropriate figures for the most critical factors in an LCA assessment.

4.5 USING THIS BOOK TO SUPPORT LIFE CYCLE ASSESSMENT CALCULATIONS

Once the research question has been appropriately constructed, the goal of the study defined, and the scope of the system established, the next step is to search out appropriate data sets and develop an appropriate life cycle inventory. This involves identifying the energy, water, and material flows associated with every stage of the bioenergy system and combining these to obtain a composite overview of the total resource, material, energy, and products of the whole system. A full life cycle inventory may track hundreds of different substances along many process steps. For greenhouse gas balance assessment, carbon dioxide is a key product, but the practitioner will also be interested in other greenhouse gases, such as methane, ammonia, and nitrous oxide. At this point in the process, the key activity is to quantify exactly how much is produced of all the substances that may impact the overall greenhouse gas balance.

This book brings together in one volume an overview of the key elements of bioenergy systems and experts in individual disciplines have contributed specific chapters which impart overview knowledge and pointers to sources of further information. Therefore, researchers can quickly gain the knowledge they need to build a life cycle inventory, carry out an assessment, and have access to sufficient depth to ensure that assessment is robust.

4.5.1 Biomass: Land-Based Production

Many bioenergy systems start with biomass production on land and it is important to realise that when a plant, crop, or tree is growing, there are many processes going on which can contribute to the overall greenhouse gas balance. The most obvious of these is that plants grow by photosynthesis and this involves carbon dioxide in the atmosphere being fixed in the form of the solid hydrocarbon biomass. However, there are fluxes of other greenhouse gases, e.g. nitrous oxide may be particularly dominant for intensively fertilised systems and it is important to be aware of how and the extent to which different crops and land use can contribute to the greenhouse gas balance. There are also exchanges between plant and soil, so that (depending on plant and management regime) there may be net increases or decreases in the amount of carbon stored in the soil, which should also be considered in greenhouse gas balances. This may be particularly affected by management regimes in the long term, e.g. whether or not straw or other crop residues are removed or ploughed into the land.

Chapter 5 covers the underpinning science associated with biomass growth, focusing on the bio-physical processes involved in carbon sequestration and plant growth. Chapter 6 then aims to help researchers decide which of the processes and fluxes are likely to be important for their bioenergy system and point them to appropriate resources to evaluate these more fully. Chapter 7 details the practices commonly used for harvest of biomass products. This will help researchers decide which assumptions they should be using for work rates and fuel consumption during harvesting.

4.5.2 Biomass: Other Production

In addition to purpose grown crops, there are two other main sources of biomass commonly encountered. The first is wood-derived from forestry processes. The dynamics of plant growth and soil behaviour are, in principle, the same as for other biomass plants, but the timescale over which sequestration occurs and the interval between harvests are very much longer. Therefore, careful consideration must be given to whether the time frame and method chosen for analysis are appropriate to the LCA research question being asked. A key issue is the temporal dynamics of carbon sequestration and release. Chapter 5 describes how carbon dioxide is sequestered from atmosphere and fixed in plants during the growth phase. However, for forestry systems, such as that described in Chapter 14, the sequestration takes place over a long time period. In principle, if we had a long period in which to deal with the problem of climate change, this would not be an issue—the carbon in the wood is removed from the atmosphere during growth, returned after release of energy, and so is effectively being 'recycled' between the ecosphere and atmosphere. This does not contribute to a net long-term increase in the atmospheric carbon dioxide concentration.

In addition, it is important to consider the impact of biomass removal on the long-term health (and yield) of the forestry system, e.g. how introduction of certain management techniques (such as changes in thinning regime) may affect biomass availability and long-term plantation yield or soil carbon content. These are areas where the current scientific evidence indicates significant variability and a number of methodological approaches are being contemplated to account for 'foregone sequestration' as well as absolute 'biomass removal'. A summary of key issues is given in Roeder and Thornley [11], but the practitioner needs to be aware that LCA results obtained are extremely sensitive to the assumptions and framing of the LCA question [12]. At present, there is not a clear consensus on how best to deal methodologically with these issues and so extreme care needs to be taken to ensure that the scope of system and time frame for assessment are consistent with the LCA objectives.

Another issue that is particularly pertinent for forestry systems is the fact that the proportion of the biomass which is used for bioenergy is often a relatively small proportion of the overall harvested or grown material. There are a number of approaches to the methodological treatment of this and they can give rise to very different results. A common response is to allocate the burdens across the combination of products, but this can still give very different results, depending on whether allocation is by a mass, price, or energy basis. A similar issue arises when we deal with material that was not necessarily produced for the purpose of bioenergy production. A good example is waste. The biogenic fraction of municipal waste can provide a significant biomass resource, but the assessment of its greenhouse gas impact is very dependent on the ideological framing of how we view waste and its treatment. This is discussed in Chapter 12, which also considers the significance of the counterfactual (how we would otherwise have treated the biomass or waste resource if not using it for energy) in the overall results obtained. It is sometimes argued that counterfactuals should have no place in life cycle assessment, since they are subjective, uncertain, and do not add information on the environmental profile of the particular product. This is logical when considering the original purpose of LCA as a technique to evaluate retrospectively the actual evidenced impact of a system. However, when using LCA prospectively to make assessments of possible future systems, the results can be affected significantly by counterfactual and other assumptions for which there may not be a robust evidence base.

4.5.3 Biomass Processing and Transport

Once harvested, biomass must be processed and transported to the point of conversion, with storage to provide appropriate buffer supply enroute. There is some discussion of processing and transportation, with indications of the most appropriate factors to use in life cycle assessments in Chapter 7. Of course, technology constantly progresses and researchers should always try to avail of the most recent figures or databases, but this gives a good indication of whether particular transport modes or logistic options are likely to make a big difference to the greenhouse gas balance for the particular bioenergy system being studied.

In addition to this, several of the case studies give good insights into the impact of different transport and processing options, e.g. the impact of sea transport is included for biodiesel from soy in Chapter 10 and, in Chapter 14, the example of North American forest residues allows exploration of the significance of potential storage impacts. Assumptions about the loss and degradation of material during storage are critical to comparison of different logistic options, e.g. whether it makes sense to torrefy or pelletise wood in advance of conversion.

4.5.4 Biomass Conversion

The biomass conversion facility is one of the key components of any bioenergy system. From a greenhouse gas balance perspective, it is critical to use an appropriate conversion efficiency. The various conversion technologies are discussed in Chapter 8 with indicative efficiences. It is also important to have a good understanding of the quality of the energy and any other products produced by the conversion process. The example chapters have been chosen to give insight into many different conversion technologies from combustion for heat (Chapter 11) to production of liquid fuel (Chapters 10 and 13) and electricity (Chapter 14).

4.5.5 Bioenergy Use

The use to which bioenergy is put can have a significant impact on the calculations associated with the greenhouse gas balance and this is explored by including a number of different end-uses from electricity (Chapter 14) to gas (Chapter 9) to heat (Chapter 11) and the contrast between first generation transport fuels (Chapter 10) and second generation (Chapter 13).

4.5.6 Other issues

One of the valuable things about bringing together so many LCA assessments in one volume is it allows identification of key cross-cutting issues and so a number of chapters near the end of this book have been devoted to examination of the agricultural, engineering, environmental, and policy lessons gleaned from the assembled body of knowledge. These are discussed in a more open way in Chapters 15–18, considering the agricultural, engineering, environmental, and policy lessons that can be extracted from some of the bioenergy LCA analyses completed. In some cases, this provides 'rules of thumb', e.g. that certain modes of transport or logistic processing generally make sense for certain feedstocks or that emissions associated with certain activities tend to be dominant. However, they also raise a number of questions where the confidence in current data may be limited, e.g. for GHG fluxes associated with some soil types/crops or where methods are still evolving, e.g. treatment of temporal aspects of GHG exchanges.

Finally, Chapter 19 recaps the key issues and learning points for practitioners in greenhouse gas assessment of bioenergy systems and lessons for implementation of sustainable bioenergy systems that maximise greenhouse gas reductions. However, there are inevitably some areas where questions or gaps remain and these are highlighted so that researchers can be aware of them and, first of all, check for the latest data sources, but secondly can think about the extent to which these parameters are likely to vary and how big an impact that variation may have on their system/results.

The accuracy of LCAs is completely dependent upon the accuracy and appropriateness of the underpinning data inputs. This is one of the reasons why it is always good practice to carry out a sensitivity test when reporting LCA results. This involves varying one or more of the parameters used in the work to assess the impact on the overall results. Many scientists will do this by examining first the sensitivity to the variation of individual parameters. Some will then combine these to examine the worst possible or best possible impact. These combinations are often not very useful as they often fail to reflect real world variations. For example, it would be absurd to examine a massive reduction in biomass plant yield alongside a massive increase in fertiliser application.

However, we must then recognise that in a real system it is very unusual for variation of a single parameter to not affect the other parameters. It is the job of the LCA practitioner to assemble a realistic set of data that is internally consistent and use that to assess the GHG balance of the bioenergy system. Sensible sensitivity studies may then be developed that vary more than one parameter, but in an internally consistent way, e.g. to replicate an alternative management technique or variation in productivity at different sites.

The most important thing is to realise that LCA results will only inform sustainable development and system improvements if they reflect realistic bioenergy systems. While it may not be possible to obtain verifiable evidence for every step in the chain, it behoves scientists to attempt to procure the best evidence available to inform their assessments.

4.6 SUMMARY

KEY POINTS

- Arguably the most important part of a bioenergy life cycle assessment is definition of the goal and scope of the assessment, with all methodological and data decisions stemming from this.
- It is important to use recent and relevant conversion data and the chapters and references in this book should provide good first indications of these.

References

[1] BSI. Environmental management life cycle assessment principles and framework. BS EN ISO 14040:2006. British Standard. Brussels; 2006. p. 32.

[2] Commission E. Directive of the European Parliament and of the Council amending Directive 98/70/EC relating to the quality of petrol and diesel fuels and amending Directive 2009/28/EC on the promotion of the use of energy from renewable sources 2012.

[3] Spath P, Mann M, Kerr D. Life-cycle assessment of coal-fired power production. Golden, CO: NREL; 1999.

[4] Directive 2009/30/EC—amendment to Directive 98/70/EC on environmental quality standards for fuel (Fuel Quality Directive). European Commission.

[5] Decision No 529/2013/EU of the European Parliament and of the Council of 21 May 2013 on accounting rules on greenhouse gas emissions and removals resulting from activities relating to land use, land-use change and forestry and on information concerning actions relating to those activities. European Parliament; 2013.

[6] OFGEM DoEaCCa, editors. Sustainability standards for electricity generation from biomass. London: HMSO; 2013.

[7] Röder M, Whittaker C, Thornley P. How certain are greenhouse gas reductions from bioenergy? Life cycle assessment and uncertainty analysis of wood pellet-to-electricity supply chains from forest residues. Biomass Bioenergy 2015;79:50–63.

[8] Upham P, Thornley P, Tomei J, Boucher P. Substitutable biodiesel feedstocks for the UK: a review of sustainability issues with reference to the UK RTFO. J Clean Prod 2009;17(Suppl. 1):S37–45.

[9] Thornley P, Gilbert P, Shackley S, Hammond J. Maximizing the greenhouse gas reductions from biomass: the role of life cycle assessment. Biomass Bioenergy 2015;81:35–43.

[10] Adams PWR, Bows A, Gilbert P, Hammond J, Howard D, Lee R, et al. Understanding greenhouse gas balances of bioenergy systems. Supergen Bioenergy Hub Report; 2013.

[11] Röder M, Thornley P. Bioenergy as a climate mitigation option within a 2°C target—uncertainties and temporal challenges of bioenergy systems. Energy Sustain Soc 2016;6.

[12] Holtsmark B. The outcome is in the assumptions: analyzing the effects on atmospheric CO_2 levels of increased use of bioenergy from forest biomass. Glob Change Biol Bioenergy 2012;5(4):467–73.

CHAPTER

5

Biogenic Carbon—Capture and Sequestration

Zoe M. Harris, Suzanne Milner†, Gail Taylor†*

*Imperial College London, London, United Kingdom †University of Southampton, Southampton, United Kingdom

5.1 BIOGENIC CARBON CAPTURE AND POTENTIAL FOR STORAGE AND SEQUESTRATION

Plants require three key inputs to grow: water, carbon dioxide, and sunlight. Through the process of photosynthesis, plants are able to convert solar radiation into chemical energy used for the growth of new tissue, a process described by the equation for photosynthesis:

$$\text{Carbon dioxide} + \text{water} \xrightarrow{\text{Sunlight+chlorophyll}} \text{glucose} + \text{oxygen}$$

The products of photosynthesis are carbohydrates and oxygen. Oxygen is released into the atmosphere and glucose is used to manufacture new cells within the plant, enabling growth and biomass production. There are two different metabolic photosynthetic pathways that lead to biomass production: C3 and C4. C3 plants tend to originate from temperate regions and fix CO_2 directly using Ribulose- 1,5-bisphosphate carboxylase oxygenase (RuBisCO) to produce two 3-carbon molecules, hence 'C3 plants' (Fig. 5.1). Examples of C3 bioenergy crops include willow, poplar, and *Arundo*. C3 plants use the Calvin cycle, but are relatively inefficient since RuBisCO can also react with O_2 in a process called photorespiration. Carbon-concentrating mechanisms have evolved in plants to make them more efficient and C4 photosynthesis is one such carbon-concentrating mechanism. In C4 plants, CO_2 is initially fixed by an enzyme with a much higher affinity for CO_2 (phosphoenolpyruvate (PEP) carboxylase) with no oxygenase activity. RuBisCO is spatially isolated in specialist cells and CO_2 is re-fixed in these bundle sheath cells (Fig. 5.1). The photosynthetic process in C4 plants is more efficient than in C3 plants, and several food crops are C4 such as maize and sorghum. They can cope with hot and dry environments since they have high water use efficiency (WUE). C4 bioenergy crops include *Miscanthus*, sugarcane, and switchgrass.

Greenhouse Gas Balances of Bioenergy Systems
https://doi.org/10.1016/B978-0-08-101036-5.00005-7

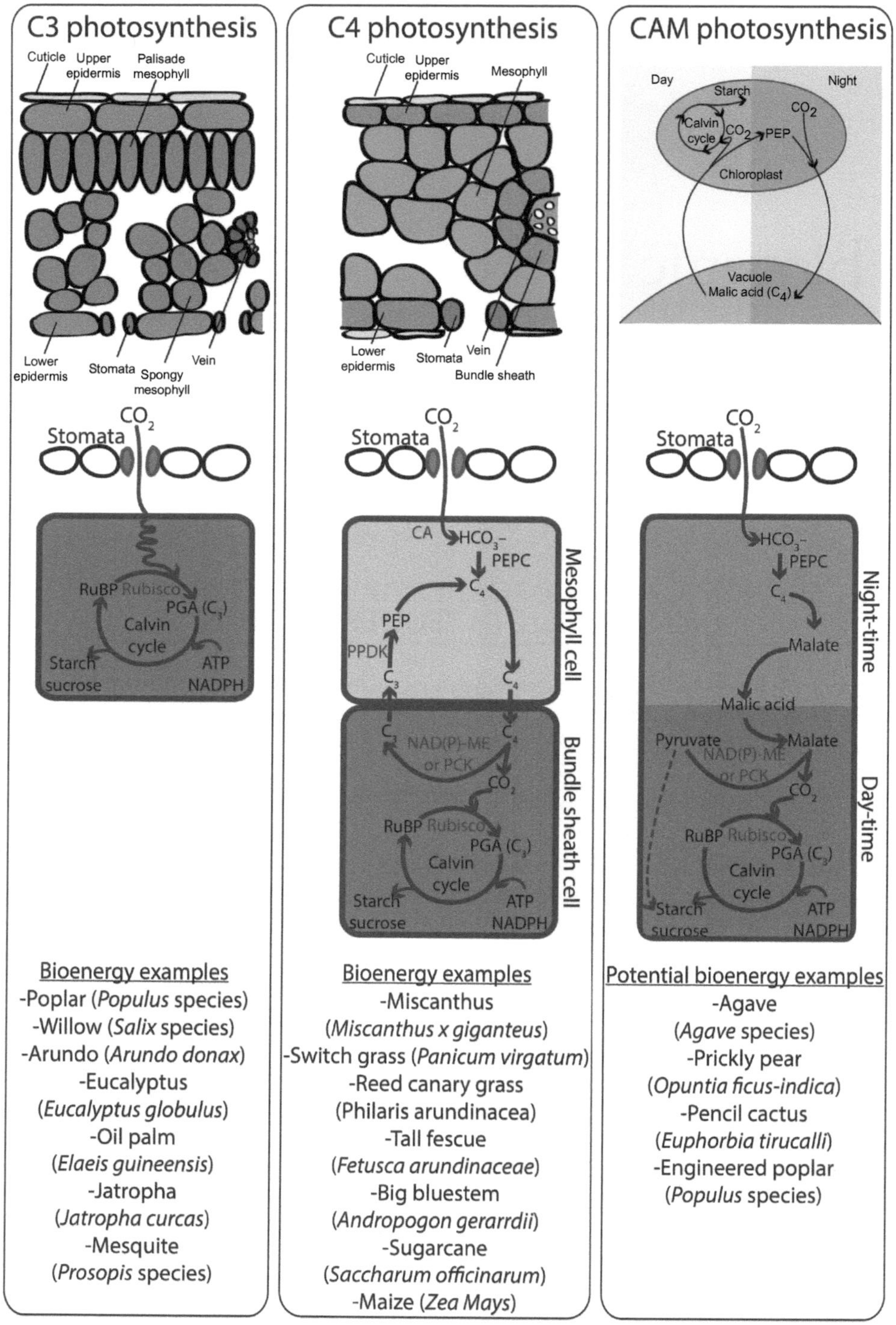

FIG. 5.1 Overview of the differences between C3, C4, and CAM photosynthesis. *Modified from Yamori W, Hikosaka K, Way DA. Temperature response of photosynthesis in C3, C4, and CAM plants: temperature acclimation and temperature adaptation. Photosynthesis Res 2014;119:101–17, and Purves WK, Orians GH, Sadava D, Heller HC. Life: the science of biology. 7th ed. Macmillan; 2004.*

A third type of photosynthesis, Crassulacean acid metabolism (CAM) photosynthesis, is much less common and is mostly only found in succulents. In CAM plants, CO_2 is fixed at night and concentrated, thus enabling CO_2 to be re-fixed by RuBisCO during the day (Fig. 5.1). CAM plants store carbon in the vacuole in the form of malic acid during the night, to be used as input into the Calvin cycle during the day. As these plants can photosynthesise with their stomata closed, they have an enhanced WUE and photosynthetic capacity. Because of this, there is currently research underway to engineer a poplar that utilises CAM photosynthesis [1]. There is also research into the potential use of the CAM plants—prickly pear (*Opuntia ficus-indica*) and the pencil cactus (*Euphorbia tirucalli*) as bioenergy sources in the form of biogas [2]. Finally, due to a high biomass production and sugar content, agave (*Agave tequilana*) has been found to produce a high yield of bioethanol [3], though at present it is not widely utilised or researched as a bioenergy crop.

Fig. 5.1 illustrates the wide diversity of bioenergy crop types that utilise differing photosynthetic machinery to fix CO_2. Using this process, plants remove carbon from the atmosphere (in the form of carbon dioxide) and fix this carbon within carbohydrates within their tissues. These carbohydrates are subsequently used for plant growth. This use of CO_2 is considered a form of carbon sequestration—biogenic carbon sequestration. Carbon sequestration can be defined as the removal of carbon from the atmosphere and storage in a long-term store or pool. Carbon sequestration can be natural where the natural processes of the carbon cycle are utilised, such as biological fixation described above, or artificial sequestration where carbon is compressed and stored, known as carbon capture and storage (CCS; discussed in more detail in Section 5.2).

In plants, once fixed, carbon is partitioned to different plant organs and the amount partitioned between different plant parts is highly dependent on the functional type of the plant. It is generally accepted that woody plants are comprised of ~50% carbon, but this varies between species with softwood species generally having a higher carbon content. During growth, biomass accumulates slightly differently depending on the type of bioenergy crop, nutrient availability, and importantly, the time of the year. Biomass partitioning, the process where plants allocate resources such as carbohydrates to regions of the plant, is integral for the growth of a crop. A comparison of total carbohydrates within energy crops (Table 5.1) found the tree species (willow and poplar), *Miscanthus* and sugarcane bagasse, contained the highest amount of cellulose. In yellow poplar (*Liriodendron tulipifera* L.), carbon was allocated evenly between leaves ($459\,mg\,C\,g^{-1}$), stem ($456\,mg\,C\,g^{-1}$), branches ($460\,mg\,C\,g^{-1}$), and roots ($446\,mg\,C\,g^{-1}$) and this relationship was maintained over time [5]. A study by Mckinley et al. [6] on biomass partitioning in Sorghum grown as a bioenergy crop found biomass, primarily leaves, leaf sheaths, and roots, accumulated predominantly until floral initiation, then stems dominated until anthesis (the opening of the flower buds), then panicles (branching cluster of flowers) until grain maturation. The biomass partitioning is also dependent on whether a crop is annual or perennial as in the winter resources are relocated in perennial plants during dormancy ready for the following season. After senescence and/or when the whole plant dies, the carbon within the plant is sequestered into the soil.

Whether bioenergy crops, forests, or arable crops, plants have the potential to sequester large amounts of carbon from the atmosphere and into the soil. Green plants are thus able to contribute to reduced CO_2 emissions, particularly when combined with CO_2 displacement through bioenergy crop utilisation.

TABLE 5.1 Carbohydrate Content Attributes of Example Energy Crops [4]

Species	Hemi-Cellulose	Cellulose	Starch	Sucrose	Total Carbohydrates
Wheat grain	3	2	70	3	82
Wheat straw	24.6	33.2			
Maize grain			74		84
Maize stover	28	35			
Sugarcane	8	24		47	
Bagasse		48			
Switchgrass	36	31.6			
Miscanthus	15.9	57.6			
Willow	14	55.9			
Poplar	23	40			

5.2 GLOBAL CARBON CYCLE AND BIOGENIC CARBON IN BIOENERGY SYSTEMS

The global carbon cycle describes the carbon pools and fluxes at a global scale (Fig. 5.2). There are six main pools holding varying amounts of carbon: the earth's crust (c. 100,000,000 PgC), the oceans (c. 38,000 PgC), fossil fuels (c. 4000 PgC), soil (c. 1500 PgC), the atmosphere (c. 750 PgC), and biomass (c. 560 PgC). Carbon moves from one pool to another through a

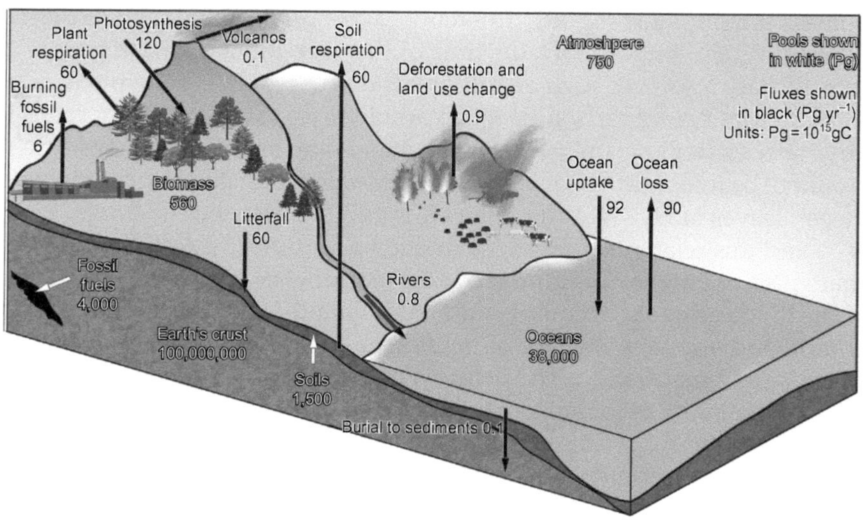

FIG. 5.2 The global carbon cycle showing where carbon can be stored in pools or where it is released as fluxes. *Modified from Sallade S, Ollinger S, Albrechtova J, et al. University of New Hampshire GLOBE Carbon Cycle Project; 2012. Available from: www.globecarboncycle.unh.edu.*

variety of mechanisms, for example, the burning of fossil fuels and release of greenhouse gas emissions (GHGs) into the atmosphere. The loss of carbon from the soil through degradation and the increase in carbon in the atmosphere (leading to warming) and the oceans (leading to acidification) are some of the key processes causing an imbalance in the global carbon cycle in the 21st century. As described above, carbon is fixed during photosynthesis into vegetation and is stored in the soil through conversion of fallen plant material into soil carbon.

Within the carbon cycle, we can define the movement of biogenic carbon as carbon which has arisen from biological sources such as that which is released from litterfall degradation or soil respiration. We can also define fossil carbon, which is carbon risen from fossil sources, which has been fixed over millions of years to the form it is in now, such as within coal and oil [7]. Biogenic carbon has a much shorter timescale for carbon cycling than fossil carbon and therefore can be utilised to aid carbon sequestration. The basic principle in this case being the conversion of land with a low carbon sequestration potential, both soil and biomass, to one with a higher carbon sequestration potential. By using bioenergy in the global carbon cycle, there is the potential to increase the carbon stored in biomass, potentially reduce the losses of carbon from land-use change (LUC) to the atmosphere, and potentially reduce the emissions from fossil fuel burning (Fig. 5.2). In recent meta-analyses, it was shown that a conversion from arable cropping to dedicated second-generation crops can increase soil carbon and reduce GHG emissions [8,9]. While this LUC may not be advocated because of the conflict with food production, it demonstrates that some LUCs are more favourable than others for biogenic carbon sequestration. One key aspect of LUC to bioenergy is the 'carbon debt' which is incurred from initial conversion and planting of the crop. This is most often reported as the number of years required for the land conversion to be able to 'pay back' the carbon to the land, lost through the LUC. For example, in one study [10], it was shown that a conversion from US grassland to corn for bioethanol would incur a carbon debt of 93 years, and from abandoned cropland to corn, a 48-year carbon debt. This is an example where a conversion from a native ecosystem leads to more negative environmental impacts than a conversion from an already anthropogenically altered ecosystem. Another more worrying estimate was one of a conversion to corn, again in the United States, presenting a 167-year payback time when indirect effects on land-use were also considered [11]. In contrast, more recently, Mello et al. [12] showed the payback time for soil C was only 8 years for a conversion from native vegetation to sugarcane ethanol, though these estimates do not consider the GHG and ecosystem C changes. Failure of studies to take into account the effects of LUCs (both direct and indirect) will give an inaccurate picture of the effects of a conversion to bioenergy crops and needs to be incorporated into all studies considering land conversions [10,11]. While these estimates are high, they are all for conversions to first-generation feedstocks with emerging research showing the payback time for second-generation crops to be shorter. For example, Mccalmont et al. [13] estimated a payback time of 8 years for a conversion from grassland to *Miscanthus*, while in another study for arable to SRC poplar, payback time was 4 years, similar to other studies for this type of conversion [14]. However, both payback times for first- and second-generation feedstocks are comparably small in terms of biogenic carbon cycling compared to carbon cycle of millions of years for fossil carbon losses through combustion. This is a key area for understanding and, as illustrated by example above, remains controversial and relatively understudied. This is important and modelling approaches are required to address these uncertainties and are emerging such as the Beac model (https://www.gov.uk/government/publications/life-cycle-impacts-of-biomass-electricity-in-2020) and ELUM metamodel (http://www.ceh.ac.uk/services/elum-model).

5.2.1 Negative Emissions Technologies

CCS involves the capture of CO_2 from industrial processes, such as energy generation or energy-intensive processes, and subsequent transport of this CO_2 into a stable long-term store. CO_2 may potentially be stored in stable geological formations where it can remain for many thousands of years [15]. For the United Kingdom, over 500 potential sites have been identified for stable off-shore storage of CO_2 [16]. An extreme example of biogenic and artificial carbon sequestration is that of bioenergy with carbon capture and storage (BECCS). Significant analysis [17,18], using IPCC Integrated Assessment Models (IAMs), shows that the probability of limiting global warming below 2°C is strongly dependent on the implementation of negative emissions technologies (NETs), defined as a technology that provides net removal of CO_2 from the atmosphere. For 101 of the 116 IPCC scenarios that hold temperature rise to below 2°C, NETs are required [17]. Although several NETs are proposed, BECCS features in most of the current assessments to limit GHG emissions and temperature rise. Of the scenarios that IPCC consider, most that hold CO_2 concentration to 420–480 ppm for a 2°C halt require NETs and half of those have BECCS exceeding 5% of primary energy supply [18]. BECCS has significant potential to deliver:

- A large net negative emission of ~3 GtC equiv. year^{-1}
- A CO_2 limit of 430–480 ppm
- <2°C temperature increase
- Provide primary energy supply of 100 EJ
- Could cost $130 billion by 2050 [18]

This potential is only matched by direct air capture (DAC), but with the considerable advantage that bioenergy and CCS technologies are already available and being deployed, unlike those for DAC. This is not an extreme BECCS scenario since other reports suggest that BECCS could provide up to 10 GtC equiv. year^{-1}, but this relies on dedicated biomass gasification or combustion routes [19]. CCS technologies work as pre-combustion, post-combustion, and oxy-combustion technologies and others in development are beginning to be utilised at varying scale, particularly for the first two technologies, but they have rarely been deployed alongside bioenergy and to date have been focussed around capture associated with fossil fuels [20].

5.2.2 Enhanced Supply of Sustainable Biomass, Resilient to Climate Change

Despite its potential, significant unresolved challenges that require step-change research are apparent to enable deployment of BECCS at appropriate scale to deliver a 3 GtC equiv. year^{-1} out to 2100. Both the CCS [21] and bioenergy components of BECCS require new insights, but significant challenges concerning the supply of sustainable biomass include:

- Land availability—finding land for adequate bioenergy crop deployment
- Yield limitations—addressing the gap between potential and actual yield
- Water use and climate resilience of bioenergy crops—water-energy-food nexus
- Sustainability with respect to ecosystem services, biodiversity, and natural capital

A second NET for carbon sequestration, which relies on capture of biogenic carbon, is the use of biochar. Biochar is black carbon produced from biomass or organic waste products that are subsequently applied to soil. Biochar is usually produced through pyrolysis of biomass, though hydro-thermal treatment of biomass and gasification are also becoming more common routes to biochar production. Biochar can be a by-product of extracting energy from biomass and returning biochar to the soil is a form of carbon sequestration using a stable form of carbon. Historically, biochar was made by burning biomass, producing charcoal. Examples of this behaviour can be found in anthrosols (man-made soil) throughout the Amazonian basin where biochar produced approximately 800 years ago are found to be visible, demonstrating the stability of biochar [22].

Biochar has the added benefit of improving soil fertility and subsequent plant growth [23], improved physical properties such as water retention [24], and altered pH, though the long-term effects of biochar into natural systems remain untested. Small amounts of labile components, partial oxidation, and biotic and abiotic surface reactions render it difficult to establish the half-life of biochar produced using modern pyrolysis techniques [25]. In terra preta soil resulting from historic burning of biomass, the nutrient content has been found to increase, particularly with respect to magnesium, manganese, and copper [22]. The physical and chemical properties of the biochar depend greatly on the feedstock and process of pyrolysis. Generally, the higher the pyrolysis temperature, the higher the proportion of aromatic carbon [26]. In terms of the feedstock, the physical properties of the biochar are determined by the surface area and porous structure of the feedstock [25]. Feedstocks that are often used include crop waste, energy crops, wood chip, municipal waste, and manure [25]. Indirect effects of using biochar as waste management include decreased methane emissions from landfill, reduction of industrial energy use and emissions due to recycling and waste reduction, and decreasing energy used in long-distance transport of waste.

The effect of biochar on the carbon flux from soil back into the atmosphere is mainly the result of microbial respiration. The biochar itself is decomposed very slowly, meaning the biochar allows carbon input into the soil more than carbon output through microbial respiration. Despite microbial respiration effects, biochar results in a net withdrawal of CO_2 from the atmosphere through photosynthetically storing carbon in the tissues of plants and the manual application of the biochar into soil and therefore holds potential as a NET.

5.3 BIOMASS CARBON AND THE GREENHOUSE GAS BALANCES OF BIOENERGY SYSTEMS

Fixed and stored biogenic carbon in the form of biomass interacts with the other carbon pools in the carbon cycle in several ways. Firstly, carbon is 'lost' from the plant system back to the atmosphere in the process of autotrophic respiration. It has been estimated that globally, the amount of carbon fixed during photosynthesis is approximately double of that which is released to the atmosphere during plant respiration (120 Pg C vs 60 Pg C, respectively, [27]). For the bioenergy crops being considered here, respiration varies between 3–8 mg CO_2 release g^{-1} dry matter h^{-1} (herbaceous crop plants) and 1–4 mg CO_2 release g^{-1} dry matter h^{-1} (deciduous trees) [28]. Biomass interfaces with the soil carbon pool predominantly through litterfall, but also through partitioning of biomass to roots and the turnover and decay of this root biomass. Significant plant carbon is found in root systems, but it is estimated that over 50% of net primary productivity (NPP) is returned to the soil through decomposition of leaf material [29]. The ability of SRC willow to effectively recycle nutrients through leaf biomass is documented within the literature [30] and is one of the reasons why these plantations do not require nutrient supplementation in the form of fertiliser. Hangs et al. [31] and Harris et al. [32] have both found that approximately 80% of SRC willow litterfall is decomposed within the first year. Hangs et al. [31] found that willow genotype and climatic variables affect the decomposability of the litterfall. SRC willow has been shown to have high quality litter (i.e. a low C:N ratio) as well as high abundance of soil fauna and microbial biomass, which may explain their rapid decomposition as observed in this experiment [33]. Litter decomposition has been shown to be very important for soil carbon sequestration and recycling of nutrients back into the soil; some studies finding that litterfall is responsible for up to 70% of annual nitrogen input [34]. In a comparison between woody bioenergy crops (willow, poplar, and black locust) and herbaceous energy crops (giant reed, *Miscanthus*, and switchgrass), it was found that there was higher soil organic carbon (SOC) accumulation in the top 10cm under woody crops which has been attributed to litterfall input [35]. While often not measured in experiments, litterfall is therefore important in carbon sequestration.

Both litter and root biomass are decomposed by a plethora of soil microbes. These communities appear to differ depending on the land-use. For example, Jenkins [36] found the species' richness of both bacteria and fungi to be higher in grassland than SRC willow. Jenkins [36] also found an increase in the fungi *Basidiomycota* (20%) of an unidentified *Agaricales* family which are usually related to tree root symbiosis and saprotrophic degradation of wood. This suggests that the shift in microbial community has occurred to accommodate the change in organic inputs that the soil will be receiving in the form of woody material. With other studies also having found increased fungal biomass in SRC soil relative to bacterial microbial biomass [37,38]. This effect may be a result of the compounds contained within SRC willow leaf litter that promote fungal growth [38]. LUC to bioenergy will require a shift in the makeup of these microbial communities to allow them to most effectively decompose root and leaf matter and requires further investigation with advances in sampling techniques.

Soil contains elemental carbon and carbonate materials (inorganic carbon) such as calcite, dolomite, and gypsum, which comprise approximately 40% of the carbon in soil [39]; however, for carbon sequestration, we are interested in the organic carbon. As organic carbon enters the soil through respiration or the breakdown of organisms, the topsoil generally has the highest concentrations. The global carbon pool is comprised of five main pools; root respiration, rhizo-microbial respirations, microbial respiration of dead plant residues, additional SOM-derived CO_2 (priming effect), and SOM-derived CO_2 (basal respiration) (Fig. 5.3). Of these, however, only SOM-derived CO_2 significantly contributes to changes in atmospheric CO_2 concentration [40]. The respiration contributions towards soil carbon dioxide flux (R_S) can be split into respiration by roots (autotrophic respiration R_A) and soil organisms (heterotrophic respiration R_H), which fix the carbon in dead organisms. Therefore, R_A utilises photosynthates, whereas R_H utilises the mineralisation of soil organic matter (SOM).

5.3.1 Impact of Management

The organic carbon content stored in the soil depends on soil type, climate, and land management. The management of the land, however, can arguably have the largest effect on SOC through tilling (or turnover rate), soil management, and plant productivity (Fig. 5.3). It is

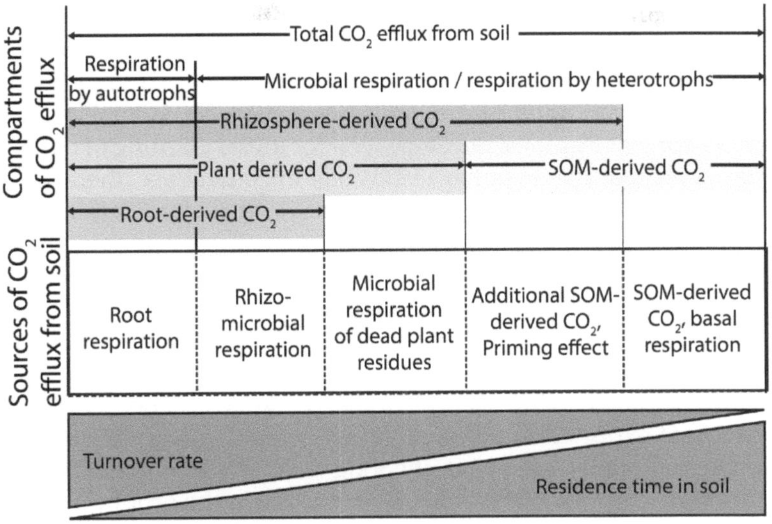

FIG. 5.3 Sources of biogenic CO_2 efflux from soil in order of turnover rates and mean residence times in the soil. *Modified from Kuzyakov Y. Sources of CO_2 efflux from soil and review of partitioning methods. Soil Biol Biochem 2006;38:425–48.*

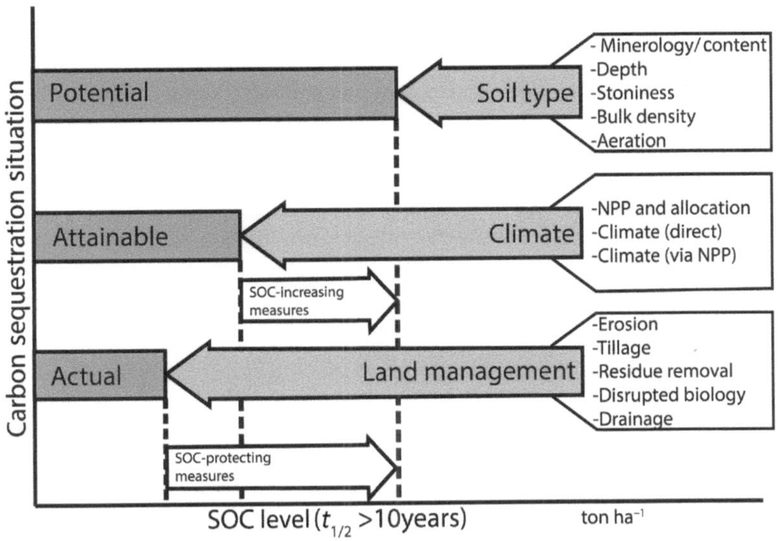

FIG. 5.4 Defining limits of carbon sequestration. *From Ingram J, Fernandes E. Managing carbon sequestration in soils: concepts and terminology. Agric Ecosyst Environ 2001;87:111–17.*

common practice to rotivate land between harvest and planting of the next crop and this enables oxidation of the SOC, causing a release of CO_2 back into the atmosphere. It is for this reason that perennial crops that involve reduced soil disturbance are often associated with higher SOC [41]. This land management usually limits the actual carbon sequestration (Fig. 5.4). In general, SOC increases as clay content of the soil increases, but traits such as bulk density are negatively related to SOC, which define the limits of the potential carbon sequestration. In terms of climatic effects, this is largely due to rainfall transferring dissolved organic carbon (DOC) into waterways, defining the attainable limit to carbon sequestration.

Management practices are important when assessing the sustainability of LUC to bioenergy. Currently, 'best practice guides' are employed for initial planting of bioenergy crops, developed for optimal yield and ease of harvest. While in general use, there is no standardised crop management practice, and thus, yield and profit often dominate agronomic practice. 'Management swing potential' describes how management can determine the ways in which bioenergy cultivation can provide negative or positive environmental impacts [42]. Several strategies can be utilised to minimise disturbance that may lead to atmospheric emissions. These include, but are not limited to, reduced or *no-tillage, retention of residues on the soil, decreased/optimised fertiliser application*, and *reduced/optimised irrigation*. The advantages and disadvantages of these practices are summarised in Table 5.2 and briefly discussed below.

Tillage (Table 5.2) is expected to occur more often for 1G crops than for 2G crops given that they are planted annually. However, the long-term effects of initial tillage during land conversion to bioenergy may be offset since 2G crops are in the ground for at least a 10-year cycle, with general agreement that reduced tillage provides less disturbance enabling soil carbon sequestration to establish more quickly compared with conventional tillage [43–46]. In a global analysis, West & Post [47] found that a switch from conventional tillage to no-till can sequester $57 \pm 14\,\mathrm{g\,C\,m^{-2}\,year^{-1}}$ with the majority of the SOC change occurring in the first

TABLE 5.2 Impacts of Different Management Regimes on Land and Their Relevance to 1G and 2G Bioenergy Crops

Management Practice	Potential Negative Impacts	Potential Positive Impacts	Relevance to 1G Bioenergy	Relevance to 2G Bioenergy
Conventional tillage (relative to reduced/no tillage)	✗ Increased disturbance ✗ Increase in soil GHG emissions ✗ Soil compaction ✗ Increased soil erosion ✗ Increased run off	✓ Favourable conditions for crop planting ✓ Incorporation of residues into soil ✓ Mechanically destroys weeds ✓ Redistribution of soil carbon deeper into soil profile	High	Low
Reduced/no tillage (relative to conventional tillage)	✗ Less favourable planting conditions	✓ Increased soil C sequestration in upper profile ✓ SOM accumulation ✓ Reduced run off ✓ Reduced soil erosion ✓ Increased yield ✓ Decreased aeration ✓ Minimised oxidation of organic matter ✓ Encourages better aggregation	High	Low

TABLE 5.2 Impacts of Different Management Regimes on Land and Their Relevance to 1G and 2G Bioenergy Crops—cont'd

Management Practice	Potential Negative Impacts	Potential Positive Impacts	Relevance to 1G Bioenergy	Relevance to 2G Bioenergy
Residue retention (relative to residue removal)	✗ Loss of potential bioethanol production from residues	✓ Reduced run off ✓ Reduced soil erosion ✓ Increases SOM ✓ Reduction in disease-producing organisms ✓ Increased carbon sequestration ✓ Increased biodiversity through habitat provision ✓ Reduced evaporation ✓ Increased yields ✓ Cost savings from not needing to remove	High	
Fertiliser	✗ N_2O emissions ✗ Run off leading to eutrophication ✗ CO_2-cost associated with fertiliser manufacture	✓ Yield increases ✓ Increased residues ✓ Use of agricultural by-product (sewage sludge)	High	High
Irrigation	✗ Depletion of water supplies ✗ Increased run off ✗ Waterlogging ✗ Stalinisation	✓ Increased yields ✓ Increased C sequestration into biomass and through residues	High	High

10–15 years. Despite these reports, there remain inconsistencies that may be related to limited sampling in the upper soil profile [48–50]; where sampling to lower depths found no significant difference in carbon storage and in many cases found that more C was stored beneath conventional systems [48,51]. There remains a lively debate occurring in the literature since the effect of a change from conventional tillage to reduced/no-tillage may merely redistribute the carbon in the soil profile than affect the amount of carbon stored [49,51–53].

In the context of land management, residues may be defined in agricultural systems as the parts of the crop which are not harvested and are left on the soil (not to be confused with residual wastes associated with bioenergy chains). Residues as co-products of some crops can themselves be used as a renewable source of energy; over 244 million tonnes of residues could be used to produce bioethanol every year in the United States [54]. Residue retention can have several positive effects on soil function as summarised in Table 5.2 [55–57]. Global retention of residues on croplands could sequester 1 billion $\mathrm{t\,C\,year^{-1}}$, which is equivalent to 30% of the annual increase in atmospheric CO_2 [58]; with other examples in the literature supporting this contention [49,59]. However, sustainable removal could allow for both soil quality gains and bioenergy production, with estimates across Europe showing 3.2% in final energy consumption could be achieved from sustainable residue removal [60].

The rise in atmospheric N_2O has occurred mostly as a consequence of fertiliser application and animal production [61] and is of particular concern as N_2O has a global warming potential of 298 [62], meaning it is 298 times more damaging to the environment as CO_2. Recent work on SRC and *Miscanthus* suggests that nitrogen fertiliser application may be the most significant management practice determining GHG balance in these plantations [63–65]. Fluxes will vary depending on crop and soil type, with some studies showing that 1G crops have higher N_2O fluxes than 2G [66]. Impacts of fertiliser use have been summarised in Table 5.2.

Several aspects of fertiliser management require further understanding:

1. The application rate of fertiliser must be clearly defined and application rate quantified after which no energy ratio benefit occurs. Effective fertiliser application would limit soil N_2O emissions and maximise economic yield benefits [67]. This type of precision agriculture is becoming routine in food cropping systems [68], but has yet to be effectively deployed for 2G crops, possibly using remote-sensing technologies and smart agriculture.
2. It is also important to understand the way in which these emissions arise and the timeframes over which they occur after application; for example, SRC showed enhanced N_2O emissions for up to 4 weeks post-application [66].
3. A fuller understanding is needed of the effects of different types of fertiliser. Addition of any type of fertiliser may lead to increased N_2O emissions; however, sewage sludge could replace more harmful chemical fertilisers as 40%–60% of the nitrogen contained is inorganic [69]. Additionally, production costs and associated emissions can be avoided as sewage sludge is an agricultural by-product (see [70] for more in-depth analysis).
4. It is important to understand the response of crops to fertilisers; this particularly applies with respect to crop variety [71].

While negative environmental effects are no doubt evident from fertiliser application and manufacture, it may be necessary for productivity and yield increases—the sustainable intensification of the landscape. Carefully planned fertiliser application could be used to increase the amount of SOC through increased aboveground biomass and residues coming into contact with the soil [66,72–74], while minimising damaging environmental impacts.

Approximately, 70% of all freshwater withdrawn globally is used for agricultural purposes [75]. Currently, only 2.3% of total irrigation water used in Europe is for bioenergy crops [76], though projected future climate changes are expected to result in increased need for irrigation to maintain yields [62]. Proper applied use of irrigation as a management strategy to reduce GHG could be effective as C is fixed into increased above- and belowground biomass [77]. 2G bioenergy crops have been found to have a higher WUE than their 1G counterparts, but an overall larger water demand [78,79]. Modelling has shown that WUE in poplar may be twice that of willow [80], suggesting that there may be room for improvement in SRC genotypes if this high WUE trait can be captured in future breeding programmes. Water availability and water use are extremely important factors when it comes to bioenergy sustainability and are further discussed in Section 5.4.

Management strategies have a large influence on the GHG balance and soil processes for bioenergy cropping systems and can determine if their cultivation will have positive or negative impacts on the landscape [42]. A full understanding is necessary of the effect of different management strategies, the varying intensities in which they can be applied, the effect of the treatment across the whole ecosystem, the effect on 1G and 2G crops, and finally the duration over which these effects will last. Appropriate and timely application of these management practices may allow for both profitable returns in yield and protection of, and ideally an increase of, soil carbon.

5.3.2 Ecosystem Carbon Accounting

On a whole ecosystem level, we can partition the movement of CO_2 into set components [81]. Starting at the plant level, plants are able to remove CO_2 from the atmosphere and fix it within their tissues in the form of new biomass, defined as *gross primary productivity* (GPP) (Fig. 5.5). This can be partitioned into the amount of carbon which has been stored in the *aboveground* portion of the plant, the stem/stalk/trunk, branches, and any leaf material, and in the *belowground biomass*, the roots of the plant. As well as fixing

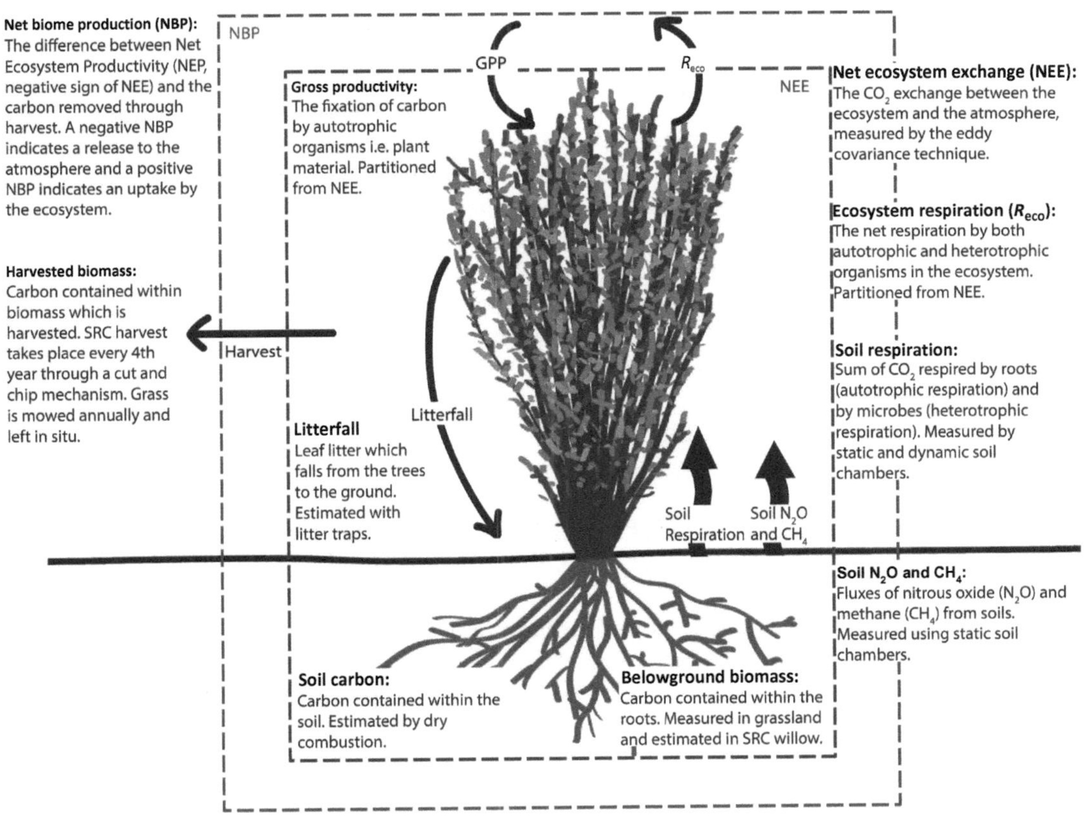

FIG. 5.5 Whole ecosystem greenhouse gas balance showing main flows of carbon through the ecosystem and examples of how they are measured in research. *Modified from Harris ZM, Alberti G, Viger M, Jenkins JR, Rowe R, Mcnamara NP, et al. Land-use change to bioenergy: grassland to short rotation coppice willow has an improved carbon balance. GCB Bioenergy 2016;9(2):469–84.*

carbon by photosynthesis, we learnt above that plants undergo respiration, releasing CO_2 into the atmosphere. Besides the respiration of living plant tissue, there is also respiration that occurs from soil micro-organisms; soil respiration (R_S), partitioned into autotropic respiration and heterotrophic respiration. *Autotrophic respiration* (R_A) is the respiration from all primary producers within the soil, namely the roots. *Heterotrophic respiration* (R_H) is the respiration of heterotrophic organisms in the soil, namely animals and micro-organism.

Plant respiration, soil respiration, and that of any other living organisms or material within the ecosystem together equate to the *ecosystem respiration* (R_{eco}). Net ecosystem exchange (NEE) is the CO_2 exchange between the ecosystem and the atmosphere which may, for example, be measured by eddy covariance techniques. Net ecosystem productivity is the sum of GPP and R_{eco} which represents the total amount of carbon in an ecosystem held within organisms (also represented as the negative portion of NEE). In agriculture, much of this carbon is removed from the ecosystem as the harvested crop, which is used elsewhere and thus net biome production (NBP) is the difference between NEP and the carbon removed via harvest. The sign of NBP is opposite to all other measures described above, where negative indicates a release to the atmosphere and a positive value indicates an uptake by the ecosystem.

The movement and partitioning of CO_2 within an ecosystem varies depending on several factors including vegetation type, climate, and management. For example, the more densely vegetated, the larger the GPP compared to more sparsely populated vegetation. Wetter climates combined with poor drainage which may result in waterlogging of the soil can create anoxic conditions and possible increases in methane emissions. Whether the land-use is native or managed will also have an impact, for example, tillage of croplands results in a loss of soil carbon through disturbance and increased soil respiration. It is important to identify and quantify which areas of the cycle are fluxing in or out of the ecosystem, as this will ultimately result in the net biome productivity. It is the balance of these variable, or imbalance, which will result in whether an ecosystem is acting as a carbon sink or a carbon source. An understanding of these flows can allow decisions to be made about facilitating management which can result in a reduction of emissions from the land-use.

5.4 APPLICATION TO DIFFERENT BIOMASS TYPES (ANNUAL CROPS, PERENNIAL CROPS, FORESTRY SYSTEMS)

Different ecosystems are able to store differing amounts of carbon dependant on the composition of the biomass and the utilisation or management of the land. Here we will discuss three land-use types and consider the differences in carbon sequestration: annual crops, perennial crops, and forestry systems.

Annual crops are able to complete their life-cycle within a year, after which time the whole plant system dies including shoots and roots. A perennial plant is one which persists for several seasons, with the aboveground soft tissues dying back, but the roots remaining alive, until the following growing season where the aboveground portion will regrow. In some cases, such as grasses, this die back will be the entire aboveground biomass. In the case of woody perennials bioenergy crops such as SRC willow and poplar, this die back is leaf loss only. Lignocellulosic grasses and woody plants survive for many years, experiencing growth and greening during the growing season, then senescence of leaves and dormancy through the winter months (there are, however, exception, such as evergreens which retain their leaves all year round). As bioenergy crops, these perennials tend to be grown from the same rootstock through harvest every 3–5 years for up to 20–30 years, at which point the regrowth efficiency reduces and the whole plant is replaced [82]. For fast growing trees in short rotation forestry (SRF), the biomass tends to be harvested every 8–20 years [83] and the rootstock remaining for approximately 100 years.

The stability of organic carbon within these ecosystems goes from the shortest to the longest term in annual and forest systems, respectively. The growth of annual plants is able to remove CO_2 from the atmosphere and fix it within both the aboveground and belowground tissues. However, after the life-cycle is complete, the entire plant dies, including the roots, resulting in release of CO_2 as a result of decomposition of these residues. This is also exaggerated by the relatively shallow depths of carbon from annual plants, combined with land management of annuals whereby turnover rate of soil is higher.

In perennial plants, where the rooting system persists belowground for several seasons, carbon is stored within larger and more extensive rooting systems compared to annual plants [84,85]. For perennial grasses such as *Miscanthus*, biomass accumulates in the summer, senesces in autumn, and over winter the crop is left to translocate nutrients belowground and reduce the moisture content of the biomass. The crop is harvested annually in early spring, and then the cycle begins again. For the woody crops, the leaves senesce in the autumn and are returned to the soil as litterfall, and nutrients are translocated into the root stock, but also carbon remains in woody biomass. In addition to the difference in wood content, the perennial grasses and SRC crops differ in their root systems with many of the perennial grasses producing rhizomes. Many bioenergy crops are propagated asexually with SRC crops commonly planted as cuttings set vertically planted in the soil with part of the cutting left above ground. Sugarcane is planted similarly with a section of the stem, sometimes called a sett, planted, but in this case the cutting is placed horizontally. *Miscanthus*, however, is planted as a rhizome which is an underground stem, or specialised root, which produce roots and shoots at regular intervals. Higher carbon accumulation rates are observed in perennial crops than annual crops across the literature [86], with the management and inputs largely influencing this difference [64,87].

Forest ecosystems are long-standing woody structures which exist for several years, with the root stock existing for approximately 100 years. Carbon is predominantly stored in the standing woody biomass of trunks and belowground in extensive root systems. Due to the length of time the rootstock is in the soil, forestry systems are associated with lower soil erosion and mechanical disturbance. Another characteristic of forestry systems is the lignin content within the plants. Lignin is a complex polymer found in the cell walls, which increases their rigidity and wood content. While providing the cellular rigidity required for increased biomass production found in woody crops, the lignin acts as a barrier against the release of sugars from the cells. This affects the extraction of bioethanol from the crops, but importantly it also affects the breakdown of the biomass and subsequent release of carbon into the soil, increasing the recalcitrance of the biomass.

Management of the landscape also plays an important role in the durability of the carbon store. Major food staple crops, such as corn and wheat, are annual crops which are cultivated on a mass scale. The constant production of these crops involves mechanical soil preparation known as tillage, planting of the crop, treatment of the crop using fertilisers or pesticides, irrigation of the crop, and finally mechanical harvest. All these processes result in disturbance of the land relative to less intensely managed ecosystems. As a result, any carbon stored in the biomass during growing season (both above ground and belowground) and any stored in the soil through litterfall/residues will be lost as a result of mechanical disturbance. Therefore, arable lands are generally referred to as carbon sources: where more carbon is released to the atmosphere than is stored in the ecosystem. Perennial crops, such as those used for bioenergy production like *Miscanthus* or SRC willow, undergo less intense management, if any at all. These crops are planted in the ground and may receive fertilisers, irrigation, or pesticides depending on the region, though often not to the scale that is used in arable cropping [42]. While these crops grow, carbon is fixed into both aboveground and belowground tissue. Once harvested, usually in winter after 2–5 years growth, the root systems of these plants remain intact and regrowth begins the following growing season [88]. Forest systems are often unmanaged or minimally managed. Managed plantations for wood products or for bioenergy are often in the ground for a minimum of 15–20 years, with forestry systems in the ground for up to 100 years [89]. During this time, carbon is stored in large aboveground and belowground biomass, as well as sequestered into the soil which will remain undisturbed for many years. In general, the less the soil is disturbed, the longer the carbon will remain in the soil because the soil carbon is relocated into deeper soil. There is still, however, a lively debate occurring since the effect of a change from conventional tillage to reduced/no-tillage may merely redistribute the carbon in the soil profile rather than affect the amount of carbon stored [8].

5.5 BRIEF DISCUSSION OF TEMPORAL ASPECTS

Carbon sequestration is a process which can happen over several timescales. The process of carbon fixation into living biomass is largely dependent on the rate of growth of the biomass and varies in the timescale of seconds to minutes. Plants which accumulate mass quickly are able to store carbon in their tissues over a shorter timeframe than those that grow less quickly, though loss to the atmosphere is often similarly over a shorter timeframe. An example where this principle is utilised is in second-generation bioenergy crops such as SRC willow or *Miscanthus*, which are favoured for their rapid accumulation of biomass. For living biomass such as trees, carbon sequestration can continue as long as growth continues; however, rate of accumulation will reach a steady state once the plant is fully grown at maturity. At maturity, fixation of carbon occurs during the growing season when greening occurs; this is balanced in the winter by senescence (the falling of leaves) and subsequent respiration. Through this seasonal variation, understanding the net movement of CO_2 is important. Fields where the fixation of carbon during the growing season is counterbalanced or higher than the respiration occurring during the plant's dormant phase are categorised as carbon sinks. This seasonal process can be observed on a global scale in the Keeling Curve. The Keeling Curve is the longest running record of global CO_2 measurements, showing an upward trend of increasing CO_2 concentration. Using this data set, we are able to observe the mass greening of temperate forests as a decrease in atmospheric CO_2, followed by a senescence and soil respiration leading to an increase in atmospheric CO_2.

Carbon sequestration into the soil carbon pool occurs over a much longer timeframe than the biomass soil pool. Though autotropic respiration will contribute towards the soil carbon pool, the heterotrophic respiration and breakdown of organic matter is the main contributor. The incorporation of carbon from the breakdown of organic matter can vary dramatically depending on their recalcitrance. For softer tissues this is months to years, but for hardwood material and humus this is the scale of centuries. On an even longer timescale, if organic matter is deeper into the soil and protected from decay by microbes, the matter is transformed into coal, oil, and natural gas and becomes what we know as fossil fuels or fossil carbon.

Due to the differences in timescale, data need to be collected over significant periods of time. On a field scale, it is possible to take spatial measurements of atmospheric CO_2 concentrations, but for carbon sequestration, temporal measurements are also needed. Commonly for this, researchers employ eddy covariance to measure the movement of atmospheric carbon. The equipment measures CO_2 concentrations in individual eddies, or parcels of air, using an infra-red gas analyser and, for this, the readings are at a very high temporal resolution of approximately 20 times per second. This can be employed on a field scale, over several hectares, but there is equipment available to analyse fluxes on a much smaller scale. These measurements, however, inform us only of the CO_2 fluxes at a given time, and not the carbon storage. As explained above, the fluxes vary depending on season and weather conditions and therefore long-term measurements are needed for a full picture of crop and management effects on fluxes. To gain a full depiction of carbon sequestration, biomass is also measured and soil cores are taken. From the soil cores, the concentration of DOC can be calculated which indicates SOC.

As research tends to be limited by time, it is a common practice to use a space-for-time substitution, where measurements are taken across multiple spatial areas with common traits but one differing trait, to substitute for measuring a single area over time. In the case of carbon sequestration, this may be, for example, measuring SOC in fields that have been converted from the same initial crop to a bioenergy crop for differing lengths of time to show the effect of LUC over time. This type of research into carbon sequestration, and other ecological research, is often termed a chronosequence.

Carbon sequestration into the soil carbon sink cannot continue indefinitely. There becomes a point where no additional carbon is able to be stored within the soil store. Land is considered at 'equilibrium' when the total carbon percentage is no longer increasing and the soil carbon capacity is saturated. A study on silty clay loam soil sown to grass from cropland at Rothamsted, United Kingdom, at differing times for differing lengths shows that the equilibrium was reached after 50–100 years, but after this time period, the soil could not take up more than a total of ca. 2.6% carbon (calculated by Smith [90] from total N values published by Johnston et al. [91]).

SUMMARY

While providing food and bioenergy, plants naturally draw carbon out of the atmosphere and convert it into biomass through the process of photosynthesis. This carbon is subsequently broken down, some is lost in the process of plant and microbial respiration and some is translocated into soils, leading to long-term carbon storage termed biogenic sequestration. The rate of biogenic sequestration compared to the release of carbon back into the atmosphere, through plant and microbial respiration is heavily reliant on land-use type, crop type and land management. In general, soil turnover rate is negatively related to the length of time the carbon remains in the soil. Green photosynthesising plants have significant potential to influence the carbon cycle and the combination of biogenic CO_2 capture and geological CO_2 storage (BECCS), may in future provide a technology able to reduce GHGs, as a negative emission technology (NET).

KEY TERMS

Aboveground biomass Any plant biomass which is above the ground this includes any trunks, stalks, branches, leaves, flowers, or seeds.

Autotropic respiration The respiration, or release of CO_2 from primary producers or plants.

Belowground biomass Any plant material which is below the ground, within the soil this includes any roots or rhizome of the plant.

Bioenergy with carbon capture and storage (BECCS) The combination of bioenergy production with carbon capture and storage technology.

Carbon capture and storage (CCS) A process in which a relatively pure stream of CO_2 from industrial and energy-related sources is separated (captured), conditioned, compressed, and transported to a storage location for long-term isolation from the atmosphere [93].

Ecosystem respiration (R_{eco}) The total respiration occurring by all organisms in an ecosystem.

Gross primary productivity (GPP) The fixation of carbon by autotrophic organisms (i.e. plant material).

Heterotrophic respiration The respiration, or release of CO_2 from organisms within the soil including bacteria, fungi, and animals.

Negative emissions technologies (NET) Technologies which allows removal of GHGs from the atmosphere such that atmospheric concentrations are reduced below the level that would have resulted without their deployment [92].

Net biome production (NBP) The difference between NEP and the carbon removed through harvest.

Net ecosystem exchange (NEE) The CO_2 exchange between the ecosystem and the atmosphere.

Net ecosystem productivity (NEP) The difference between the GPP and the total ecosystem respiration.

Soil respiration The sum of CO_2 respired by the roots (autotropic respiration) and by soil microbes or biota (heterotrophic respiration).

References

[1] Yang X, Cushman JC, Borland AM, et al. A roadmap for research on crassulacean acid metabolism (CAM) to enhance sustainable food and bioenergy production in a hotter, drier world. New Phytol 2015;207:491–504.

[2] Mason PM, Glover K, Smith JC, Willis KJ, Woods J, Thompson IP. The potential of CAM crops as a globally significant bioenergy resource: moving from 'fuel or food'to 'fuel and more food'. Energy Environ Sci 2015;8:2320–9.

[3] Borland AM, Griffiths H, Hartwell J, Smith JC. Exploiting the potential of plants with crassulacean acid metabolism for bioenergy production on marginal lands. J Exp Bot 2009;60:2879–96.

[4] Karp A, Shield I. Bioenergy from plants and the sustainable yield challenge. New Phytol 2008;179:15–32.
[5] Wullschleger S, Norby R, Love J, Runck C. Energetic costs of tissue construction in yellow-poplar and white oak trees exposed to long-term CO_2 enrichment. Ann Bot 1997;80:289–97.
[6] Mckinley B, Rooney W, Wilkerson C, Mullet J. Dynamics of biomass partitioning, stem gene expression, cell wall biosynthesis, and sucrose accumulation during development of *Sorghum bicolor*. Plant J 2016;88:662–80.
[7] Berner RA. The long-term carbon cycle, fossil fuels and atmospheric composition. Nature 2003;426:323–6.
[8] Harris Z, Spake R, Taylor G. Land use change to bioenergy: a meta-analysis of soil carbon and GHG emissions. Biomass Bioenergy 2015;82:27–39.
[9] Qin Z, Dunn JB, Kwon H, Mueller S, Wander MM. Soil carbon sequestration and land use change associated with biofuel production: empirical evidence. GCB Bioenergy 2016;8:66–80.
[10] Fargione J, Hill J, Tilman D, Polasky S, Hawthorne P. Land clearing and the biofuel carbon debt. Science 2008;319:1235–8.
[11] Searchinger T, Heimlich R, Houghton RA, et al. Use of US croplands for biofuels increases greenhouse gases through emissions from land-use change. Science 2008;319:1238–40.
[12] Mello FF, Cerri CE, Davies CA, et al. Payback time for soil carbon and sugar-cane ethanol. Nat Clim Chang 2014;4:605–9.
[13] Mccalmont JP, Mcnamara NP, Donnison IS, Farrar K, Clifton-Brown JC. An inter-year comparison of CO_2 flux and carbon budget at a commercial scale land-use transition from semi-improved grassland to Miscanthus x giganteus. GCB Bioenergy 2017;9(1):229–45.
[14] Price D, Orchansky A, Jassal R, Arevalo C, Thomas B. Soil respiration in a hybrid poplar plantation located in central Alberta. In: Alberta soil science workshop; 2009.
[15] Haszeldine RS. Carbon capture and storage: how green can black be? Science 2009;325:1647–52.
[16] Pale Blue Dot Energy, Axis Well Technology, Costain et al. A summary of results from the strategic UK CO_2 storage appraisal project. Progressing development of the UK's strategic carbon dioxide storage resource; 2016.
[17] Fuss S, Canadell JG, Peters GP, et al. Betting on negative emissions. Nat Clim Chang 2014;4:850–3.
[18] Smith P, Davis SJ, Creutzig F, et al. Biophysical and economic limits to negative CO_2 emissions. Nat Clim Chang 2016;6:42–50.
[19] IEAGHG. ECOPHYS: an ecophysiological growth process model for juvenile poplar. In: Potential for biomass with carbon capture and storage, IEA Greenhouse Gas R&D Programme; 2011.
[20] Boot-Handford ME, Abanades JC, Anthony EJ, et al. Carbon capture and storage update. Energy Environ Sci 2014;7:130–89.
[21] Haszeldine RS. Can CCS and NET enable the continued use of fossil carbon fuels after CoP21? Oxf Rev Econ Policy 2016;32:304–22.
[22] Glaser B, Birk JJ. State of the scientific knowledge on properties and genesis of Anthropogenic Dark Earths in Central Amazonia (terra preta de Índio). Geochim Cosmochim Acta 2012;82:39–51.
[23] Viger M, Hancock RD, Miglietta F, Taylor G. More plant growth but less plant defence? first global gene expression data for plants grown in soil amended with biochar. GCB Bioenergy 2015;7:658–72.
[24] Abel S, Peters A, Trinks S, Schonsky H, Facklam M, Wessolek G. Impact of biochar and hydrochar addition on water retention and water repellency of sandy soil. Geoderma 2013;202:183–91.
[25] Sohi S, Lopez-Capel E, Krull E, Bol R. Biochar, climate change and soil: a review to guide future research. CSIRO land and water science report 5; 2009. p. 17–31.
[26] Sohi SP. Carbon storage with benefits. Science 2012;338:1034–5.
[27] Sallade S, Ollinger S, Albrechtova J, et al. University of New Hampshire GLOBE carbon cycle project, 2012Available from: www.globecarboncycle.unh.edu; .
[28] Larcher W. Physiological plant ecology: ecophysiology and stress physiology of functional groups. Berlin, Heidelberg: Springer-Verlag; 2003.
[29] Wardle DA, Bardgett RD, Klironomos JN, Setälä H, Van Der Putten WH, Wall DH. Ecological linkages between aboveground and belowground biota. Science 2004;304:1629–33.
[30] Ericsson T. Nutrient cycling in energy forest plantations. Biomass Bioenergy 1994;6:115–21.
[31] Hangs RD, Schoenau JJ, Van Rees KC, Bélanger N, Volk T. Leaf litter decomposition and nutrient-release characteristics of several willow varieties within short-rotation coppice plantations in Saskatchewan, Canada. BioEnergy Res 2014;7:1074–90.
[32] Harris ZM, Alberti G, Viger M, Jenkins JR, Rowe R, Mcnamara NP, et al. Land-use change to bioenergy: grassland to short rotation coppice willow has an improved carbon balance. GCB Bioenergy 2017;9(2):469–84.

[33] Heděnec P, Novotný D, Usťak S, et al. The effect of native and introduced biofuel crops on the composition of soil biota communities. Biomass Bioenergy 2014;60:137–46.

[34] Bauer G, Persson H, Persson T, et al. Linking plant nutrition and ecosystem processes. Carbon and nitrogen cycling in european forest ecosystems. Berlin, Heidelberg: Springer; 2000.

[35] Chimento C, Almagro M, Amaducci S. Carbon sequestration potential in perennial bioenergy crops: the importance of organic matter inputs and its physical protection. GCB Bioenergy 2016;8(1):111–21.

[36] Jenkins J. Using next generation sequencing to assess soil microbial communities associated with drought, biochar and land use change to bioenergy. [Unpublished thesis for the degree of Doctor of Philosophy]. University of Sotuhampton; 2016.

[37] Liang C, Jesus EDC, Duncan DS, Jackson RD, Tiedje JM, Balser TC. Soil microbial communities under model biofuel cropping systems in southern Wisconsin, USA: impact of crop species and soil properties. Appl Soil Ecol 2012;54:24–31.

[38] Stauffer M, Leyval C, Brun J-J, Leportier P, Berthelin J. Effect of willow short rotation coppice on soil properties after three years of growth as compared to forest, grassland and arable land uses. Plant Soil 2014;377:423–38.

[39] Lal R. Soil carbon sequestration impacts on global climate change and food security. Science 2004;304:1623–7.

[40] Kuzyakov Y. Sources of CO_2 efflux from soil and review of partitioning methods. Soil Biol Biochem 2006;38:425–48.

[41] Kahle P, Möller J, Baum C, Gurgel A. Tillage-induced changes in the distribution of soil organic matter and the soil aggregate stability under a former short rotation coppice. Soil Tillage Res 2013;133:49–53.

[42] Davis SC, Boddey RM, Alves BJ, et al. Management swing potential for bioenergy crops. GCB Bioenergy 2013;5:623–38.

[43] Chen H, Marhan S, Billen N, Stahr K. Soil organic-carbon and total nitrogen stocks as affected by different land uses in Baden-Württemberg (southwest Germany). J Plant Nutr Soil Sci 2009;172:32–42.

[44] Lennon M, Nater EA. Biophysical aspects of terrestrial carbon sequestration in Minnesota. St. Paul: University of Minnesota; 2006.

[45] Paustian K, Andrén O, Janzen HH, et al. Agricultural soils as a sink to mitigate CO_2 emissions. Soil Use Manag 1997;13:230–44.

[46] Van Groenigen KJ, Hastings A, Forristal D, Roth B, Jones M, Smith P. Soil C storage as affected by tillage and straw management: an assessment using field measurements and model predictions. Agric Ecosyst Environ 2011;140:218–25.

[47] West TO, Post WM. Soil organic carbon sequestration rates by tillage and crop rotation. Soil Sci Soc Am J 2002;66:1930–46.

[48] Baker JM, Ochsner TE, Venterea RT, Griffis TJ. Tillage and soil carbon sequestration—what do we really know? Agric Ecosyst Environ 2007;118:1–5.

[49] Dolan MS, Clapp CE, Allmaras RR, Baker JM, Molina JE. Soil organic carbon and nitrogen in a Minnesota soil as related to tillage, residue and nitrogen management. Soil Tillage Res 2006;89:221–31.

[50] Six J, Bossuyt H, Degryze S, Denef K. A history of research on the link between (micro)aggregates, soil biota, and soil organic matter dynamics. Soil Tillage Res 2004;79:7–31.

[51] Blanco-Canqui H, Lal R. No-tillage and soil-profile carbon sequestration: an on-farm assessment. Soil Sci Soc Am J 2008;72:693–701.

[52] Angers DA, Bolinder MA, Carter MR, et al. Impact of tillage practices on organic carbon and nitrogen storage in cool, humid soils of eastern Canada. Soil Tillage Res 1997;41:191–201.

[53] Vandenbygaart AJ, Yang XM, Kay BD, Aspinall JD. Variability in carbon sequestration potential in no-till soil landscapes of southern Ontario. Soil Tillage Res 2002;65:231–41.

[54] Tally S. Making biofuel from corn stover. Biocycle 2002;43(11):62.

[55] Franzluebbers AJ. Soil organic matter stratification ratio as an indicator of soil quality. Soil Tillage Res 2002;66:95–106.

[56] Lal R. Crop residues as soil amendments and feedstock for bioethanol production. Waste Manag 2008;28:747–58.

[57] SQNTD Team. In: Service USDOaaNRC, editor. Crop residue removal for biomass energy production: effects on soils and recommendations. North Carolina: USDA; 2006.

[58] Karlen DL, Lal R, Follett RF, et al. Crop residues: the rest of the story. Environ Sci Technol 2009;43:8011–5.

[59] Buford MA, Stokes BJ. Incorporation of biomass into forest soils for enhanced productivity, restoration and biostorage: a modelling study to evaluate research needs. N Z J For Sci 2002;30:130–7.

[60] Scarlat N, Martinov M, Dallemand J-F. Assessment of the availability of agricultural crop residues in the European Union: potential and limitations for bioenergy use. Waste Manag 2010;30:1889–97.
[61] IPCC. In: IPOC Change, editor. Contribution of working group III to the fourth assessment report of the intergovernmental panel on climate change. Cambridge/New York, NY: Cambridge University Press; 2007.
[62] IPCC. In: IPOC Change, editors. Climate change 2007: working group I: the physical science basis. Cambridge: Cambridge University Press; 2007.
[63] Behnke G, David M, Voigt T. Greenhouse gas emissions, nitrate leaching, and biomass yields from production of Miscanthus × Giganteus in Illinois, USA. BioEnergy Res 2012;5:801–13.
[64] Drewer J, Finch JW, Lloyd CR, Baggs EM, Skiba U. How do soil emissions of N_2O, CH_4 and CO_2 from perennial bioenergy crops differ from arable annual crops? GCB Bioenergy 2012;4:408–19.
[65] Nikiema PR, Miller DE, Raymond O. Initial greenhouse gas emissions and nitrogen leaching losses associated with converting pastureland to short-rotation woody bioenergy crops in northern Michigan, USA. Biomass Bioenergy 2012;39:413–26.
[66] Hellebrand HJ, Scholz V, Kern J. Fertiliser induced nitrous oxide emissions during energy crop cultivation on loamy sand soils. Atmos Environ 2008;42:8403–11.
[67] Sehy U, Ruser R, Munch JC. Nitrous oxide fluxes from maize fields: relationship to yield, site-specific fertilization, and soil conditions. Agric Ecosyst Environ 2003;99:97–111.
[68] Mulla DJ. Twenty five years of remote sensing in precision agriculture: key advances and remaining knowledge gaps. Biosyst Eng 2013;114:358–71.
[69] DEFRA. Fertiliser recommendations for agricultural and horticultural crops. UK: Department for Environment FRA; 2003.
[70] Gilbert P, Thornley P, Riche AB. The influence of organic and inorganic fertiliser application rates on UK biomass crop sustainability. Biomass Bioenergy 2011;35:1170–81.
[71] Zalesny RS, Jr. Wiese AH, Bauer EO, Riemenschneider DE. Ex situ growth and biomass of Populus bioenergy crops irrigated and fertilized with landfill leachate. Biomass Bioenergy 2009;33:62–9.
[72] Fronning BE, Thelen KD, Min D-H. Use of manure, compost, and cover crops to supplant crop residue carbon in corn stover removed cropping systems. Agron J 2008;100:1703–10.
[73] Izaurralde RC, Mcgill WB, Robertson JA, Juma NG, Thurston JJ. Carbon balance of the breton classical plots over half a century. Soil Sci Soc Am J 2001;65:431–41.
[74] Kaur T, Brar BS, Dhillon NS. Soil organic matter dynamics as affected by long-term use of organic and inorganic fertilizers under maize–wheat cropping system. Nutr Cycl Agroecosyst 2008;81:59–69.
[75] Agriculture CaOWMI. In: Molden D, editor. Water for food, water for life: a comprehensive assessment of water management in agriculture. London/Colombo: Earthscan/International Water Management Institute; 2007.
[76] Dworak T, Elbersen B, Van Diepen K, et al. Assessment of inter-linkages between bioenergy development and water availability. Berlin/Vienna: Ecologic Institute for International and European Environmental Policy; 2009.
[77] Henriksen CB, Hussey K, Holm P. Exploiting soil-management strategies for climate mitigation in the european union: maximizing 'Win-Win' solutions across policy regimes. Ecol Soc 2011;16:22–38.
[78] Vanloocke A, Twine TE, Zeri M, Bernacchi CJ. A regional comparison of water use efficiency for miscanthus, switchgrass and maize. Agric For Meteorol 2012;164:82–95.
[79] Zeri M, Hussain MZ, Anderson-Teixeira KJ, Delucia E, Bernacchi CJ. Water use efficiency of perennial and annual bioenergy crops in central Illinois. J Geophys Res Biogeosci 2013;118:581–9.
[80] Tallis MJ, Casella E, Henshall PA, Aylott MJ, Randle TJ, Morison JIL, et al. Development and evaluation of ForestGrowth-SRC a process-based model for short rotation coppice yield and spatial supply reveals poplar uses water more efficiently than willow. GCB Bioenergy 2013;5:53–66.
[81] Chapin FS, Woodwell GM, Randerson JT, et al. Reconciling carbon-cycle concepts, terminology, and methods. Ecosystems 2006;9:1041–50.
[82] Aylott MJ, Casella E, Tubby I, Street N, Smith P, Taylor G. Yield and spatial supply of bioenergy poplar and willow short-rotation coppice in the UK. New Phytol 2008;178:358–70.
[83] Hastings A, Tallis MJ, Casella E, et al. The technical potential of Great Britain to produce ligno-cellulosic biomass for bioenergy in current and future climates. GCB Bioenergy 2014;6:108–22.
[84] Baum C, Leinweber P, Weih M, Lamersdorf N, Dimitriou I. Effects of short rotation coppice with willows and poplar on soil ecology. Landbauforschung 2009;59:183–96.

[85] Mann L, Tolbert V. Soil sustainability in renewable biomass plantings. AMBIO J Hum Environ 2000;29:492–8.
[86] Grandy AS, Robertson GP. Land-use intensity effects on soil organic carbon accumulation rates and mechanisms. Ecosystems 2007;10:59–74.
[87] Lal R. Soil carbon sequestration to mitigate climate change. Geoderma 2004;123:1–22.
[88] Lewandowski I, Clifton-Brown J, Scurlock J, Huisman W. Miscanthus: European experience with a novel energy crop. Biomass Bioenergy 2000;19:209–27.
[89] Hardcastle P, Calder I, Dingwall C, Garrett W, Mcchesney I, Mathews J, et al. A review of the potential impacts of short rotation forestry. Penicuik: LTS International; 2006.
[90] Smith P. Do grasslands act as a perpetual sink for carbon? Glob Chang Biol 2014;20:2708–11.
[91] Johnston AE, Poulton PR, Coleman K. Soil organic matter: its importance in sustainable agriculture and carbon dioxide fluxes. Adv Agron 2009;101:1–57.
[92] Mclaren D. A comparative global assessment of potential negative emissions technologies. Process Saf Environ Prot 2012;90:489–500.
[93] Allwood JM, Dubash NK, Gómez-Echeverri L, Von Stechow C. Glossary. In: Edenhofer O, Pichs-Madruga R, Sokona Y, Farahani E, Kadner S, Seyboth K, Adler A, Baum I, Brunner S, Eickemeier P, Kriemann B, Savolainen J, Schlömer S, von Stechow C, Zwickel T, Minx JC, editors. Climate change 2014: mitigation of climate change. Contribution of working group III to the fifth assessment report of the Intergovernmental Panel on Climate Change. Cambridge/New York, NY: Cambridge University Press; 2014.

Further Reading

[1] Ingram J, Fernandes E. Managing carbon sequestration in soils: concepts and terminology. Agric Ecosyst Environ 2001;87:111–7.
[2] Purves WK, Orians GH, Sadava D, Heller HC. Life: the science of biology. 7th ed. New York: Macmillan; 2004.
[3] Yamori W, Hikosaka K, Way DA. Temperature response of photosynthesis in C3, C4, and CAM plants: temperature acclimation and temperature adaptation. Photosynth Res 2014;119:101–17.

CHAPTER

6

Greenhouse Gas (GHG) and Biogenic Volatile Organic Compound (bVOC) Fluxes Associated With Land-Use Change to Bioenergy Crops

Julia Drewer, David Howard†, Niall McNamara†*

*NERC Centre for Ecology & Hydrology, Penicuik, United Kingdom †NERC Centre for Ecology & Hydrology, Lancaster, United Kingdom

6.1 INTRODUCTION

Carbon (C) is the chemical basis for all life on earth. It is combined with oxygen (O_2), hydrogen, and other elements to form a myriad of organic compounds that take gaseous, liquid, and solid forms. Through photosynthesis, plants capture C from carbon dioxide (CO_2) in the atmosphere and use sunlight as an energy source to produce complex energy-rich organic material. All organisms respire to release the energy in organic material for their benefit and, as a consequence, release CO_2. In the atmosphere, CO_2 is relatively rare (0.04% or 400 parts per million), but, unlike nitrogen and oxygen that form around 99% of the air, it absorbs, holds, and releases energy from infrared radiation coming from the earth after it has been warmed by the sun's shorter wavelength radiation. Along with other much scarcer gases, it acts as a blanket trapping heat and raising the temperature of our planet; the process is known as the greenhouse effect and the gases are called greenhouse gases (GHGs). The process is important in creating and maintaining the temperatures that support life; without it the earth would be around 30°C cooler on average. The low concentration of CO_2 in the atmosphere means that relatively small releases of previously fixed C into the atmosphere will have a noticeable impact on global temperature. Different gaseous compounds have different capacities to absorb radiation; each gas's total contribution to the greenhouse effect is the product of the amount of energy individual gas molecules can absorb and the number of molecules. Quantifying the effect is a little more complicated as the global warming potential

Greenhouse Gas Balances of Bioenergy Systems
https://doi.org/10.1016/B978-0-08-101036-5.00006-9

(GWP) of a gas depends not simply on the amount of infrared radiation it absorbs, but from which part of the electromagnetic spectrum it absorbs the radiation and how long the gas remains in the atmosphere. The different parts of the spectrum are important as the gases overlap in their capacities to absorb energies; some parts of the spectrum already have more than enough capacity to absorb all the energy that is transmitted, so adding more capacity at that wavelength would not alter the greenhouse effect.

Water (H_2O) is the most common compound to show a greenhouse effect in the atmosphere and has the highest impact contributing about 50% of the total effect in vapour form and 25% as clouds; it is the main reason why the earth's temperature is hospitable to life. However, in terms of importance it is usually ignored since the individual molecules will usually only spend a very short period (a matter of days) in the atmosphere, evaporating from the surface (sea) when the temperature is high enough, condensing to form clouds (as liquid droplets) that in turn are deposited back to the surface through precipitation. The amount and location of water in the atmosphere are both variable and dynamic and are major components of our weather systems and, although man influences this (e.g. with vapour trails from aircraft), it does not have the same legacy issues related to the other GHGs. As CO_2 is the second most common GHG in the atmosphere, after water vapour, but is persistent in the environment once released, its impact is considered to be of the highest magnitude.

The GHG balance in the atmosphere does vary naturally over millennia, but our use of fossil fuels to release energy from organic matter that has been geologically petrified is upsetting the balance. Once in the atmosphere, CO_2 molecules will persist for around a 100 years, so the impact of releases is cumulative and, if we stopped releasing any, the effects will continue to be seen for decades. Bioenergy, recycling carbon that has recently been in the atmosphere, is seen as a strategy to provide energy without significantly upsetting the GHG balance.

Plants are also responsible for the generation of compounds that are directly contributing to the greenhouse effect, most notably CO_2, methane (CH_4), and nitrous oxide (N_2O). Even though CH_4 and N_2O are not emitted in the quantity of CO_2, their GWP per molecule are 28–34 and 265–298 times higher than CO_2, respectively, on a 100-year time horizon which highlights their importance [1]; the range of GWP values reflects the differences in methods of estimation in this case being estimated without and with climate-carbon feedbacks and gives an idea of the level of confidence in the effects, but the relative GWP in relation to CO_2 remain consistent. Bioenergy from crops is expected to make a considerable contribution to climate change mitigation because of the energy sources it replaces and the potential to be 'carbon negative' [2]. However, with some crop production and management techniques, bioenergy is not only not C or GHG neutral, but may emit the equivalent GHG total as the substituted fossil fuels [3]. Plants also produce biogenic volatile organic compounds (bVOCs), for example, hemiterpenes such as isoprene and monoterpenes such as α-pinene and β-pinene which are predominantly involved in growth and defence. They can also function as communication media between plants and insects or plant communities [4]. bVOCs are important atmospheric trace gases as they directly and indirectly influence the concentrations and lifetimes of air pollutants and GHGs. For example, oxidation of bVOCs generates tropospheric ozone [5,6], which is also an important GHG and secondary organic aerosols (SOA) [7], both of which have adverse human (and plant) health impacts and affect the Earth's radiative balance. The effect of bVOC photochemistry on the oxidising capacity of the atmosphere also affects the lifetime of CH_4. BVOC emissions from crops have previously been demonstrated to have a significant impact on atmospheric composition [8]. Therefore, direct emissions from cultivation of bioenergy plants can impact GHG balances and overall life cycle analyses (LCAs, see Chapters 3 and 4). However, in this chapter we are only assessing GHG balances or C neutrality up to the farm gate, which means emissions from combustion and transport, for example from farm machinery, are usually not considered.

6.2 KEY GREENHOUSE GASES AND BIOGENIC VOLATILE ORGANIC COMPOUNDS RELEASED DURING PLANT GROWTH AND BIOLOGICAL PROCESSES

This section covers the processes and environmental conditions that lead to greenhouse gas (GHG) and biogenic volatile organic compound (bVOC) fluxes.

6.2.1 Carbon Dioxide (CO_2) Emission and Uptake

Put simply, plants trap C from CO_2 while growing, which is then released when the biomass is burned. The process of burning is equivalent to plant respiration returning the recently captured CO_2 to the atmosphere, thereby retaining the C balance. Using bioenergy avoids using ancient stored C from fossil fuels. Even though bioenergy crops are not always C neutral with the use of nitrogen (N) fertilisers (leading to emissions of N_2O with higher GWP) and fossil fuels in the agricultural practices and processing, they usually have a better C balance than traditional arable crops and are clearly better than using fossil fuels to provide the energy. Nonetheless, this does not take into account CO_2 respiration from soils or all the other activities associated with their growth, harvesting and management. Numerous studies have been carried out investigating CO_2 emission and uptake from bioenergy crops [3,9,10]. Most studies report CO_2 to be the dominant GHG compared with CH_4 and N_2O.

In most soil types and under most situations, the main losses of C from the soil are as CO_2 from microbial respiration and the mineralisation of soil organic matter [11,12]. During this enzymatic oxidation, microorganisms transform C compounds into CO_2, water, energy, and decomposer biomass. Carbon dioxide emissions are highly variable from heterogeneous soil microsites. They are controlled by a host of factors including microbial processes, root activity, litter and crop residue content, catalytic properties of clay colloids, and on microclimate [13]. Usually soils under perennial bioenergy crops are ploughed or tilled less often than annual arable crops and therefore have smaller amounts of physical release of CO_2 due to land management operations/soil disturbance. Perennial crops are, once established, permanent crops for about 25–30 years and harvested annually or every 2–3 years depending on crop. However, during the establishment or removal of perennial bioenergy crops, the physical release of CO_2 can play an important part.

6.2.2 Methane (CH_4) Emission and Consumption

Methane (CH_4) emissions are more important in peat and wetland soil under anaerobic conditions than agricultural mineral soils that are usually aerobic. In contrast, microbiological consumption of CH_4 in soil has been identified as the main terrestrial CH_4 sink, and N inputs have been demonstrated to weaken this sink-strength substantially over the short to medium term [14]. Tillage and compaction during ploughing not only disrupts the soil, but can also affect the soil CH_4 sink [15,16]; therefore, untilled bioenergy crops may be a greater sink of CH_4. Don et al. [3] reported that of the bioenergy sites where CH_4 was measured in Denmark, Germany, the United Kingdom, and Ireland, CH_4 uptake rates were small with between 2 and 17 kg CO_2 equiv. $ha^{-1} year^{-1}$. The CH_4 uptake compensates for around 3% (up to 5.2% in annual crops and 12.7% in SRC) of the N_2O emissions calculated as CO_2 equivalents. Methane is not usually included in bVOCs from plant origin, but is important in anaerobic decomposition and may be generated by poorly stored biomass for fuel [17–19].

6.2.3 Nitrous Oxide (N_2O) From Nitrification and Denitrification

Nitrous oxide (N_2O) is a product of nitrification and denitrification, two contrasting microbial processes in the soil. The former is the stepwise oxidation of NH_4^+ to nitrite (NO_2^-) and then to nitrate (NO_3^-), the latter being one of the most important oxidising agents [20]. Heterotrophic bacteria and fungi are also known to perform nitrification processes [21,22]. If concentrations of O_2 are low, nitrifying bacteria can use NO_2^- as an electron acceptor, reducing it to nitric oxide (NO) and N_2O, which are emitted to the atmosphere [22]. The NO and N_2O production increases when partial pressure of O_2 decreases. However, the effect of soil aeration is difficult to estimate as the efficiency of NH_4^+ oxidation decreases at the same time [22].

Denitrification is the reduction of NO_3^- to molecular nitrogen (N_2) under anaerobic conditions, performed by heterotrophic facultative anaerobic bacteria. In this reaction, N_2O is an intermediate electron acceptor. The amount of N_2O emitted to the atmosphere from denitrification depends on soil structure and water content. Anaerobic soils with large saturated clods produce high amounts of N_2O. In contrast, well-aerated soils can present anaerobic microsites that are due to localised high respiration rates. Nitrification and denitrification and the subsequent release of N_2O are processes controlled by soil management, such as fertilisation and tillage, and abiotic factors such as climate, frost and thaw frequency, soil porosity, moisture content, pH, and organic C availability [23]. Mineral fertilisers for agricultural production are the largest single global N_2O source. Perennial energy crops, however, provide the opportunity to produce bioenergy with low associated N_2O emissions due to low N fertilisation demand and higher N-use efficiency [3].

6.2.4 Biogenic Volatile Organic Compounds (bVOCs-Isoprene, Monoterpenes, etc.)

The reasons why plants produce bVOCs are complex and remain uncertain. Isoprene has no clear function [24], though it is thought to potentially have a role in protecting plants from thermal damage [25]. The function of monoterpenes is also largely unknown; however, they are thought to have a general ecological role, providing functions as diverse as attracting pollinators, competitive phytotoxicity, and defending against herbivores [26].

Terpenes have a carbon skeleton composed of characteristic C5 units. According to the number of C5 units, they are sub-divided into hemiterpenes (C5, e.g. isoprene), monoterpenes (C10, e.g. α-pinene), sesquiterpenes (C15, e.g. β-caryophyllene), and other sub-groups. Isoprene is the most common bVOC and is estimated to account for around 500 Tg C $year^{-1}$ (43% of total bVOC emissions) [27]. Besides playing an important role in atmospheric chemistry, monoterpenes also are economically significant, providing essential oils, carotenoid pigments, and natural rubber. Sesquiterpenes are thought to be less important in terms of impact on atmospheric chemistry, although some studies have reported emission rates on a scale similar to monoterpenes [28,29]. The rate and composition of bVOC emissions varies widely between and within plant species, so it is important to make measurements to build up an inventory of all terrestrial sources. However, there is still considerable uncertainty regarding the magnitude and seasonality of fluxes from different environments. Compounds emitted and emission rates vary widely between plant species as well as within species. Better knowledge of the characteristics and impacts of emissions from all vegetation is needed to say whether extensive planting of bioenergy crops will contribute significantly to the budgets of these trace gases.

6.3 FLUXES OF GHGS AND bVOCs FROM THE DIFFERENT TYPES OF BIOENERGY CROPS

This section describes the seasonal differences and emissions from specific life cycle stages (e.g. the initial planting replacing an existing land-use through to the establishment of a different use for the land once it is no longer cropped for bioenergy). Examples of the three major energy crop types are included along with a cross-tabulation of crops and key GHG/bVOCs.

6.3.1 Perennial Energy Crops (*Miscanthus* and Reed Canary Grass)

Perennial grasses such as *Miscanthus*, Reed Canary Grass (*Phalaris arundinacea*), and sorghum (*Sorghum bicolor* (L.) *Moench*) are typically harvested once a year with an estimated life span of around 25–30 years. Usually, they are not fertilised as they have a high N-use efficiency.

Drewer et al. [10] reported that perennial bioenergy crops such as *Miscanthus* in the United Kingdom are not 'GHG neutral', but appear to emit less N_2O than annual crops during cultivation because they did not receive N fertiliser. However, experiments using fertilisers have shown that if perennial bioenergy crops are fertilised, N_2O emissions will probably increase substantially which could offset the GHG balance more, and the GHG savings in comparison with annual crops would be even smaller. Low emission rates of isoprene and α-pinene were measured from *Miscanthus* in 2010 of the same field (limit of detection (<lod) to 6.42 $\mu g\, g^{-1} h^{-1}$ and <lod to 20.8 $\mu g\, g^{-1} h^{-1}$, respectively), but were not detected in subsequent years [30]. The measured emission rates followed a clear seasonal trend.

A study on sorghum investigated the effects of N fertilisation (unfertilised control vs 280 kg urea-N ha^{-1}), residue management (no biomass return vs 50% of sorghum biomass returned), crop sequence (corn [*Zea mays* L.]-*sorghum* vs *sorghum*-sorghum), and their associated CO_2 and N_2O emissions on a silty clay loam soil in central Texas [31]. Generally, CO_2 and N_2O fluxes were relatively higher than observed by others in the United States, despite drought conditions throughout much of the study period. During the growing season, highest fluxes of both gases were measured, especially following a precipitation-irrigation event and shortly after N fertilisation. Residue return increased cumulative CO_2 emissions each year; highest emissions of both gases were observed likely due to increased heterotrophic microbial activity. Nitrogen addition in the form of fertiliser significantly increased cumulative emissions of N_2O in both years, but only impacted cumulative CO_2 emissions in one of the study years. While crop rotation (a management strategy employing crops in specific sequences) impacted biomass yield, it had no significant effect on cumulative CO_2 or N_2O emissions [31].

6.3.2 High Yielding Forest Bioenergy

Short Rotation Forestry (SRF) and Short Rotation Coppice (SRC) are methods of growing high yielding energy forestry crops. SRF involves fast growing tree species over short rotations (between 10 and 20 years) using single stem management with conventional forest establishment and harvesting techniques. SRF have a higher density of trees (>2500 ha^{-1}) than traditional forests. SRC consist primarily of densely planted (~15,000 stools ha^{-1}), high yielding varieties of willow, and occasionally poplar, harvested on a 2–5 year cycle, although typically every 3 years. The rootstock or stool is left in the ground and after each harvest new shoots emerge, as in traditional coppice. It is currently expected that an SRC plantation should be viable for 20–30 years before it needs replanting, although there are currently no fixed guidelines (e.g. in the United Kingdom the Forestry Commission suggests that productivity will decline after 20 years, while the Department of the Environment, Food and Rural affairs (DEFRA) state that a plantation could be viable for up to 30 years) [32].

Drewer et al. [33] report GHG fluxes from several land-use systems on adjacent fields under the same soil and climatic conditions; in particular, the transition from arable agricultural land and fallow to perennial bioenergy crops SRC willow and SRF. Fertiliser-induced N_2O fluxes were an order of magnitude higher from the arable field than from the perennial crops. For the bioenergy crops, CH_4 and N_2O fluxes were negligible compared with CO_2 fluxes. This makes SRC and SRF viable options as low GHG-emitting bioenergy crops.

In a UK study by Morrison et al. [30], the largest bVOC emission rates were measured from SRC willow in England and varied between years. Isoprene emission rates varied between <lod and 1960 $\mu g\, g^{-1} h^{-1}$. Of the monoterpenes detected from willow, α-pinene emission rates were highest (<lod to 803 $\mu g\, g^{-1} h^{-1}$), followed by <lod to 268 $\mu g\, g^{-1} h^{-1}$ for δ-3-carene, <lod to 125 $\mu g\, g^{-1} h^{-1}$ for β-pinene, and <lod to 80.4 $\mu g\, g^{-1} h^{-1}$ for limonene. BVOC emission rates measured in Scotland were much lower. Crude extrapolations based solely on data gathered in that study indicate that isoprene emissions from willow could correspond to 0.004%–0.03% (United Kingdom) and 0.76%–5.5% (Europe) of current global isoprene if 50% of all land potentially available for bioenergy crops is planted with willow [30].

An Italian study investigated the consequences of land-use change from a cropland to a short rotation coppice plantation by comparing the GHG budgets of these two systems over a single rotation (24 months) in Viterbo, Italy [34]. The reference site was a crop rotation between grassland and winter wheat, which had been the management of the SRC site before its conversion to short rotation coppice. Reported CO_2 uptake was $3512 \pm 224\,g\,CO_2\,m^{-2}$ at the SRC site in 2 years, and $1838 \pm 107\,g\,CO_2\,m^{-2}$ at the reference site. The SRC site could offset $379.7 \pm 175.1\,g\,CO_2$ equiv. m^{-2} from fossil fuel displacement. Soil CH_4 and N_2O fluxes were negligible. Overall, the reference site was GHG neutral with $156 \pm 264\,g\,CO_2$ equiv. m^{-2}, while the SRC site was a net sink of $2202 \pm 792\,g\,CO_2$ equiv. m^{-2}. In conclusion, the experiment led to a positive evaluation from a GHG viewpoint of the conversion of cropland to bioenergy SRC. The study is a good example of how simple snapshots are taken and interpreted; the paired site approach is effective, but the observations from a single rotation may not properly represent the full cycle. Carbon may be lost from the system during transition through ploughing, aeration, and the energy used converting; the initial phase may show rapid sequestration and rootstock become established followed by a period of more consistent gains in the standard rotation. The figures do suggest a clear benefit in reducing GHG emissions, but the absolute values may not be suitable for translation into other locations or with other crops.

6.3.3 Annual Rotational/Traditional Agricultural Crops (e.g. Oilseed Rape, Wheat)

The difference in GHG emissions, especially N_2O from perennial bioenergy crops compared with annual arable crops, is predominantly due to fertiliser-induced N_2O emissions [10,33]. As long as bioenergy crops are not fertilised, their net GHG balance will be better than that of traditional fertilised crops.

Oilseed rape is one of the leading feedstocks for biodiesel production in Europe. The climate change mitigation effect of rape methyl ester (RME) is counterbalanced by the GHG emissions during crop production, mainly as N_2O from soils. Oilseed rape requires high levels of N fertiliser and crop residues are also rich in N, both can lead to enhanced N_2O emissions. GHG emissions of oilseed rape production are often estimated using emission factors that account for crop type specifics only with respect to crop residues. A metaanalysis by Walter et al. [35] assessed annual N_2O emissions from winter oilseed rape, to compare them to those of cereals and to explore the underlying reasons for differences. For the identification of the most important factors, linear mixed effects models were fitted with 43 N_2O emission data points derived from 12 different field sites. N_2O emissions increased exponentially with N fertilisation rates, but interyear annual and site-specific variability were high and climate variables or soil parameters did not improve the prediction model. Annual N_2O emissions from winter oilseed rape were 22% higher than those from winter cereals fertilised at the same rate. Fertiliser is commonly applied at a rate of $200\,kg\,N\,ha^{-1}\,year^{-1}$; the mean fraction of fertiliser N lost as N_2O-N was 1.27% for oilseed rape compared to 1.04% for cereals. The risk of high yield-scaled N_2O emissions increased after a critical N surplus of about $80\,kg\,N\,ha^{-1}\,year^{-1}$. The difference in N_2O emissions between oilseed rape and cereal cultivation was especially high after harvest due to the high N contents in oilseed rape's crop residues [35]. No significant emission rates of bVOCs were measured from oilseed rape in the United Kingdom [30].

6.4 NET ANNUAL AND ENTIRE LIFESPAN FLUXES TO ATMOSPHERE OF CROPS

This section examines the contribution of GHG fluxes at different temporal and spatial scales. Confidence in national and global estimates varies, but must be taken into account when making assessments.

6.4.1 UK Crops

According to the UK DEFRA [36], which released statistics in 2015, there were 122 thousand hectares of agricultural land used for bioenergy in the United Kingdom in 2014. The area of crops grown for bioenergy equated to 2% of all arable land in the United Kingdom in that year. Just over 68% of land used for bioenergy in 2014 was for biofuel for the UK road transport market, which means that 1.2 million tonnes of UK crops were produced. Maize grown for use in anaerobic digestion accounted for 19% of the total maize area in England in 2015 and 0.7% of England's total arable area.

To date, no experimental studies have captured the whole life cycle of GHG emissions from planting to end of life cycle, nor have they followed in detail the implications of land-use change to another use such as re-establishment of the previous crop or cultivation of a second rotation crop. A systematic review and metaanalysis by Harris et al. [37] evaluated the current state of knowledge and quantified the effects of land-use change (LUC) to second generation (2G), non-food bioenergy crops on soil organic carbon (SOC) and GHG emissions of relevance to temperate zone agriculture. Across the range of transitions to 2G, the mean age of crops varied from 5.4 to 7.1 years old (range 1–16 years), thus not capturing the end of life cycle. Following this analysis of 138 original studies, transitions from arable to short rotation coppice (SRC, poplar or willow) or perennial grasses (mostly *Miscanthus* or switchgrass) resulted in increased SOC (+5.0 ± 7.8% and +25.7 ± 6.7%, respectively). Transitions from grassland to SRC were broadly neutral (+3.7 ± 14.6%), while grassland to perennial grass transitions and forest to SRC both indicated a decrease in SOC (−10.9 ± 4.3% and −11.4 ± 23.4%, respectively). There were insufficient paired data to conduct a strict metaanalysis for GHG emissions, but summary figures of general trends in GHGs from the studies revealed increased and decreased soil CO_2 emissions following transition from forests and arable to perennial grasses. However, the study identified significant knowledge gaps surrounding the effects of land-use change to bioenergy on greenhouse gas balance, particularly for CH_4. There is also large uncertainty in quantifying transitions from grasslands and transitions to short rotation forestry. A striking finding of this review is the lack of empirical studies that are available to validate modelled data. Given that models are extensively used in the development of bioenergy LCA and sustainability criteria, this is an area where further long-term data sets are required [37]. Using data from Harris et al. [37] and from detailed experimental studies [38–41], the effect of LUC to 1G and 2G crops was assessed spatially in the United Kingdom to 2050 using the ECOSSE soil process model [42]. Conversion of traditional agricultural rotation crops to *Miscanthus*, SRC, and SRF and conversion of permanent grass to SRF show that, over a significant area, the flux of GHG moves towards the sequestration of C in the soil with SOC increasing. Conversion of permanent grass to *Miscanthus*, permanent grass to SRF, and forest to SRF shows detrimental changes in soil GHG balance over a significant area. Conversion of permanent grass to wheat, oilseed rape, sugar beet, and SRC and all conversions from forest show large detrimental changes in soil GHG balance over most of the United Kingdom, largely due to moving from uncultivated soil to regular cultivation. Overall, the dominant driver of the net GWP were changes in SOC with the relative net contribution of N_2O and CH_4 on a CO_2 equivalent basis being small to negligible, respectively.

6.4.2 European Crops

The European Union (EU) has Targets and Directives for reducing GHG emissions and increasing the use of biofuels; more land across the EU will be dedicated to the production of bioenergy. The plant species selected to be grown across the EU fit into the same categories described above, traditional crops (first generation, 1G and second generation, 2G) and 2G perennial grasses and woody plants, although their productivity and GHG and bVOC production will differ in different climates. The range of species, their potential distribution, and impact on GHG have been modelled under different climate scenarios [43]. Renewable energy statistics of the European Union [44] report that the primary production of renewable energy within the EU-28 in 2013 was 192 million tonnes of oil equivalent (toe) of 24.3% share of total primary energy production from all sources. The quantity of renewable energy produced within the EU-28 increased overall by 84.4% between 2003 and 2013, equivalent to an average increase of 6.3% per year. Among renewable energies, the most important source in the EU-28 was biomass and renewable waste, accounting for just under two thirds (64.2%) of primary renewables production in 2013.

There are different ways of increasing the production of bioenergy which include spatial expansion (replacing existing land-uses), increasing intensity of management (e.g. use of fertilisers, pesticides, and mechanisation), and changing the species planted. The European Environment Agency has modelled different scenarios across Europe focussing on the land use [44a]. Their models constrained bioenergy planting from the perspective of the existing land-use and estimated the yields using different crop mixes; they examined the impacts using a number of metrics covering water, soil, biodiversity as well as GHG.

Increasing the use of additional conventional biofuels up to 2020 on the scale anticipated in the 23 European national renewable energy action plans would lead to between 81% and 167% more GHG emissions than meeting the same need through fossil fuel use [45]. Specific emissions due to land-use change are especially high for bioenergy systems with low yield per hectare. Technological developments along the supply chain and improved dedicated energy crop types and crop management will reduce the impact of the bioenergy feedstock on the GHG balance. Improving crop yields per hectare, fostered by increased crop prices and improved conversion efficiencies, will decrease the energy-specific C debt [3].

6.4.3 Global

In other parts of the world, bioenergy is equally important. Recent statistics [46] show that bioenergy accounts for roughly 10% (50 EJ) of world total primary energy supply today. However, in developing countries the use of bioenergy is much more diverse dominated by activities such as cooking and heating, using very inefficient open fires or simple cook stoves. There are considerable impacts on health (smoke pollution, particulates, dioxins, etc.) and environment (deforestation). Modern bioenergy supply, while still very small in the developing world, has been growing steadily in the last decade. For example, a total of 370 TWh of bioenergy electricity was produced in 2012 corresponding to 1.5% of world electricity generation. Subsistence style bioenergy exploits material from the local environment including wood and agricultural waste, but new feedstocks are being developed including Jatropha and palm.

Recent work has focussed also on oil palm plantations as palm oil can either be used for biofuels or substitute oils like rapeseed oil that are used for biofuels and no longer available as food oils. Planting of palm oil plantations has come under scrutiny as the land-use change from primary or even secondary rainforest to oil palm plantations also results in elevated GHG emissions in a similar manner that palm oil plantations are fertilised and can potentially generate high N_2O emissions [47]. Fertilisation of palm oil plantations can be very inefficient and over-use of fertiliser is a common occurrence.

A study in the US (southcentral Wisconsin and southwestern Michigan) also reports higher N_2O emissions from traditional crops compared with bioenergy crops [48]. Cumulative 3-year N_2O emissions from annuals were 142% higher than from perennials, with fertilised perennials 190% higher than unfertilised perennials. Emissions ranged from 3.1 to 19.1 kg N_2O-N ha^{-1} $year^{-1}$ for the annuals with continuous corn > corn-soybean-canola rotation and 1.1–6.3 kg N_2O-N ha^{-1} $year^{-1}$ for perennials. Nitrous oxide peak fluxes typically were associated with precipitation events that closely followed fertilisation.

Soils had an almost neutral C budget with a slight emission of 10.0 Tg CO_2 equiv. by 2030. However, the ethanol C offset by displacing fossil fuels could payback that C deficit and ensures the environmental benefits of sugarcane ethanol. Their results show an increase of C reservoirs (biomass and soil) through conversion of arable and pastoral lands into sugarcane and a decrease of C reservoirs when citrus, plantation forest, and natural forest are converted to sugarcane [49].

6.5 SIGNIFICANCE OF CROP MANAGEMENT ON GHGS

Different crops emit different levels of GHG, but there are interactions with timing of management and geographic location.

6.5.1 Soil Improvement (Biochar, Drainage, Aeration, etc.)

Biochar is charcoal, a C-rich substance produced from biomass, which is applied to soils. It is being promoted as a climate change mitigation tool as it has the potential to increase soil

C sequestration and reduce soil GHG emissions when applied as a soil amendment. For this reason, combining bioenergy cultivation with biochar application to improve the GHG balance of bioenergy crops is an attractive proposition. Biochar is created by heating biomass in a low-oxygen environment (a process called pyrolysis, typically heated to between 350°C and 600°C) [50]. In the study in the United Kingdom, soil N_2O emissions were not significantly suppressed with biochar amendment, although they were generally low. Soil CH_4 fluxes were below minimum detectable limits in both experiments. These findings demonstrate that biochar amendment has the potential to suppress net soil CO_2 equiv. emissions in bioenergy crop systems for up to 2 years after addition, primarily through reduced CO_2 emissions. Suppression of soil CO_2 emissions may be due to a combined effect of reduced enzymatic activity, the increased carbon-use efficiency from the co-location of soil microbes, soil organic matter and nutrients, and the precipitation of CO_2 onto the biochar surface [50].

A full accounting of net greenhouse gas balance (NGHGB) and greenhouse gas intensity (GHGI) affected by biochar amendment combined with or without N-fertiliser application was examined in an annual coastal reclaimed Jerusalem artichoke bioenergy cropping system in China [51]. The net ecosystem exchange of CO_2 (NEE) was determined by the difference between soil heterotrophic respiration (R-H) and net primary production (NPP) using the static chamber method. Biochar amendment raised the seasonal R-H, but without suppressing the NPP during the Jerusalem artichoke cropping season. Soil CH_4 emissions were 72% and 80% lower in the biochar amended than unamended plots when combined with N-fertiliser application during the Jerusalem artichoke cropping and noncropping seasons, respectively. The biochar-induced soil N_2O mitigation efficiency was weakened by N-fertiliser input over the annual cycle. Annual NGHGB and GHGI were negative for all the field treatments and were significantly lower in biochar amended than in unamended soils, suggesting that Jerusalem artichoke cropping system served as a net sink of GHGs due to net ecosystem CO_2 and biochar-induced C sequestration exceeding CO_2-equivalents released as CH_4 and N_2O emissions. On average, biochar amendment significantly enhanced GHGs sink capacity by resulting in almost four- to fivefolds decrease in annual NGHGB or GHGI when combined with N-fertiliser application or not. Therefore, higher biomass gain as potential alternative source of biofuels but lower climatic impacts of bioenergy production would be reconciled by biochar use in southeast coastal China.

Despite being a long established method of soil improvement in tropical soils, there is little understanding of the processes involved in different soil types. The observational evidence described above and elsewhere is strongly suggestive of a reduction in GHG emissions from soil following treatment, but further research is needed before it can be widely exploited as a mitigation measure. The key issue is that the mechanism by which the GHGs are suppressed is not fully understood and it remains the question of longevity of this method. Therefore, this limits the extent to which biochar can be relied on as a mitigation measure.

6.5.2 Straw or Forest Residue (Include Comment on Value of Co-products Depending on Residue Management)

A significant fraction (30% in Europe) of bioenergy feedstock is derived from waste and crop by-products such as straw, industry residues, and manure [52]. These have a high potential to contribute to future bioenergy production, as they do not induce any land-use change. However, in future the need to mitigate GHG emissions in agriculture by increasing the soil organic carbon (SOC) of croplands may create a competitive market for these 'waste products' as soil amendments to improve SOC and soil fertility and this will increase the value of such waste. Non-harvested by-products such as crop and forest residues left on the ground contribute to SOC sequestration in the ecosystems for a limited transitional period. Whether their harvest and energetic use as a substitute for fossil fuel or their being left on the site results in higher CO_2 savings depends on an array of parameters such as the C saturation level of the ecosystem and the energy investments for collection and transport of residues [53].

Residues are often viewed by land managers as a problem to be solved at lowest cost and with as little effort as possible. Straw and stubble, agricultural residues, are still burnt in fields in many parts of the world (including United States, France, and India) having the benefit of killing pests and weeds and it is also argued that it releases N. However, it is also recognised to increase nutrient run off and create smoke; the practice was banned in the United Kingdom in 1993 as it is in China, but the C balance benefits of a ban are disputed [54]. Where straw is used as a feedstock in bioenergy power plants, it needs to be replaced by other organic soil amendments to maintain a positive SOC balance. Not only is SOC important directly for the GHG balance, but also for soil fertility, erosion protection, and water and nutrient retention in soils, all of which indirectly influence the GHG balance. It has been argued that financially adroit farmers will sell their straw to biorefineries and purchase mineral fertiliser with the proceeds having a significant detrimental impact on GHG emissions, especially N_2O [55].

6.5.3 Sustainable Removal Rate

The removal or replanting phase is a critical step in the life cycle of perennial crops and depends on the agricultural techniques used. Perennial plantations are only viable for around 20–30 years and these numbers are not set in stone. It varies by crop and environmental conditions and mainly depends on how long a viable yield can be maintained.

It is estimated that *Miscanthus*, for example, would need to be replanted every 25 years and may take 4–5 years for the maximum yield to be established. Due to biomass decay on the soil surface, the removal caused extra GHG emissions of 1.5 t CO_2 equiv. ha^{-1} compared to a rotation of annual crops or the continuation of *Miscanthus*, which should be considered in life cycle assessment (LCA). In one study, planting an annual crop to replace *Miscanthus* caused an additional emission of 1.4 t CO_2 equiv. ha^{-1} relative to a rotation of annual arable crops, while shifting to set-aside actually reduced GHG emissions by 5.1 t CO_2 equiv. ha^{-1} relative to the *Miscanthus* control. This occurred despite a net increase in N_2O emissions amounting to 147 kg CO_2 equiv. ha^{-1} [56].

6.6 POTENTIAL, FUTURE, AND WIDER IMPACTS ON GHG FLUXES

This section describes how the effects may be mitigated by change or indirectly contribute to additional releases.

6.6.1 Fertiliser Input

Fertiliser application rates are directly linked to GHG emissions for both conventional and energy cropping systems. The emissions arise from N_2O release from soils, along with additional emissions associated with fertiliser production and transport [3]. Fertiliser application practice depends on soil conditions and agro-economic constraints, which are highly variable throughout Europe and globally. However, the N-fertiliser demand of perennial crops such as *Miscanthus* and SRC was always much lower when compared with annual crops. Perennial energy crops have a higher N-use efficiency and thus need less fertiliser with reduced N loss as N_2O or NO_3. Several studies have now concluded that GHG fluxes from bioenergy crops are only lower than traditional crops when not fertilised [3,10,51]. Some field trials of *Miscanthus* have, however, shown that yield gains after nitrogen fertilisation offset any additional N_2O emissions [57]. Other work did not observe any yield benefit through fertilisation of *Miscanthus*, suggesting the benefits of any additional inputs may be context-dependent [58].

6.6.2 Climate Change and Scenarios

A systematic review of the life cycle components of first- and second-generation bioenergy crops and their emissions driven by different uses (heat and power or transport) identified three distinct sources of variation; real (e.g. cultivation), methodological (data capture and processing), and statistical uncertainty [59].

The mitigation of climate change is a major driver for the deployment of bioenergy, but environmental conditions also change the performance of different crops and their yield. Bioenergy crops appear to be more resilient than grain crops to climate change, but the GHGs arising from management change may well increase [60]. Modelling future scenarios at national and global levels for different points in the future and under different constraints and drivers is relatively common (e.g. UK's Committee on Climate Change Bioenergy Review Technical Paper 2 Global and UK bioenergy supply scenarios of December 2011), but the GHG implications and confidence in impacts are often ignored. Research [43] is providing data for productivity, emissions, and distribution under different climate scenarios, but with a complex, poorly understood system, interacting with other complex imperfectly understood components (e.g. agriculture), the significant effects are difficult to separate.

6.6.3 Indirect Land-Use Change (LUC, Both Land and Intensity)

Searchinger et al. [61] raised the issue of indirect land-use change (iLUC) associated with bioenergy describing how displaced land-use increased GHG emissions. The process is understood to create impacts in two ways, first there are categorical changes in land-use where the material produced prior to the land being converted to bioenergy displaces another land-use that in turn can displace other land-use with a cascade effect. The second impact pathway is through increasing the intensity of land management; the production lost to bioenergy is made up by increasing the yield of land already producing that commodity through greater use of systems such as fertilisers that can add to the GHG load. The concept appears easy to understand and has been considered by authorities to create more comprehensive assessments of policy; the United Nations Environment Programme's World Conservation Monitoring Centre (UNEP-WCMC) has created a framework to assess the effects. However, the assessments have been fiercely criticised; there is no widely accepted method of assessing in detail the impacts of specific land-use change. While there is strong support for the case of significant global impacts, the uncertainties associated with it seem to still be growing [62]. It is now recognised that the level of change is strongly influenced by economics, world trade, value systems, material and monetary flows, and above all producer and consumer behaviour [63].

The importance of indirect effects, especially when they are difficult to quantify and are likely to be impacted by a diversity of other drivers, may seem trivial. However, it is important to consider the objectives of an analysis before deciding the approach to be used and defining its domain. In this case, the goal of examining GHG emissions is to solve a global problem of climate change, clearly an examination of the consequential impacts of a series of complex processes. The value of actions to reduce emissions locally may vary from psychological comfort to improved political and commercial image, but if the consequence at a global scale is trivial or even an increase of emissions, then no long-term benefit is gained. For LCA, since robust figures are not readily to hand, it may seem reasonable to define a domain that excludes iLUC, especially if the approach is attributional, but the goal and scope definition stage should always force the opposite. That said, better accounting for iLUC is rapidly needed, so it can be used in making better decisions.

6.7 SUMMARY

Our knowledge of the relationship between bioenergy, GHG, and bVOCs is incomplete, and shows the system is complex. It is not possible to put absolute or generic values on the different components, but the sections above have shown that from observations and estimates the relative status can be described. In most cases in the developed world, bioenergy production is relatively new and expanding; it is replacing a range of existing land-uses (agriculture, forestry, and marginal land), meaning that at the moment case by case studies are needed for different situations.

From the studies described above, it appears that for bioenergy feedstock production CO_2 is the dominant GHG in most cases with strong interactions with SOC. Through the different phases of the life cycle, SOC acts as either a sink or source and shows different capacities depending on the soil type, crop cultivated, and management practices.

Nitrogen, usually in the form of fertiliser, has a much more variable presence in bioenergy feedstock GHG emissions than normal agriculture. Many bioenergy crops are seen as low (or even zero) input crops which makes them financially attractive. However, some traditional crops (including wheat) are grown as 1G bioenergy crops (usually for the production of biofuels) and these commonly still have considerable fertiliser input (see Fig. 6.2). For 2G crops, the careful application of fertiliser will also increase yield and, as the material

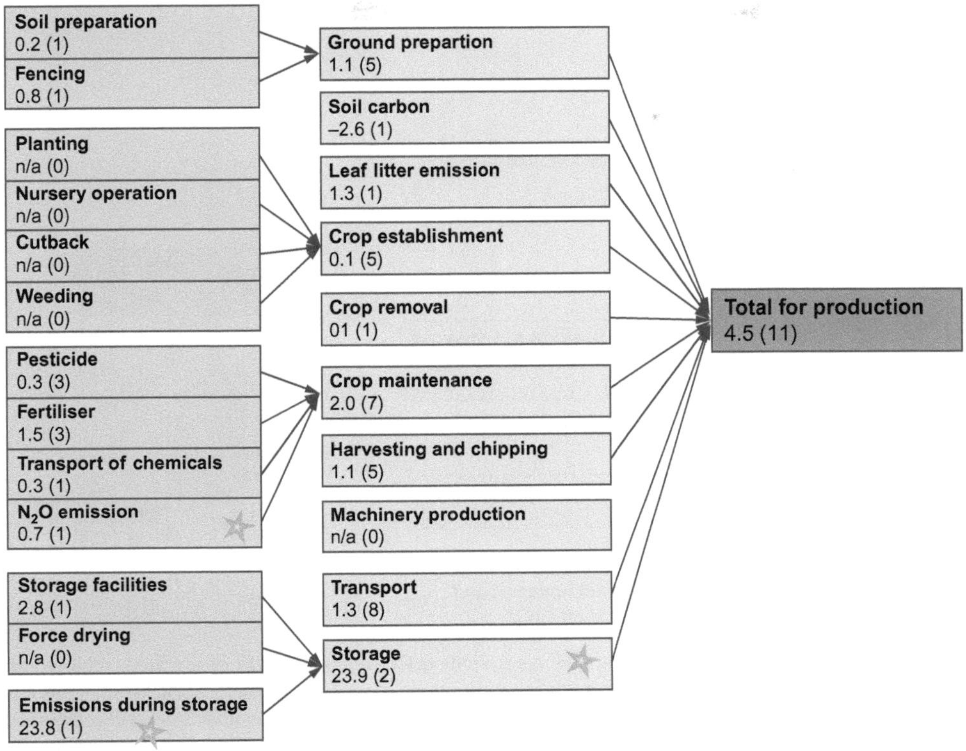

FIG. 6.1 Life cycle compartments showing GHG emissions (g CO_2 equiv. MJ^{-1}) in production of woody crop feedstock. Values are arithmetic means derived from different numbers of scenarios (given in brackets); n/a is where figures are not available. *Stars* indicate major contributors. *Based on Rowe RL, et al. Potential benefits of commercial willow Short Rotation Coppice (SRC) for farm-scale plant and invertebrate communities in the agri-environment. Biomass Bioenergy 2011;35(1):325–36.*

is not for human consumption, the sources of fertiliser may be less constrained (e.g. more indiscriminate use of slurry). The notion that bioenergy crops can produce good returns from marginal or poor land has been exposed with the example of Jatropha. It was seen as a plant that will survive extreme arid situations where other crops will fail and did receive a lot of support for planting in the tropics, but, while it survives in harsh conditions, it is not productive and as with all plants has its own requirements for water and nutrients to maximise its yield. There is still hope that Jatropha will extend our geographical range of cultivation, but will depend on research into new genetic varieties and sustainable management practices.

Figs 6.1–6.3 give an example of how the different compartments compare in their contribution to GHG emissions for energy feedstocks. They are from a systematic review of the GHG and energy balances comparing liquid biofuels with biomass used for heat and power [59,64].

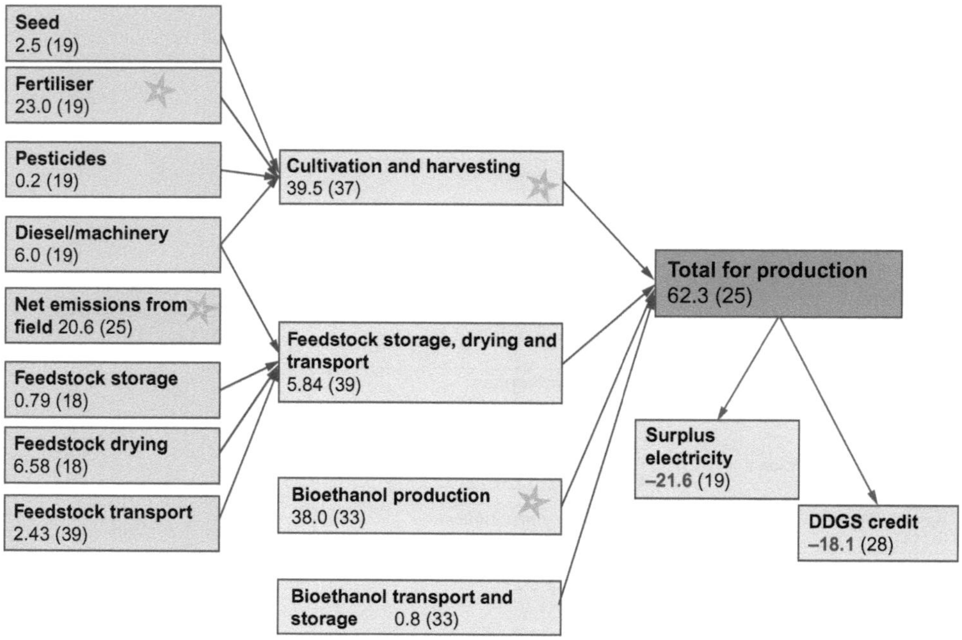

FIG. 6.2 Life cycle compartments showing GHG emissions (g CO_2 equiv. MJ^{-1}) in production of bioethanol from wheat. Values are arithmetic means derived from different numbers of scenarios (given in brackets); n/a is where figures are not available. *Stars* indicate major contributors. *Based on Rowe RL, et al. Potential benefits of commercial willow Short Rotation Coppice (SRC) for farm-scale plant and invertebrate communities in the agri-environment. Biomass Bioenergy 2011;35(1):325–36.*

The review constructed a life cycle for different forms of bioenergy and then attempted to populate the compartments from published literature. A number of inconsistencies and uncertainties were reported including the use of different units, the omission of different life cycle stages, and different methodologies to credit co-products. Woody crops and energy grasses show similar categories and levels of emission that are considerably lower than using fossil fuels. One surprisingly high emission was associated with the storage of wood, which had been stacked wet and stored in near anaerobic conditions leading to CH_4 generation; the system is not typical, indeed reports often describe the unusual rather than the norm.

The implication arising from these figures is that management decisions are a major factor in determining the GHG emissions. The yields of the 2G crops could be improved by the use of fertiliser, although not necessarily to the level applied to wheat. The current low levels of application do not represent an evangelical striving to combat climate change, but a gradual learning of how to handle a new crop and the belief of myths such as '*Miscanthus* needs no fertiliser'. The agronomists will eventually make economic decisions about use of fertiliser, following new trendsetters.

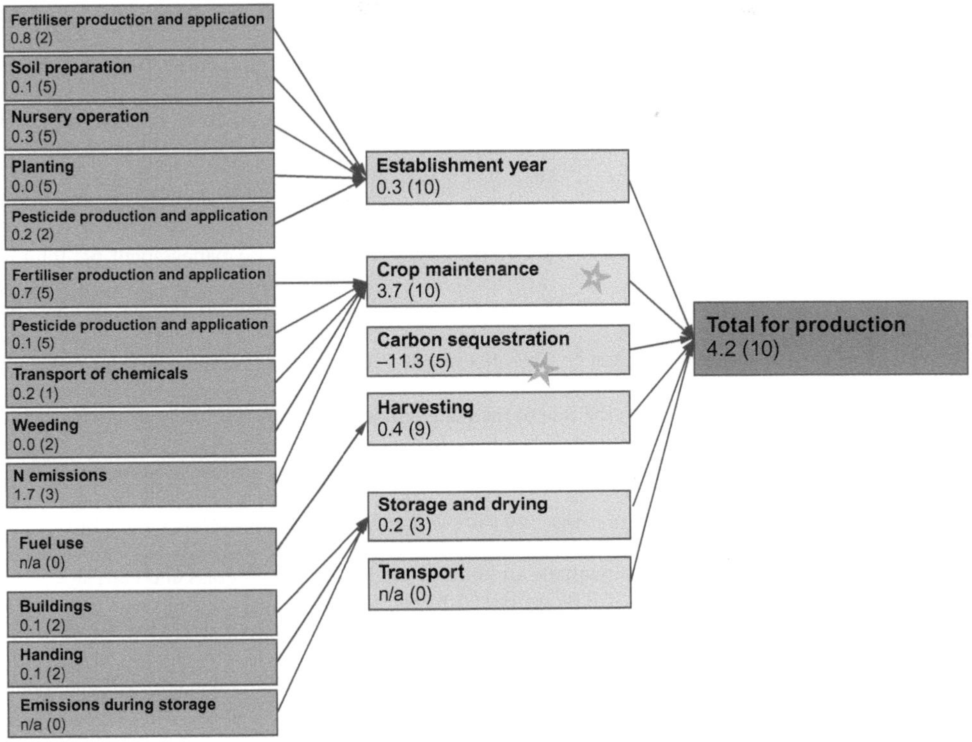

FIG. 6.3 Life cycle compartments showing GHG emissions (g CO_2 equiv. MJ^{-1}) in production of energy grass feedstock. Values are arithmetic means derived from different numbers of scenarios (given in brackets); n/a is where figures are not available. *Stars* indicate major contributors. *Based on Rowe RL, et al. Potential benefits of commercial willow Short Rotation Coppice (SRC) for farm-scale plant and invertebrate communities in the agri-environment. Biomass Bioenergy 2011;35(1):325–36.*

References

[1] Myhre G, et al. Anthropogenic and natural radiative forcing. In: Stocker TF, Qin D, Plattner G-K, Tignor M, Allen SK, Boschung J, Nauels A, Xia Y, Bex V, Midgley PM, editors. Climate change 2013: the physical science basis. contribution of working group I to the fifth assessment report of the intergovernmental panel on climate change. Cambridge/New York, NY: Cambridge University Press; 2013.

[2] Tilman D, Hill J, Lehman C. Carbon-negative biofuels from low-input high-diversity grassland biomass. Science 2006;314(5805):1598–600.

[3] Don A, et al. Land-use change to bioenergy production in Europe: implications for the greenhouse gas balance and soil carbon. GCB Bioenergy 2012;4(4):372–91.

[4] Laothawornkitkul J, et al. Biogenic volatile organic compounds in the Earth system. New Phytol 2009;183(1):27–51.

[5] Fehsenfeld F, et al. Emissions of volatile organic compounds from vegetation and the implications for atmospheric chemistry. Glob Biogeochem Cycles 1992;6(4):389–430.

[6] Pierce T, et al. Influence of increased isoprene emissions on regional ozone modeling. J Geophys Res Atmos 1998;103(D19):25611–29.

[7] Claeys M, et al. Formation of secondary organic aerosols from isoprene and its gas-phase oxidation products through reaction with hydrogen peroxide. Atmos Environ 2004;38(25):4093–8.

[8] Hewitt CN, et al. Quantification of VOC emission rates from the biosphere. Trends Analyt Chem 2011;30(7):937–44.

[9] Dondini M, et al. Simulation of greenhouse gases following land-use change to bioenergy crops using the ECOSSE model: acomparison between site measurements and model predictions. Glob Change Biol Bioenergy 2016;8(5):925–40.

[10] Drewer J, et al. How do soil emissions of N_2O, CH_4 and CO_2 from perennial bioenergy crops differ from arable annual crops? GCB Bioenergy 2012;4(4):408–19.

[11] Dawson JJC, Smith P. Carbon losses from soil and its consequences for land-use management. Sci Total Environ 2007;382(2–3):165–90.

[12] Singh BK, et al. Microorganisms and climate change: terrestrial feedbacks and mitigation options. Nat Rev Microbiol 2010;8(11):779–90.

[13] Muñoz C, Paulino L, Monreal C, Zagal E. Greenhouse gas (CO_2 and N_2O) emissions from soils: a review. Chilean J Agric Res 2010;**70**(3):485–97.

[14] Dobbie KE, et al. Effect of land use on the rate of methane uptake by surface soils in northern Europe. Atmos Environ 1996;30(7):1005–11.

[15] Teepe R, et al. Nitrous oxide emission and methane consumption following compaction of forest soils. Soil Sci Soc Am J 2004;68(2):605–11.

[16] Yamulki S, Jarvis S. Short-term effects of tillage and compaction on nitrous oxide, nitric oxide, nitrogen dioxide, methane and carbon dioxide fluxes from grassland. Biol Fertil Soils 2002;36(3):224–31.

[17] Emery I, Mosier N. Direct emission of methane and nitrous oxide from switchgrass and corn stover: implications for large-scale biomass storage. GCB Bioenergy 2015;7(4):865–76.

[18] Jämsén M, et al. Measurement methodology for greenhouse gas emissions from storage of forest chips—a review. Renew Sust Energ Rev 2015;51:1617–23.

[19] Wihersaari M. Evaluation of greenhouse gas emission risks from storage of wood residue. Biomass Bioenergy 2005;28(5):444–53.

[20] Šimek M, Hynšt J, Šimek P. Emissions of CH_4, CO_2, and N_2O from soil at a cattle overwintering area as affected by available C and N. Appl Soil Ecol 2014;75:52–62.

[21] Prosser JI. Nitrogen in soils | nitrification A2. In: Hillel D, editor. Encyclopedia of soils in the environment. Oxford: Elsevier; 2005. p. 31–9.

[22] Smith KA. Greenhouse gas emissions A2. In: Hillel D, editor. Encyclopedia of soils in the environment. Oxford: Elsevier; 2005. p. 145–53.

[23] Skiba U, Smith KA. The control of nitrous oxide emissions from agricultural and natural soils. Chemosphere Global Change Sci 2000;2(3–4):379–86.

[24] Fall R. Biogenic emissions of volatile organic compounds from higher plants A2. In: Hewitt CN, editor. Reactive hydrocarbons in the atmosphere. San Diego: Academic Press; 1999. p. 41–96 [chapter 2].

[25] Loreto F, Velikova V. Isoprene produced by leaves protects the photosynthetic apparatus against ozone damage, quenches ozone products, and reduces lipid peroxidation of cellular membranes. Plant Physiol 2001;127(4):1781–7.

[26] Monks PS. Tropospheric photochemistry. In: Hewitt CN, Jackson AV, editors. Handbook of atmospheric science: principles and applications. Malden, MA, USA: Blackwell Science Ltd; 2003. p. 156–87. https://doi.org/10.1002/9780470999318. [chapter 6].

[27] Guenther A, et al. Estimates of global terrestrial isoprene emissions using MEGAN (Model of emissions of gases and aerosols from nature). Atmos Chem Phys 2006;6(11):3181–210.

[28] Schuh G, et al. Emissions of volatile organic compounds from sunflower and beech: dependence on temperature and light intensity. J Atmos Chem 1997;27(3):291–318.

[29] Winer AM, et al. Emission rates of organics from vegetation in California's Central Valley. Atmos Environ Part A 1992;26(14):2647–59.

[30] Morrison EC, Drewer J, Heal MR. A comparison of isoprene and monoterpene emission rates from the perennial bioenergy crops short-rotation coppice willow and Miscanthus and the annual arable crops wheat and oilseed rape. GCB Bioenergy 2016;8(1):211–25.

[31] Storlien JO, et al. Carbon dioxide and nitrous oxide emissions impacted by bioenergy sorghum management. Soil Sci Soc Am J 2014;78(5):1694–706.

[32] DEFRA. Growing short-rotation coppice; 2002. p. 32.

[33] Drewer J, Yamulki S, Leeson SR, Anderson M, Perks MP, Skiba UM, McNamara NP. Difference in soil methane (CH_4) and nitrous oxide (N_2O) fluxes from bioenergy crops SRC willow and SRF scots pine compared with adjacent arable and fallow in a temperate climate. Bioenergy Res 2017;10(2):575–82.

[34] Sabbatini S, et al. Greenhouse gas balance of cropland conversion to bioenergy poplar short-rotation coppice. Biogeosciences 2016;13(1):95–113.

[35] Walter K, et al. Direct nitrous oxide emissions from oilseed rape cropping–a meta-analysis. GCB Bioenergy 2015;7(6):1260–71.

[36] DEFRA. 2015. https://www.gov.uk/government/uploads/system/uploads/attachment_data/file/483812/nonfood-statsnotice2014-10dec15.pdf.

[37] Harris ZM, Spake R, Taylor G. Land use change to bioenergy: a meta-analysis of soil carbon and GHG emissions. Biomass Bioenergy 2015;82:27–39.

[38] Harris ZM, Alberti G, Viger M, Jenkins JR, Rowe R, McNamara NP, Taylor G. Land-use change to bioenergy: grassland to short rotation coppice willow has an improved carbon balance. GCB Bioenergy 2017;9:469–84.

[39] Keith AM, Alberti G, Viger M, Jenkins JR, Rowe R, McNamara NP, Taylor G. Implications of land-use change to short rotation forestry in Great Britain for soil and biomass carbon. Glob Change Biol Bioenergy 2015;7(3):541–52.

[40] McCalmont JP, McNamara NP, Donnison IS, Farrar K, Clifton-Brown JC. An interyear comparison of CO_2 flux and carbon budget at a commercial-scale land-use transition from semi-improved grassland to *Miscanthus* x *giganteus*. GCB Bioenergy 2017;9:229–45. https://doi.org/10.1111/gcbb.12323.

[41] Rowe RL, et al. Evaluating ecosystem processes in willow short rotation coppice bioenergy plantations. GCB Bioenergy 2013;5(3):257–66.

[42] Richards M, et al. High-resolution spatial modelling of greenhouse gas emissions from land-use change to energy crops in the United Kingdom. GCB Bioenergy 2017;9(3):627–44.

[43] Tuck G, et al. The potential distribution of bioenergy crops in Europe under present and future climate. Biomass Bioenergy 2006;30(3):183–97.

[44] EU.2015.Availablefrom:http://ec.europa.eu/eurostat/web/environmental-data-centre-on-natural-resources/natural-resources/energy-resources/energy-from-biomass.

[44a] EEA. EU bioenergy potential from a resource efficiency perspective. EEA Report No 6/2013 ISSN. Luxembourg: Publications Office of the European Union; 2013. ISBN 978-92-9213-397-9.

[45] Bowyer C. Anticipated indirect land use change associated with expanded use of biofuels and bioliquids in the EU—an analysis of the national renewable energy action plans. London, UK: IEEP; 2010.

[46] IEA. 2015. Available from: https://www.iea.org/topics/renewables/subtopics/bioenergy/.

[47] Fowler D, et al. Effects of land use on surface–atmosphere exchanges of trace gases and energy in Borneo: comparing fluxes over oil palm plantations and a rainforest. Philos Trans R Soc Lond B Biol Sci 2011;366(1582):3196–209.

[48] Oates LG, et al. Nitrous oxide emissions during establishment of eight alternative cellulosic bioenergy cropping systems in the North Central United States. Glob Change Biol Bioenergy 2016;8(3):539–49.

[49] Bordonal RD, et al. Greenhouse gas balance from cultivation and direct land use change of recently established sugarcane (*Saccharum officinarum*) plantation in south-central Brazil. Renew Sustain Energy Rev 2015;52:547–56.

[50] Case SDC, et al. Can biochar reduce soil greenhouse gas emissions from a Miscanthus bioenergy crop? GCB Bioenergy 2014;6(1):76–89.

[51] Zhang YJ, et al. Annual accounting of net greenhouse gas balance response to biochar addition in a coastal saline bioenergy cropping system in China. Soil Tillage Res 2016;158:39–48.

[52] AEBIOM. European biomass statistics. Brussels: E.B. Association; 2010. p. 75.

[53] Lal R. Soil and environmental implications of using crop residues as biofuel feedstock. Int Sugar J 2006;108(1287):161–7.

[54] Chan KY, Heenan DP. The effects of stubble burning and tillage on soil carbon sequestration and crop productivity in southeastern Australia. Soil Use Manag 2005;21(4):427–31.

[55] Thornley P, Chong K, Bridgwater T. European biorefineries: implications for land, trade and employment. Environ Sci Pol 2014;37:255–65.

[56] Dufossé K, et al. Effects of a 20-year old Miscanthus × giganteus stand and its removal on soil characteristics and greenhouse gas emissions. Biomass Bioenergy 2014;69:198–210.

[57] Roth B, et al. Are the benefits of yield responses to nitrogen fertilizer application in the bioenergy crop Miscanthus × giganteus offset by increased soil emissions of nitrous oxide? GCB Bioenergy 2015;7(1):145–52.

[58] Behnke GD, David MB, Voigt TB. Greenhouse gas emissions, nitrate leaching, and biomass yields from production of Miscanthus x giganteus in Illinois, USA. Bioenergy Res 2012;5(4):801–13.

[59] Whitaker J, et al. Sources of variability in greenhouse gas and energy balances for biofuel production: a systematic review. GCB Bioenergy 2010;2(3):99–112.
[60] Kyle P, et al. Meeting the radiative forcing targets of the representative concentration pathways in a world with agricultural climate impacts. Earth's Future 2014;2(2):83–98.
[61] Searchinger T, et al. Use of U.S. croplands for biofuels increases greenhouse gases through emissions from land-use change. Science 2008;319(5867):1238–40.
[62] Plevin RJ, et al. Greenhouse gas emissions from biofuels' indirect land use change are uncertain but may be much greater than previously estimated. Environ Sci Technol 2010;44(21):8015–21.
[63] Meyfroidt P, et al. Globalization of land use: distant drivers of land change and geographic displacement of land use. Curr Opin Environ Sustain 2013;5(5):438–44.
[64] Rowe RL, et al. Potential benefits of commercial willow Short Rotation Coppice (SRC) for farm-scale plant and invertebrate communities in the agri-environment. Biomass Bioenergy 2011;35(1):325–36.

Further Reading

[1] Lal R. World crop residues production and implications of its use as a biofuel. Environ Int 2005;31:575–84.
[2] Monks PS. Tropospheric photochemistry. In: Hewitt CN, Jackson A, editors. Handbook of atmospheric science. Oxford: Blackwell Science Ltd; 2003.
[3] Muñoz C, Paulino L, Monreal C, Zagal E. GHG (CO_2 and N_2O) emissions from soils: a review. Chil J Agric Res 2010;70(3):485–97.
[4] Roeder M, Whittaker C, Thornley P. How certain are greenhouse gas reductions from bioenergy? Life cycle assessment and uncertainty analysis of wood pellet-to-electricity supply chains from forest residues. Biomass Bioenergy 2015;79:50–63.

CHAPTER

7

Biomass Harvesting, Processing, Storage, and Transport

Carly Whittaker, Ian Shield
Rothamsted Research, Harpenden, United Kingdom

7.1 INTRODUCTION

Biomass supply chains involving forestry or perennial crops consist of cultivation, harvesting, processing, storage, and transportation. This chapter focuses on stages of the supply chain that occur from the point at which the crop is ready to harvest. Some main biomass systems are used to illustrate typical processes and fuel requirements for harvesting, though in some cases this may be influenced by the biomass yield per hectare. These are usually variable costs/requirements, whereas site establishment and termination can be seen as a 'fixed' parameter in life cycle assessments (LCAs). Some biomass types, like wastes, do not have such a phase attributed to them, rather their 'production' system begins at processing. Once harvested, or collected, there are various options for processing biomass into a form that is more suitable for transportation, storage, and final use. The key of processing is to produce a more homogenous material that offers end users a fuel whose behaviour and quality are easy to predict, monitor, and feed into combustion systems in a reliable manner. Finally, transport is an unavoidable component of biomass supply chains, which must also be included in LCA studies.

7.2 HARVESTING OPTIONS IN BIOENERGY SUPPLY CHAINS

7.2.1 Harvesting Options in Willow

Willow can be harvested as chips, billets, and rods, each requiring a specialist machine. A forage harvester is typically used to harvest as chip, for which farmers are recommended to budget £30/h for fuel requirements alone. This equates to approximately 45l/h, assuming average yields of 9 oven-dried tonnes (ODT)/h/year. Two tractors are required to carry the crop offsite using 50l/h and a loader tractor forming the storage pile at 4l/h. There is very little data available on billet harvesting, which involves cutting the willow stem into roughly

Greenhouse Gas Balances of Bioenergy Systems
https://doi.org/10.1016/B978-0-08-101036-5.00007-0

20 cm pieces, but the author's currently unpublished estimate places fuel use at 37 l/h with similar carting and store management as above, but requiring subsequent chipping. Rod harvesting requires around 30 l/h and forms small dispersed heaps around the field, delaying the need for carting and chipping until later in time. It should be considered that these costs are incurred only every third year.

7.2.2 Harvesting Options in Miscanthus

For harvesting Miscanthus, a mower conditioner [1] or forage harvester may be used to cut the crop at the base for later to be baled by a conventional baler. The forage harvester breaks the stems more than the mower and may chip the material. Chipped Miscanthus has a very low bulk density, and this is only usually carried out for equestrian and animal bedding end uses at short distance from the field. In normal crop (yield = 12 tonne/h DM), the fuel requirement is 15 l/h for cutting, 12 l/h for baling, and 27 l/h for loading, carting, and unloading trailers [2].

7.2.3 Harvesting Options in Forestry

Biomass supply chains in the forestry sector will handle a number of forms of material, including whole tree thinnings, roundwood, and brash bales. Whole trees are typically the first two thinnings in a stand, after which the stems are large enough (<15 cm) to be harvested for pulpwood [3]. This is typically performed by a forest harvesting head with an integrated chainsaw that cuts the tree at the base and then de-limbs and tops the trees onsite. More valuable stemwood (>30 cm) are typically harvested in this manner; however, the stems are 'bucked' into lengths determined by the consumer and the diameter of the wood. The remaining brash on the site can be collected into brash bales by using modified brash bundle heads [4], though this is not conventionally carried out in current biomass supply chains. This material is not only important for nutrient recycling, but also needed for the protection of soil from compaction by passing machinery. It has been suggested that removing stem tips and branches from the forest floor may lead to a decreased flux of carbon into the forest litter, resulting in a reduction of the carbon storage in the forest soil [5]. It is instead suggested that low value roundwood will enter the biomass market. After cutting, all material must be carried from the site using either a skidder or forwarder. The difference between them is that a skidder drags the biomass along the forest floor, but a forwarder carries the material on a trailer. Some fuel consumptions of these processes are listed below in Table 7.1.

TABLE 7.1 Fuel Consumption for Harvesting Forestry Biomass Resources

	Bulk Density (tonnes/m^3) (30% MC)	Fuel Consumption (l/tonne)		
		Felling/Collecting	Forwarding	Total
Roundwood	0.306	3.92	2.94	6.86
Whole trees	0.283	4.24	3.18	7.42
Brash bales	0.330	6.06	2.73	8.79

From Whittaker C, Mortimer N, Murphy R, Matthews R. Energy and greenhouse gas balance of the use of forest residues for bioenergy production in the UK. Biomass Bioenergy 2011;35:4581.

7.3 PROCESSING OPTIONS FOR BIOMASS

Once harvested, there are various options for processing biomass into a form that is more suitable for transportation, storage, and final use. Two somewhat opposed processes are detailed here: comminution and densification: other more advanced processing options are covered in other chapters of this book.

7.3.1 Comminution

Nearly all biomass types will need to be comminuted to increase drying efficiencies or further processing. In woody biomass supply chains, such processing is usually carried out by a wood chipper or less commonly by a tubgrinder [7]. Chippers are high-speed rotary devices operating with either a disc, drum, or screw design, the difference depending on where the cutting knives are positioned [8]. Mobile chippers are often diesel-powered and tractor or self-propelled, though the latter is more powerful and best suited for larger operations [9]. Stationary chippers are often much larger and powerful than mobile chippers, being restricted to the plant where the biomass is received, and are powered using the electricity supply as supplied to the plant [4]. From a review of the literature, the fuel consumption and hence throughput of chippers are dependent on the feed in bulk dimensions, chipper type, and moisture content of wood [8]. Despite these factors, the fuel consumption of chipping tends to range between 1 and 2 l, or 2.5–5 kg CO_2 equiv./tonne of wood processed, with a throughput of around 7 tonnes/h [8]. Stationary chippers have a much higher throughput, of around 200 tonnes/h [4], which translates to a delivered energy of 30–45 MJ/tonne, although the GHG emissions depend on the energy mix that comprises the electricity supply. This will be country-specific and dependent on whether self-generated or grid electricity is utilised.

Tubgrinders, or shredders, are semimobile units used for processing particularly larger items of biomass at a high rate (up to 100 tonne/h [7]) and can be seen as a preprocessing stage for chipping. This may be utilised in wood waste processing due to the heterogeneous nature of the material. The process is diesel-powered and typically consumes around 4 l, or 10 kg CO_2 equiv./tonne shredded (Dalkia pers. Com. 2013).

7.3.2 Densification

7.3.2.1 Wood Pellets

Imported wood pellets are currently the dominant biomass resource utilised in the United Kingdom. They are made from finely powdered and dried untreated biomass that is passed through a pellet dye at high temperature and pressure. The act of compression and the high temperature causes the lignin within the biomass to melt, glueing the particles together to then re-solidify as a pellet after cooling. The result is a high-value and high-density pellet that is consistently shaped and thus more efficiently transported compared to wood chips or straw bales [10].

The process of pelleting requires a significant amount of energy: a moisture content of 8%–12% is required to produce adequately durable pellets [11], which cannot be achieved

TABLE 7.2 Typical Energy Requirements and GHG Emissions From Production of 1 tonne (wood) Pellets (tp) Based on [17], and [18] With Natural Gas as a Heat Source

Stage	Amount (MJ/tonne pellets)	GHG Emissions (kg CO_2 equiv./tonne pellets)	Contribution (%)
Drying	93	11.51	5
Drying (heat)	2880	147.23	69
Hammermilling	120	14.85	7
Pelleting	270	33.41	16
Cooling	6	0.74	0
Auxiliary power	43	5.32	2
Total		213.06	

without artificial drying. A great deal of energy is required to hammermill the material to a fine enough particle size to ensure sufficient bonding between them [12]. Finally, the pellets must be placed under a high pressure (300 MPa) in the pellet mill [13]. Table 7.2 shows typical energy requirements for pelleting. Other types of biomass pellets (e.g. agricultural residues) may have differences; for example, in good weather conditions artificial drying may not be required [14]. The energy requirements for drying could be provided from conventional means (natural gas) or from biomass or waste sources, which could have a large impact on the total GHG emissions. Assuming natural gas is used, the production of one tonne of wood pellets adds an additional 213 kg CO_2 equiv./tonne of biomass feedstock. Although this additional processing decreases the GHG emission savings achievable from a given biomass type, they cannot undermine the potential for it to cost-effectively mitigate GHG emissions in energy applications [15,16].

7.3.2.2 Bales

Densified biomass also includes lower density biomass types, such as bales, though this form presents a more viable option to transport such dry feedstocks. Producing bales is estimated to require 3.5 l/tonne baled and forwarded to the road side [2].

7.4 STORAGE OPTIONS FOR BIOMASS

Storage is an unavoidable stage of any supply chain. With biomass, there is a need to provide a buffer between the supply of fuel and its demand. This is particularly an issue in heat supply chains, where the demand will mainly occur during the winter [19]. The supply of biomass may depend on seasonally limited harvesting windows, or in waste supply chains, the sporadic and dispersed nature of the supply means that aggregation points are needed to regulate the flow of material.

Though the GHG emissions associated with handling and storing biomass are small [15], it can influence the final GHG emission results in a number of ways. Storage can be used to season biomass, particularly wood, which ultimately increases its net calorific value [20]. Also, one of the most challenging tasks in biomass supply chains is storing the material to reduce dry matter (DM) losses and unwanted quality changes due to degradation [21]. DM losses not only increase the GHG emissions for providing a unit of available fuel, but also increase the land or resource potential required to supply a given amount of energy. Changes in fuel quality also introduce losses through rejection and can affect the lower heating value (LHV) or combustion properties. For example, higher ash contents can decrease the LHV, and fuel-bound nitrogen can influence emissions of NO_x during combustion [22]. Fines, or particles less than 3.15mm, tend to burn rapidly and generate very high temperatures and can lead to ash melting and slagging [23].

DM losses of storing fresh wood in this form will depend on a number of factors, including the form and moisture content of the material, ambient temperatures, rainfall, ground conditions, exposure, and how densely the material is packed. In general, comparing the results from storage studies is difficult because they are often performed under different conditions, with different species and for differing periods of times. The following subsections focus on storage options for woody biomass, baled biomass, and higher value densified biomass.

7.4.1 Wood

7.4.1.1 Small Logs, Brash, and Coppice Rods

As mentioned previously, seasoning wood is necessary where drier feedstocks are required. This involves leaving wood in piles outside or under covered storage where it can dry in ambient conditions rather than through more expensive active drying processes. The DM losses of storage of whole logs are believed to be small, or around 2% after a year [24], though in reality few studies have examined this. Likewise, rod storage is not well-documented, though in-house experimentation found average DM losses of 5% across 4 heaps stored for 9 months. A number of studies show that DM losses during brash bale storage range between 5% and 21%, with some reductions when stored under cover (Table 7.3). Sometimes the dry matter losses of storing brash can appear to be higher in conifer species due to needle drop [25,26].

TABLE 7.3 Range of DM Losses From Brash Bale Storage From Literature

Storage Conditions	DM Losses (%)	Storage Period (Months)	References
Outside piles	21	9	[25]
Covered storage	12	9	
Outside and covered piles	14	10	[27]
Outside piles	8	12	[28]
Outside piles	17	12	
Covered storage	12	12	
Covered storage	14	12	
Outside piles	13	6	[20]
Outside piles	5	12	
Outside piles	10	10	Estimate based on [29]

7.4.1.2 Wood Chips

It is common practice to store wood chips outside in piles, aiming to redistribute moisture away from the chips to the edge of the heap where it can evaporate [30]. It was previously suggested that the chips can dry from 50% to 25% MC by the end of the summer [19], though more recent studies suggest the outside 'crust' remains wet and the final lifting of the heap means the stack only dries by about 10% [31,32].

It is typical for the temperature to rise very rapidly after heap establishment; indicating microbial decomposition [19,29,33,34]. In the literature, DM losses from wood chip storage range widely, between 1% and 27% [35], though two recent studies showed whole heap dry matter losses 21%–23% in willow chip after 3–5 months of storage in the United Kingdom [31,32]. Similarly, over 9 months DM losses of 21%, 22% were found in covered heaps of poplar chips [36]. That study showed the loss plateaued after 4–5 months, suggesting the losses are due to decay fungi that operate at cooler temperatures, rather than thermophilic bacteria. Covering the storage heaps with plastic could slightly help, but not enough to compensate for the increased costs of this operation [37]. A sensitivity analysis examining the GHG emission savings from willow found that a DM loss of 20% increases the GHG emissions (per MWh) by 11% [32].

It is suggested that the decay process can lead to the release of methane [21], but there is some uncertainty over the extent at which this occurs [38]. A recently published sensitivity analysis showed that 1% loss of carbon in the form of methane would triple the GHG emissions from willow-based heat [35]. Experimental data for willow heaps showed that if any methane is produced within the stack, it occurs during a very short peak after the initial heating phase [31], and there is uncertainty to whether gases leave the stack [32] or, like in landfill, are oxidised as they migrate outwards by methanotrophic bacteria [34].

7.4.2 Bales

Baled biomass typically includes agricultural residues and energy grasses, though willow can also be baled. As this technique usually involves drier feedstocks, issues of self-heating and degradation are not as large of a concern, though there is evidence that patches of moisture can support microbial growth and lead to losses [39]. Uncovered storage can incur losses of 13% DM [40], or in wrapped or stored bales, it is more like 3% [41]. Therefore, in order to preserve their dry state, bales are best stored under cover, in either in an open barn or under tarpaulin.

7.4.3 Densified Biomass

A substantial amount of processing is required to produce densified pellets. As a result, losses during storage would have a higher impact on the net GHG balance of the system compared to upstream losses. Pellets rapidly disintegrate in humid conditions [42]; therefore, it is vital that storage facilities can adequately protect pellets from rain, condensation, or ground water [43,44].

There are reports of self-heating in pellet stores, though here it is due to chemical oxidation of the biomass, or from physical forces of condensation and adsorption [45,46]. The effect of this on DM losses has not been tested, though it is believed that this can lead to the condensation of water causing pockets of damp, usually in a centralised 'chimney' [47], which could lead to pellet disintegration. This is, however, believed to be a slow process that only occurs at large scales, where it is best practiced to flood the storage areas with nitrogen gas [48].

7.5 TRANSPORTING BIOMASS

Due to its dispersed nature and low energy density, biomass must be collected from a wide area. The most cost-effective method of land-based transportation is via rail; however, many biomass resources are not located close to rail hubs, and loading between trucks and rail will increase costs. Therefore, many supply chains will rely on transportation via roads. Exceptions for this are where imports are received at ports, which have supporting rail infrastructures.

7.5.1 Road Transport

There are many sources of GHG emissions from road transport in literature; however, the data is usually provided as consolidated numbers that are based on many assumptions [49]. Emission factors for road transport provided by DEFRA and DECC usually suffice for most feedstocks, though there may be requirements to adjust the factors for transporting particularly low-density feedstocks. With these feedstocks, the issue is that of volume, where the full payload capacity of the vehicle is not reached. Having less cargo in the vehicle also increases the cost and GHG emissions per tonne transported. In most large articulated trucks (with a volume of around 70 m^3), any feedstock under 400 kg/m^3 in density will be volume-limited. According to size and height restrictions, a flat-bed trailer can fit 36 Heston bales, or a total weight of 18 tonnes, so the best payload (86% full) achievable in theory is achieved using a 32 gross vehicle weight (GVW) truck with a payload of around 21 tonnes [4]. Due to differences in bulk density, different feedstocks will have different road transport GHG emission factors (Fig. 7.1).

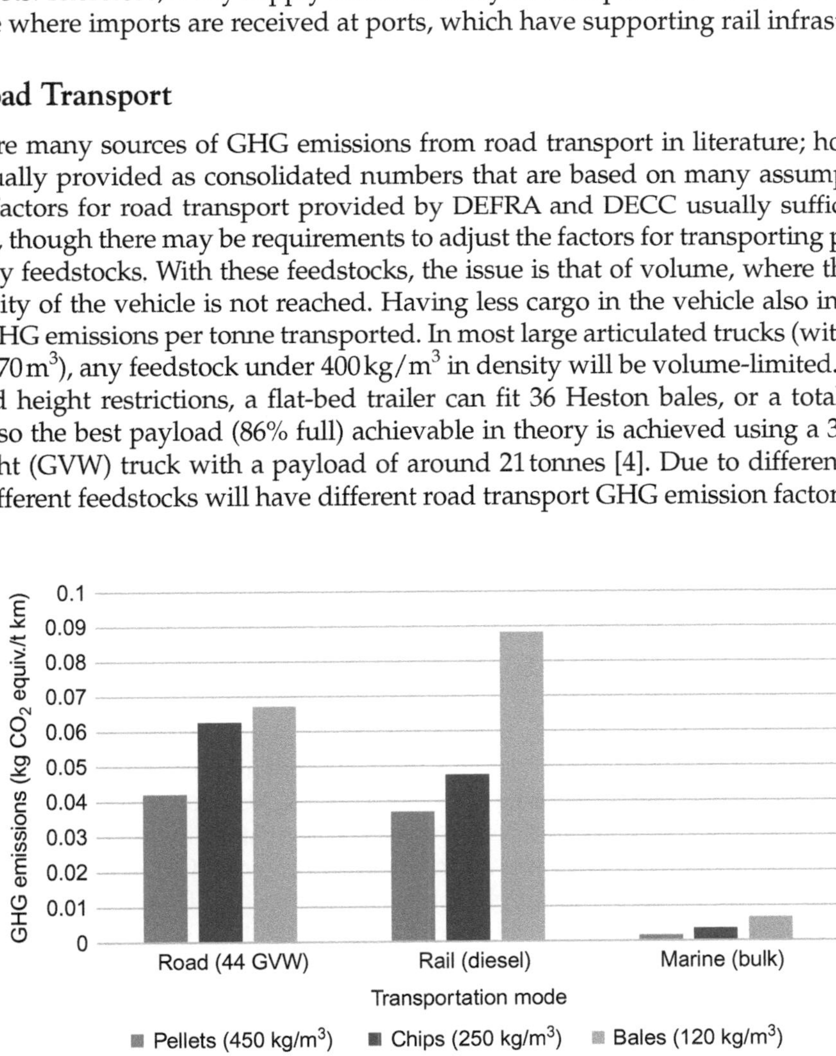

FIG. 7.1 Transport emissions for three different biomass forms via road, rail, and marine transport. *The figures are based on DEFRA, DECC. 2013 Guidelines to DEFRA/DECC's GHG conversion factors for company reporting: methodology paper for emission factors. London; 2013 to account for an empty return journey and the volume of the biomass.*

7.5.2 Rail Transport

Estimating the volume of train cargo is difficult as trains are flexible with the mass of cargo carried and the number of train hoppers used to contain the load. The emission factors shown in Fig. 7.1 are based on the following assumptions; that a max of 1600 tonnes can be carried by a train without dramatically affecting speed, the maximum train length is 700 m [50], and the typical volume of a biomass-specific train hopper is 116 m^3 [51]. Freight trains can be hauled by diesel or electric locomotives, but the former represents 93% of the total [18].

7.5.3 Marine Transport

Wood pellets are favoured in large-scale biomass supply chains as the densified nature of the feedstock reduces costs throughout the supply chain in regard to handling, storage, and transport. The trade in wood chips is predominantly limited between European countries, although there are some instances where longer transoceanic supply chains exist [52]. Marine transport of biomass occurs in bulk carriers that can carry 80,000 tonnes or even more. As the vessels are so large, the feedstock will always be weight-limited. The GHG emissions from marine transport are generally small, about 40 times lower compared to road transport.

7.5.4 Transport Costs

There are limited ways to calculate costs of a hypothetical transport operation, without getting a quote direct from a haulage company for a specific delivery. Many freight operations will have agreed contracts to deliver goods at a given time [53], which cost according to distances required, time of delivery, and speciality of the cargo. Typical road transportation costs can be found [54], which are based on a number of fixed and mileage-related costs. It must be stressed that quotes must be based on the whole journey rather than a 'per hour or km' basis. In this case, a quotation method, used in the retail industry, can be used where costs are based on the number of return-journeys that can be made in 1 day.

7.6 SUMMARY AND TRADE-OFFS

There are a number of options available to bioenergy growers, producers, collectors, and processors. Most automated screw-fed combustion systems require that the biomass is supplied in chip or pellet form to ensure a consistent and reliable fuel feed. Producing pellets may require extra investment in capital costs, as well as a greater input of primary energy and result in higher GHG emissions; however, they can still provide GHG savings compared to fossil fuels. Also, further down the supply chain, GHG emissions from transport and storage losses can be reduced by having a denser fuel compared to chips or bales. Harvesting wood as chip may be far cheaper, but unless the supply chain is equipped to dry the chip, it will be subject to losses due to decay in storage. Harvesting as logs or coppice stems may require rarer specialist machinery, but can reduce dry matter losses during storage, though incurring additional costs from handling and then from chipping or pelleting. Therefore, the selections of how biomass is harvested and processed will depend on the supply chain structure, machine availability, and the final end use demand.

References

[1] Bullard M, Metcalf P. Estimating the energy requirements and CO_2 emissions from production of the perennial grasses Miscanthus, Switchgrass and Reed Canary Grass; 2001.

[2] Whittaker C. The importance of life cycle assessment methodology in the regulation of biofuels. Ph.D. thesis. Bath: Mechanical Engineering, The University of Bath; 2013.

[3] Matthews RW, Mortimer ND. Estimation of carbon dioxide and energy budgets of wood-fired electricity generation systems. Farnham and Sheffield: Mensuration Branch, Forestry Commission Research Agency and Resources Research Unit Sheffield Hallam University; 2000.

[4] Whittaker C, Mortimer N, Murphy R, Hillier J, Smith P. The life cycle analysis of using biomass for heat and electricity production in the UK. London: Imperial College London; 2009.

[5] Schlamadinger B, Apps M, Bohlin F, Gustavsson L, Jungmeier G, Marland G, et al. Towards a standard methodology for greenhouse gas balances of bioenergy systems in comparison with fossil energy systems. Biomass Bioenergy 1997;13:359.

[6] Cummer KR, Brown RC. Ancillary equipment for biomass gasification. Biomass Bioenergy 2002;23:113.

[7] Webster P. Internal Project Information Note 19/06: Large Chippers. Forest Research; 2007.

[8] Spinelli R, Hartsough B. A survey of Italian chipping operations. Biomass Bioenergy 2001;21:433.

[9] Evans G. Techno-economic assessment of biomass "densification" technologies. York: NNFCC; 2008.

[10] Lehtikangas P. Quality properties of pelletised sawdust, logging residues and bark. Biomass Bioenergy 2001;20:351.

[11] Hamelinck CN, Suurs RAA, Faaij APC. International bioenergy transport costs and energy balance. Biomass and Bioenergy 2005;29:114.

[12] Nielsen NPK, Gardner DJ, Poulsen T, Felby C. Importance of temperature, moisture content, and species for the conversion process of wood residues into fuel pellets. Wood Fiber Sci 2009;41:414.

[13] Obernberger I, Thek G. Physical characterisation and chemical composition of densified biomass fuels with regard to their combustion behaviour. Biomass Bioenergy 2004;27:653.

[14] DEFRA, DECC. 2013 Guidelines to Defra/DECC's GHG conversion factors for company reporting: methodology paper for emission factors. London; 2013.

[15] Sultana A, Kumar A. Development of energy and emission parameters for densified form of lignocellulosic biomass. Energy 2011;36:2716.

[16] AEA Technology, Associates NE. Biomass Environmental Assessment Tool. 2.1 ed. Oxford and Sheffield; 2010.

[17] Wilson TO, McNeal FM, Spatari SG, Abler D, Adler PR. Densified biomass can cost-effectively mitigate greenhouse gas emissions and address energy security in thermal applications. Environ Sci Technol 2011;46:1270.

[18] Thörnqvist T. Drying and storage of forest residues for energy production. Biomass 1985;7:125.

[19] Garstang J, Weekes A, Poulter R, Bartlett D. Identification and characterisation of factors affecting losses in the large-scale, nonventilated bulk storage of wood chips and development of best storage practices. London: Department of Trade and Industry; 2002.

[20] Wihersaari M. Greenhouse gas emissions from final harvest fuel chip production in Finland. Biomass Bioenergy 2005;28:435.

[21] Baxter XC, Darvell LI, Jones JM, Barraclough T, Yates NE, Shield I. Miscanthus combustion properties and variations with Miscanthus agronomy. Fuel 2014;117(Part A):851.

[22] Francescato V, Antonini E, Metschina C, Schnedl C, Krajnc N, Koscik K, et al. Wood fuels handbook. Legnaro: AIEL-Italian Agriforestry Energy Association; 2008.

[23] Nurmi J. Changes in volumetric energy densities during storage of whole-tree feed stocks from silvicultural thinnings. Biomass Bioenergy 2014;61:114.

[24] Jirjis R. Storage and drying of wood fuel. Biomass Bioenergy 1995;9:181.

[25] Eriksson L, Gustavsson L. Comparative analysis of wood chips and bundles—costs, carbon dioxide emissions, dry-matter losses and allergic reactions. Biomass Bioenergy 2010;34:82.

[26] Pettersson M, Nordfjell T. Fuel quality changes during seasonal storage of compacted logging residues and young trees. Biomass Bioenergy 2007;31:782.

[27] Filbakk T, Høibø OA, Dibdiakova J, Nurmi J. Modelling moisture content and dry matter loss during storage of logging residues for energy. Scand J For Res 2011;26:267.

[28] Nurmi J. The storage of logging residue for fuel. Biomass Bioenergy 1999;17:41.

[29] Noll M, Jirjis R. Microbial communities in large-scale wood piles and their effects on wood quality and the environment. Appl Microbiol Biotechnol 2012;95:551.
[30] Whittaker C, Yates NE, Powers SJ, Misselbrook T, Shield I. Dry matter losses and greenhouse gas emissions from outside storage of short rotation coppice willow chip. Bioenergy Res 2016;9:288.
[31] Whittaker C, Yates NE, Powers SJ, Donovan N, Misselbrook TH, Shield I. Dry matter losses and greenhouse gas emissions from a willow wood chip storage heap. Biomass Bioenergy [submitted for publication].
[32] Hellebrand HJ. Emission of nitrous oxide and other trace gases during composting of grass and green waste. J Agric Eng Res 1998;69:365.
[33] Barlaz MA. Forest products decomposition in municipal solid waste landfills. Waste Manag 2006;26:321.
[34] Whittaker C, Macalpine W, Yates NE, Shield I. Dry matter losses and methane emissions during wood chip storage: the impact on full life cycle greenhouse gas savings of short rotation coppice willow for heat. Bioenergy Res 2016;9:288.
[35] Lenz H, Idler C, Hartung E, Pecenka R. Open-air storage of fine and coarse wood chips of poplar from short rotation coppice in covered piles. Biomass Bioenergy 2015;83:269.
[36] Manzone M, Balsari P, Spinelli R. Small-scale storage techniques for fuel chips from short rotation forestry. Fuel 2013;109:687.
[37] BTG. In: PCFPlus Research, editors. Methane and nitrous oxide emissions from biomass waste stockpiles. Enschede: Biomass Technology Group BV; 2002.
[38] Smith WA, Bonner IJ, Kenney KL, Wendt LM. Practical considerations of moisture in baled biomass feedstocks. Biofuels 2012;4:95.
[39] Sanderson MA, Egg RP, Wiselogel AE. Biomass losses during harvest and storage of switchgrass. Biomass Bioenergy 1997;12:107.
[40] Shinners KJ, Binversie BN, Muck RE, Weimer PJ. Comparison of wet and dry corn stover harvest and storage. Biomass Bioenergy 2007;31:211.
[41] Hartley ID, Wood LJ. Hygroscopic properties of densified softwood pellets. Biomass Bioenergy 2008;32:90.
[42] DEPV. Recommendations for storage of wood pellets. Berlin: German Wood Fuel and Pellet Association; 2012.
[43] Alakangas E, Paju P. Wood pellets in Finland—technology, economy and market: OPET Report 5. Jyvaskyla: VTT Processes; 2002.
[44] Svedberg URA, Hogberg H-E, Hogberg J, Galle B. Emission of hexanal and carbon monoxide from storage of wood pellets, a potential occupational and domestic health hazard. Ann Occup Hyg 2004;48:339.
[45] Ferrero F, Malow M, Noll M. Temperature and gas evolution during large scale outside storage of wood chips. Eur J Wood Wood Prod 2011;69:587.
[46] Larsson SH, Lestander TA, Crompton D, Melin S, Sokhansanj S. Temperature patterns in large scale wood pellet silo storage. Appl Energy 2012;92:322.
[47] Dafnomilis I, Schott D, Lodewijks G. Current Practices in Solid Biomass Terminals in the Netherlands. In: 23rd European biomass conference and exhibition, Vienna; 2015.
[48] The Neutral G. Freight emissions—Neutral Group UK; 2011.
[49] Beth M. Intermodal freight operations. Railway technical web pages; 2016.
[50] Drax. Drax unveils high-volume biomass wagon. Railway Gazette; 2013.
[51] Lamers P, Junginger M, Marchal D, Shouwenberg PP, Cocchi M. Global wood chip trade for energy. Utrecht: IEA Bioenergy; 2012.
[52] Cole S. Applied transport economics: policy, management & decision making. 3rd ed. London/Sterling: Kogan Page; 2005.
[53] LTD DFFI. RHA cost tables 2010. Bristol, UK; 2010.

Further Reading

[1] Whittaker C, Mortimer N, Murphy R, Matthews R. Energy and greenhouse gas balance of the use of forest residues for bioenergy production in the UK. Biomass Bioenergy 2011;35:4581.

CHAPTER

8

Biomass Conversion Technologies

Paul Adams, Tony Bridgwater†, Amanda Lea-Langton‡, Andrew Ross§, Ian Watson¶*

*University of Bath, Bath, United Kingdom †Aston University, Birmingham, United Kingdom ‡University of Manchester, Manchester, United Kingdom §University of Leeds, Leeds, United Kingdom ¶University of Glasgow, Glasgow, United Kingdom

8.1 INTRODUCTION

Biomass can be converted into several useful forms of energy using different processes. Fig. 8.1 portrays a number of different biomass (to bioenergy) conversion pathways. Various factors affect the choice of conversion process; these can include: the type, quantity, and characteristics of biomass feedstock, the desired form of the energy or end-use requirements, environmental standards, policy, economic conditions, and project-specific factors [1]. In most situations, it is the form in which the energy is required and the feedstocks which are available that determine the appropriate process route.

Bioenergy is the term used to describe energy derived from biomass feedstocks. A number of stages (such as harvesting, drying, storage, transportation, processing, etc.) are required to convert biomass into a useful energy source. Conversion of biomass to energy is undertaken using three main process technologies: bio-chemical, thermo-chemical, and physio-chemical. Bio-chemical conversion encompasses two main process options: anaerobic digestion (AD) (production of biogas, a mixture of mainly methane and carbon dioxide) and fermentation (production of ethanol) [3]. Within thermo-chemical conversion, the four main process options are combustion, pyrolysis, gasification, and liquefaction [1]. Physio-chemical conversion consists principally of extraction (with esterification) where oilseeds are crushed to extract oil. An overview of each of the main energy conversion options is presented in the following sub-sections to provide an understanding of the technologies in the context of greenhouse gas (GHG) emissions.

The information in this chapter should provide sufficient background for scientists and analysts to be able to outline and understand the steps in a typical bioenergy conversion scheme. Sufficient numerical detail is given of yields, efficiencies, etc. to support life cycle assessment (LCA) practitioners in devising initial process schemes and to check data

Greenhouse Gas Balances of Bioenergy Systems
https://doi.org/10.1016/B978-0-08-101036-5.00008-2

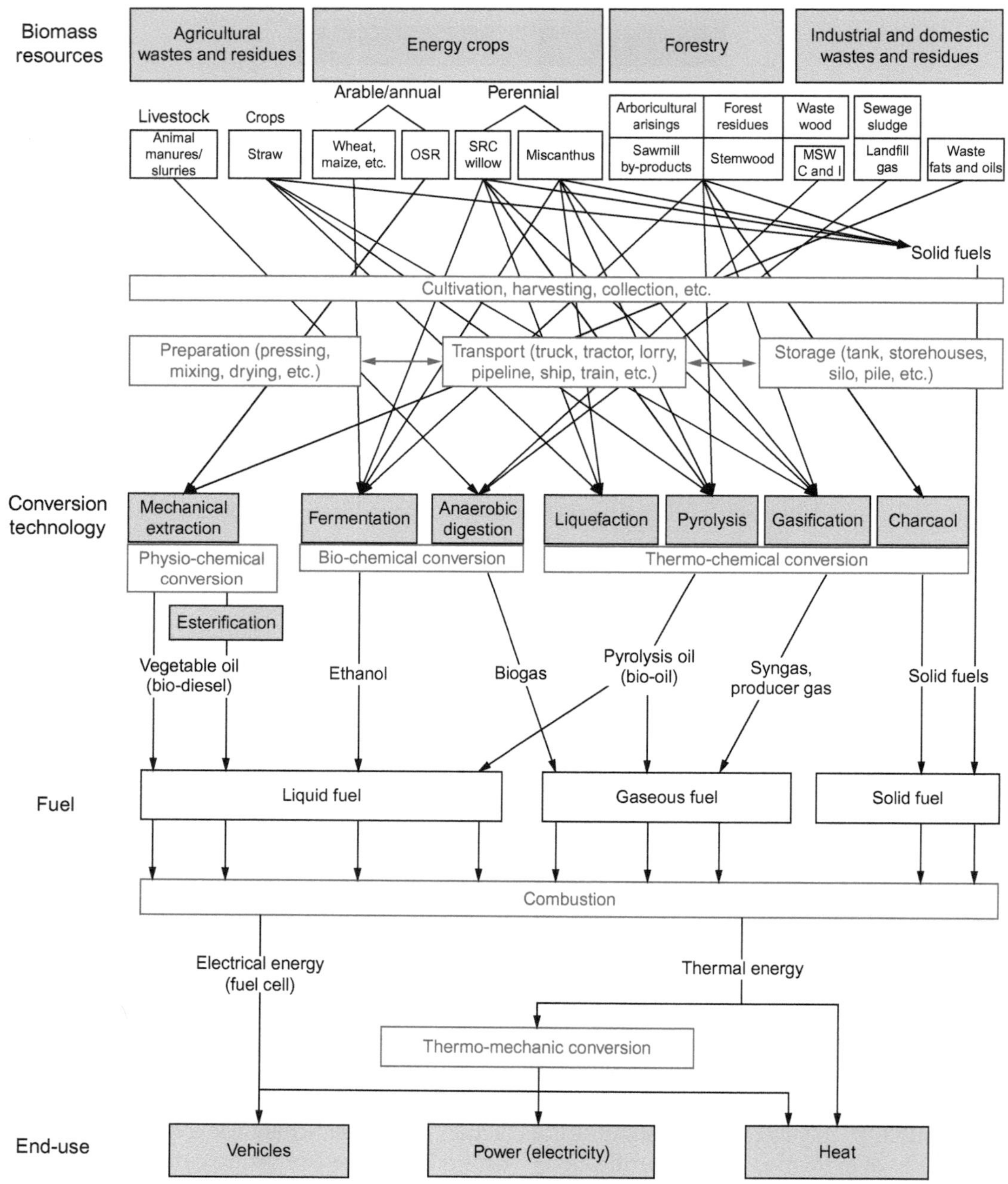

FIG. 8.1 Schematic representation of biomass conversion pathways [1–3].

robustness. However, accurate LCA assessments require accurate underpinning mass-energy balances and this is usually best achieved with a combination of engineering systems modelling and validation with experimental data, reported results, or operational plant data.

For some processes, such as combustion, the conversion process directly releases the desired energy end-product. For other processes (such as fermentation) conversion results in an intermediate vector (in this case a liquid) which is then converted to provide energy. For some prime-movers, e.g. aero-engines, significant upgrading of the fuel may be required to meet appropriate standards and prevent damage. Those final conversion steps are often in boilers, engines, and turbines that have been adapted to the characteristics of the biomass-derived intermediate. It is important to note that the efficiency and performance of that final conversion step often varies for different vectors and it cannot be assumed, e.g. that pyrolysis oil will achieve the same combustion efficiency as fossil-derived fuel oil or that the airborne emissions from syngas combustion will be the same as natural gas combustion. Wherever possible, primary data relating to the performance of the specific option being pursued should be used. If that is not available, analysts should carefully consider what factors might make a difference to the GHG balance of the overall system before using 'best-fit' data from a similar system and carrying out appropriate sensitivity checks.

8.2 BIO-CHEMICAL CONVERSION

Bio-chemical (or biological) conversion processes include both AD and fermentation. AD utilises bacteria to transform the organic matter into gaseous products, offering reasonable economics and abundant applications worldwide. By comparison, fermenting the feedstock, yeasts are used to convert the contained sugar into ethanol. This produces a diluted alcohol which then needs to be distilled and thus suffers from a lower overall process performance and high plant cost. The science of both technologies is well-understood and has been adapted to several feedstocks, environments, and purposes globally [3,4].

8.2.1 Anaerobic Digestion

AD is the conversion of organic material directly to a gas, known as biogas, which is a mixture of mainly methane and carbon dioxide with small quantities of other gases such as hydrogen sulphide [4]. Organic non-lignocellulosic (non-woody) material, the feedstock (also known as substrate) is converted by micro-organisms in the absence of oxygen. This conversion process produces stable and commercially useful compounds and is similar to composting except that composting is aerobic (involving oxygen) in its breakdown of organic matter [5]. The biomass is converted by bacteria in an anaerobic environment, producing a gas with an energy content of about 20%–40% of the lower heating value (LHV) of the feedstock [1]. AD is a commercially proven technology and is widely used for treating high moisture content (m.c.) organic wastes, i.e. 80%–90% m.c.

Feedstocks for AD include organic wastes and residues (such as animal manures or slurry) or energy crops (such as maize silage) grown specifically for feeding the AD plant. Anaerobic digesters produce conditions that encourage the natural breakdown of organic matter by bacteria in the absence of air. The two main products from an AD plant are:

- Biogas. This is a mixture of ~50%–60% methane, ~40%–50% carbon dioxide, and traces of other 'contaminant' gases [4];
- Bio-fertiliser (digestate). This is an inert and sterile wet product with valuable plant nutrients and organic humus. It can be separated into liquid and solid fractions for application to land or secondary processing [5].

AD produces biogas which can be combusted to generate electricity and/or heat or upgraded to biomethane and injected into the gas grid or used as a road fuel. It also provides a useful organic fertiliser as a by-product, making it an extremely promising technology with diverse applications. The generation of biogas is based on a complex process in various stages. Many factors influence the process from the cultivation of biomass, operating conditions, through to the end utilisation of the biogas (see Chapter 9). A schematic of a typical on-farm AD system showing the main on-site processes is presented in Fig. 8.2.

8.2.1.1 Manure Storage

The most common manure storage systems are cellars, silos, and manure bags. Manure contains bacteria that will produce methane as soon as it is excreted. Methane production in

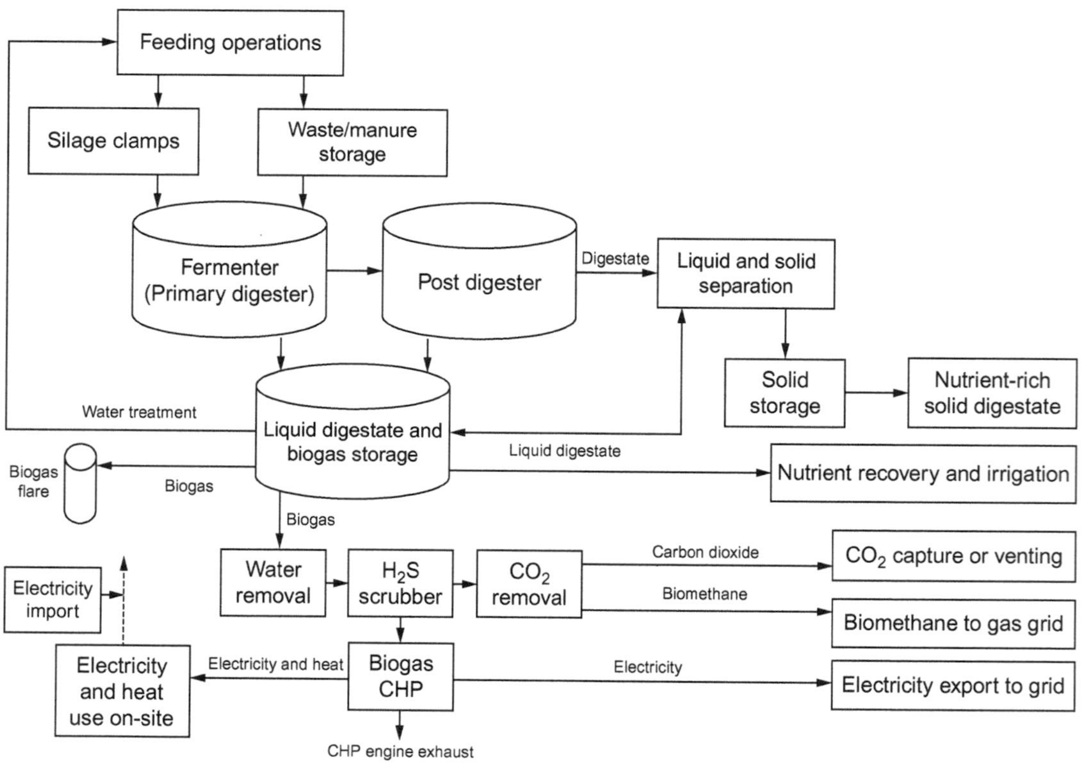

FIG. 8.2 Schematic overview of a typical farm-based AD system.

storage is a source of GHG emissions and will lower the biogas yield of the digester. Therefore, it is important to transport manure from storage to the digester as soon as possible, which is generally done by means of a pump.

8.2.1.2 Crop Storage

The addition of other biomass (called a co-substrate) with a higher energy density than manure can substantially increase the biogas yield. Crop silage is generally stored in large clamps which store enough feedstock from the harvest to supply the digester for a year. Well-produced silage should only produce very negligible or zero GHG emissions. However, poorly made silage can suffer from losses and deteriorate, resulting in a higher requirement for feedstock and therefore increased GHG emissions from cultivation and harvesting.

8.2.1.3 Pre-Treatment

Essentially, there are three different methods for pre-treatment, depending on the type of co-substrate and the size characteristics as delivered to the AD plant: mechanical treatment, pre-heating, and thermal treatment. Some co-substrates require a size reduction, such as grinding or chopping, to prevent pieces of co-substrate that are too large for the pumps and mixers of the installation [6]. Size reduction also increases the surface area for the bacteria, which will accelerate biogas production. Other types of co-substrate, such as fats, may require pre-heating in order to improve the flow characteristics. Certain co-substrates need thermal treatment in order to fulfil sanitation requirements [7].

8.2.1.4 Digester

In the digester, the substrates are heated and the AD process takes place. Contents of the digester are stirred periodically to mix new substrate with the old substrate to improve the penetration of bacteria with the fresh substrate, to realise an even temperature in the digester, to prevent and disturb the build-up of sedimentary layers, and to improve the metabolism of the bacteria by removing the gas bubbles and replacing them with fresh feedstock [6]. GHG emissions can arise from the digester only if gas can leak; generally modern digesters are gas-tight.

8.2.1.5 Post-Digestion Storage and Use

Digested substrates (digestate) are normally stored in a post-digestion storage tank for subsequent use as organic fertiliser or recirculation in the AD system. Post-digesters also allow for additional biogas to be generated and captured. Liquid digestate is often used as a fertiliser and for irrigation, and there are several nutrient-recovery technologies available to increase the nutrient concentration. The solid digestate fraction can be used as a nutrient-rich soil improver.

Biogas produced in the digester requires storage until it can be used. The storage type will depend on the selected construction, but is usually stored either in the digester or an external gas storage unit. Modern larger biogas facilities have the ability to store biogas in both the post-digester and combined digestate and biogas storage tanks. Emissions from digestate or biogas storage can arise depending on the plant design and how the storage systems are managed; this is further assessed in Chapter 9.

8.2.1.6 Gas Engine

The gas engine, functioning as a combined heat and power (CHP) unit, uses the biogas to generate electricity and heat. Biogas electricity usually supplies the parasitic load of the AD facility with additional electricity generation exported to the grid. Produced heat will often be used to heat the digester and for pasteurisation with the remaining heat used for heating buildings, greenhouses, drying biomass, or for other industrial processes. Biogas can be used for generating heat only or can be processed for use as a transportation fuel or for supply to a natural gas grid.

8.2.1.7 Biomethane

An alternative to combustion of raw biogas in CHP engines or boilers is the upgrading to a purer methane form known as biomethane. Upgrading biogas involves the removal of most of the carbon dioxide and further purification of other trace gases. Biomethane can then be used directly in vehicles with gas engines or injected into the gas grid. In terms of GHG emissions, biogas upgrading requires additional energy (usually electricity) and can require other inputs such as chemicals, water, or membranes depending on the upgrading technology employed. Methane emissions can also arise in the upgrading process. This increased energy requirement and methane loss can increase the GHG emissions from biomethane compared to biogas; however, it gives a more versatile high value fuel and can be used for a wider range of applications. Additionally, the carbon dioxide stream can be captured and used in industrial processes such as in greenhouses or the food and drinks industry. Methane slip (which increases fugitive GHG emissions) can be reduced to almost zero through the use of a thermal oxidiser. Production of biomethane can also have a nominally higher efficiency than biogas CHP since the electricity generation may waste heat; hence the use of LCA to assess and compare different conversion pathways is important.

8.2.2 Fermentation

Fermentation is another anaerobic biological process with the simple sugars from biomass feedstock converted to alcohol and carbon dioxide by the action of a different set of micro-organisms, usually yeasts. Ethanol (C_2H_5OH) is then separated from other components using heat to distill the mixture, so that the ethanol boils off and can then be cooled and condensed back to liquid [3]. Fermentation is used commercially on a large-scale in various countries to produce ethanol from sugar crops (e.g. sugar cane, sugar beet) and starch crops (e.g. maize, wheat). The biomass is ground down and the starch converted by enzymes to sugars, with yeast then converting the sugars to cellular energy and thereby producing ethanol. Purification of ethanol by distillation is an energy-intensive step which can result in relatively low net energy balances [8] with about 450 L of ethanol being produced per ton (t) of dry corn [1]. Solid residues produced from the fermentation process can be used as cattle-feed, and in the case of sugar cane, the bagasse can be used as a fuel for boilers or for subsequent gasification [1].

Bio-based ethanol (bioethanol) is most commonly used as an extender in petrol (gasoline) containing a percentage of ethanol. Higher blends (typically 85% ethanol, 15% gasoline—designated E85) are used in 'flexible-fuel vehicles' (FFVs) in countries such as Australia, Sweden, the United States, and Brazil. Perhaps the best known example of fermentation for fuel is the production of ethanol from sugar cane in Brazil where many cars run on 100% bioethanol [9]. Sugar cane stalks contain sufficiently high amounts of sugar and the plant is currently the lowest cost source of producing ethanol. Maize (also known as corn) is the second largest source of bio-fuel feedstock today, primarily because of its dominance in the United States for ethanol production [10]. Producing ethanol from grain starches requires more land than sugar cane, because the crops have lower fuel yields per hectare. The main sources of feedstock for fermentation in the Europe are wheat and sugar beet. Both of these crops are grown for food purposes and there are not significant surpluses available; hence fermentation is not anticipated to be used for bioenergy production on a large-scale in many parts of the world. However, fermentation can also be used as a pathway to a wide variety of products in a biorefinery, which can replace fossil fuel hydrocarbons in conventional refineries and there are significant research efforts to progress advanced technologies to enable fermentation of lingocellulosic feedstocks by pre-treatment and/or identification of appropriate enzymes (see Chapter 13).

8.3 THERMO-CHEMICAL CONVERSION

Thermo-chemical conversion of biomass can occur via pyrolysis, gasification, liquefaction, and combustion [11]. Pyrolysis is the preliminary stage of all thermo-chemical processes since it comprises chemical reactions to form solid, liquid, and gaseous products without oxygen. Technologies that produce gaseous or liquid intermediate products that can be upgraded to higher value energy (transportation fuels, electricity, etc.) are particularly valuable [11]. The main processes, the intermediate energy carriers, and the final energy products resulting from thermo-chemical conversion are illustrated in Fig. 8.3.

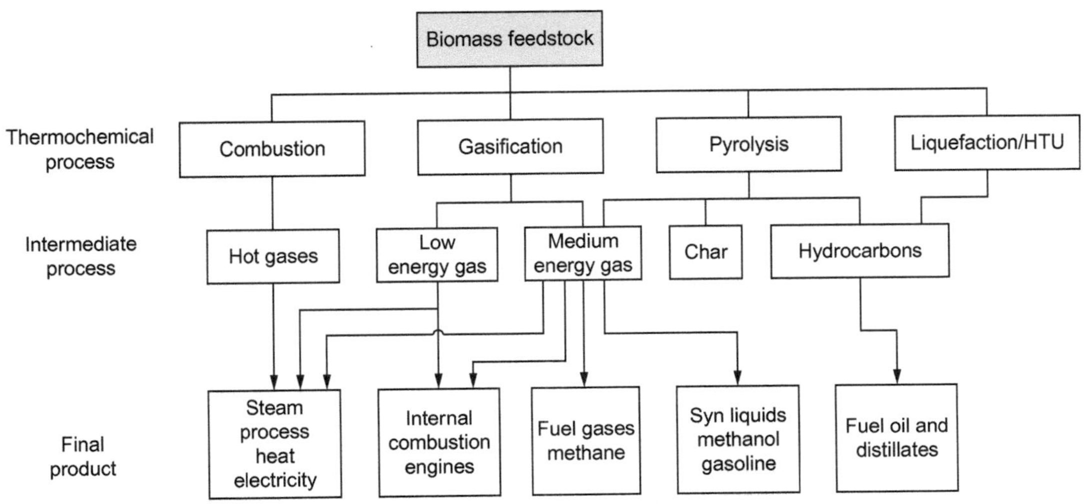

FIG. 8.3 Main processes, intermediate energy carriers, and final energy products from the thermo-chemical conversion of biomass. *Adapted from McKendry P. Energy production from biomass (part 2): conversion technologies. Bioresour Technol 2002;83(1):47–54. Available from: http://linkinghub.elsevier.com/retrieve/pii/S0960852401001195 [Cited 6 March 2017].*

In general, thermo-chemical processes have higher efficiencies than bio-chemical/biological processes, quicker reaction times (a few seconds or minutes for thermo-chemical processes compared to several days, weeks, or longer for bio-chemical processes) [12], and the superior ability to destroy most of the organic compounds, e.g. lignin materials in biomass are typically considered to be non-fermentable and consequently cannot be completely decomposed via biological approaches, whereas they are decomposable via thermo-chemical approaches [13,14].

As shown in Figs 8.1 and 8.3, the stored energy contained within biomass can be released either directly as heat via combustion/co-firing, or could be transformed into solid (e.g. charcoal), liquid (e.g. bio-oils), or gaseous (e.g. synthetic gas) fuels via pyrolysis, liquefaction, or gasification with various utilisation purposes [15]. Table 8.1 summarises the different types of thermo-chemical conversion processes, their classification, process conditions, reactors used, and typical product yields [11].

TABLE 8.1 Overview of Biomass Thermo-Chemical Conversion Technologies

			Product Yield			
Conversion Technology	**Process Condition**	**Reactor Types**	**Liquid (wt.%)**	**Gas (wt.%)**	**Solid (wt.%)**	**Reference**
PYROLYSIS (SECTION 8.3.4)						
Fast Pyrolysis	Atmospheric pressure, small particle size (<3 mm), short residence time (0.5–2 s), moderate temperature (400–550°C) in absence of oxygen	Fixed bed reactor, Tubular reactor, Bubbling fluidised bed reactor, Circulating fluidised bed reactor, Ablative Pyrolyzer, Rotating Cones reactor, Auger reactor, Cyclone reactor (N.B. the reactor type that can be used will depend on the pyrolysis method)	65–75	13–25	12–19	[16–18]
Slow Pyrolysis	Low heating rate, moderate temperature (350–750°C), atmospheric pressure, long residence time in absence of oxygen		30–50 in 2 phases	15–30	30–60	[19, 20]
Intermediate Pyrolysis	Moderate temperature (<500°C), moderate vapor residence time (4–10 s) and atmospheric pressure		45–55 in 2 phases	25–35	15–25	[21]
Flash Pyrolysis	Rapid heating (<0.5 s), very small particle sizes (<0.5 mm), temperature (400–1000°C)		60–70	10–15	15–25	[22, 23]
Vacuum Pyrolysis	Moderate temperature (300–500°C), pressure below atmospheric (<50 kPa)		45–60	17–27	19–27	[24–26]
Ablative Pyrolysis	Moderate temperature (450–600°C), atmospheric pressure, particle size <3.5 mm		60–80	6–10	12–20	[27]

TABLE 8.1 Overview of Biomass Thermo-Chemical Conversion Technologies—cont'd

Conversion Technology	Process Condition	Reactor Types	Product Yield			Reference
			Liquid (wt.%)	Gas (wt.%)	Solid (wt.%)	
Gasification (Section 8.3.5)	Moderate to high temperature (600–1200°C), presence/absence of catalyst, small particle size desirable, gasifying agent	Fixed bed, moving bed, fluidised bed and entrained flow gasifier	Gas yield (m^3 gas/kg biomass): 1–2.6			[28]–[30]
Combustion (Section 8.3.6)	Maximum temperature: 740–1300°C, Air mass flow: 0.1–0.5 kg/m^2s	Fixed bed, fluidised bed, circulating bed and entrained flow bed combustor	Power and heat			[31]
Co-firing (Section 8.3.6.8)	wt.% of biomass: 5–20	Boiler	Power and heat			[32]
Hydrothermal Liquefaction (Section 8.3.7)	Temperature: moderate to high (250–550°C), pressure: 5–25 MPa, solvent type, heating rate: 5–140°C/min	High pressure reactor	60–75	15–20	8–20	[33]
Hydrothermal Carbonisation (Section 8.3.7)	Temperature: low to high: (400–1200°C), heating rate: 4–5°C, high pressure	Stainless steel container inside a furnace	Bio-char (wt.%): 20–35			[34]

Modified from Patel M, Zhang X, Kumar A. Techno-economic and life cycle assessment on lignocellulosic biomass thermochemical conversion technologies: a review. Renew Sustain Energy Rev 2016;53:1486–99.

8.3.1 Feedstock Considerations

The low sulphur content of biomass and its lower carbon footprint than fossil alternatives has led to considerable interest in thermo-chemical conversion of biomass and co-conversion with coal or oil to reduce the environmental impact of fossil fuel use. Prins et al. provide a detailed comparison of the thermodynamic efficiencies of coal and biomass feedstocks in terms of the atomic H/C and O/C ratios [35]. However, it should be realised that the fundamentally different chemical composition of the feedstocks and their combination may result in constraints on conversion plant design or operation, ultimately impacting GHG emissions and the production efficiency of the syngas or liquid fuels and consequently its end utilisation. For example, combustion of waste material is often carried out at lower steam temperatures to avoid boiler corrosion and fouling; gasification of woody biomass produces a syngas with significant quantities of tar that must be removed or converted to avoid damage to downstream conversion plant and liquid bio-oil must be deoxygenated to be used satisfactorily in conventional combustion engines and trace impurities removed prior to successful catalytic conversion. Feedstock selection therefore significantly impacts the process scheme configuration and reliability/maintainability of conversion plant.

8.3.2 Scale of Operation

It is also important to consider scale of operation. Biomass conversion plants are often much smaller than traditional fossil fuel-fired facilities owing to the level of local feedstock availability. For example, many downdraft gasifiers or small scale pyrolysis units operate in rural areas, often on locally sourced feedstock from farms. LCA of feedstock production parameters (i.e. inventory for cultivation, harvesting, and delivery) for different fuels, logistics of supply, and satisfying demand requirements are relatively well-understood. McKendry provides an overview of biomass as a fuel source [36]. The fuel type and quality significantly influence the process scheme, maintenance requirements, contaminant/by-product levels and associated upgrading requirements, and airborne emissions. These have material and energy resource and output implications and so it is essential, therefore, that the feedstock type, harvesting, and delivery to the plant are fully considered in the LCA. Further information to support this is given in Chapters 5–7, with Chapter 7 particularly focusing on the supply and logistic issues. Where possible, primary data on feedstocks should be sought that explicitly connects the processing requirements to the production regime. Where new, untested feedstocks are being considered, great care needs to be taken to ensure the overall process scheme contains all steps and accurately reflects the likely conversion performance for the specific feedstock being considered. The case studies in Chapters 9–14 cover a variety of different feedstock types to give examples of how such issues are addressed: farm wastes/residues (manure/slurry—Chapter 9); agricultural food crops (soy—Chapter 10); purpose grown energy crops (willow—Chapter 11); waste arisings (waste wood—Chapter 12); forest products (harvesting residues—Chapter 14).

The choice of the most appropriate technology for a bioenergy development is therefore driven by a range of factors: available feedstock and its properties, scale of operation, and desired energy end-vector. This results in different technologies having different market niches, e.g. use of pyrolysis to overcome logistic costs with small units providing a liquid fuel input to larger scale upgrading facilities to transport fuels, small-medium scale gasification for gas grid injection or CHP, and large-scale waste combustion for power or CHP.

8.3.3 Pre-Processing Technologies

Different conversion processes require different feedstock pre-processing. Downdraft gasifiers can operate with chipped or pelletised solid fuels, fluidised beds require fairly consistently sized feedstocks, and feedstocks for entrained flow gasifiers generally need to be ground or slurried, depending on the gasifier type. This places different demands on resources and GHG emissions for pre-treatment. The energy requirements are substantially greater for producing powders or sub-millimetre particulates than for larger chipped size feedstock.

Pelletising may offer some advantages as it results in a homogenised, uniform feedstock, improving process controllability (a particular advantage for waste and refuse-derived feedstocks, with high levels of feedstock variability). Pelletising may also minimise feedstock losses during storage and pellets can sometimes be conveniently integrated into existing fossil fuel infrastructure (e.g. co-firing in mills at large coal-fired utility plants). However, pelletising requires drying, grinding, and pressing of the material, exerting sufficient pressure to melt the lignin in the wood. Sometimes a binder may also be added to improve pellet physical properties. After pelletising, the feedstock should be more robust to mechanical handling, resulting in lower levels of losses along the supply chain.

Torrefaction (low temperature thermal treatment under low oxygen conditions) has also been considered to improve the cost-effectiveness or resource efficiency of overall supply chain logistics since it can reduce degradation of stored material and reduce the energy costs for grinding and producing small particulate sizes. It is also attracting increasing attention for combustion processes because of its potential to reduce flue gas emission levels (especially particulates). It can also have a significant effect on gasification efficiency as many of the volatiles can be removed in this pre-treatment, reducing tar content further downstream but generally at the cost of a reduced syngas higher heating value (HHV). As with all bioenergy systems and processing options, the key to accurate LCA is to ensure that appropriate data for the relevant feedstocks, technologies, and scales is consistently incorporated at every stage to allow a full evaluation of the extent to which penalties associated with processing may be offset by conversion process benefits.

8.3.4 Pyrolysis

Pyrolysis has become of major interest due to the flexibility in operation, versatility of the technology, and adaptability to a wide variety of feedstocks and products. Pyrolysis operates in anaerobic conditions where heat is usually provided externally and the constituents of biomass are thermally cracked to gases and vapours which usually undergo secondary reactions, thereby giving a broad spectrum of products. The conditions and circumstances that can have a major impact on the products and the process performance include feedstock, technology, reaction temperature, additives, catalysts, hot vapour residence time, solids residence time, and pressure.

The word pyrolysis is derived from the Greek words 'pyro' meaning fire and 'lysis' meaning decomposition or breaking down into constituent parts [37]. Pyrolysis has been applied for thousands of years for charcoal production, but it is only in the last 35 years that fast pyrolysis for liquids has been developed. This operates at moderate temperatures of around 500°C and very short reaction times and hot vapour residence times of less than 2 s. Fast pyrolysis is of considerable interest because this directly gives high yields of liquids of up to 75 wt.% on dry biomass, which can be used directly in a variety of applications [17], used as an efficient energy carrier, and/or used as a precursor for bio-fuels. Intermediate and slow pyrolysis focus on production of solid char as the main product with the liquids and gases usually as by-products, although increasing attention is being paid to maximising the value of the products. Slow pyrolysis has been used for over 70 years to reduce quantities of waste that require disposal as well as making the residues less harmful to the environment.

8.3.4.1 Science of Pyrolysis

Pyrolysis is the thermal decomposition of biomass occurring in the absence of oxygen, converting biomass to liquid (usually called bio-oil), solid, and gaseous fractions with various chemical reactions of feedstock. Lower process temperatures and longer hot vapour residence times favour the production of charcoal. Higher temperatures and longer residence times increase biomass conversion to gas, and moderate temperatures and short hot vapour residence time are optimum for producing liquids. Three products are always produced, but the proportions can be varied over a wide range by adjustment of the process parameters. Table 8.2 and Fig. 8.4 indicate typical product distribution obtained from different modes of pyrolysis, showing the considerable flexibility achievable by changing process conditions. Fast pyrolysis for liquids production is currently of particular interest commercially as the liquid can be stored and transported and used for energy, transport fuels, chemicals, or as an energy carrier.

8.3.4.2 Fast Pyrolysis

In fast pyrolysis, biomass decomposes very quickly to generate mostly vapours and aerosols and some charcoal and gas. After cooling and condensation, a dark brown homogenous mobile liquid is formed which has a heating value about half of that of conventional fuel oil.

TABLE 8.2 Typical Product Weight Yields (Dry Wood Basis) Obtained by Different Modes of Pyrolysis of Wood [38]

Mode	Conditions	Liquid	Solid	Gas
Fast	~500°C. Short hot vapour residence time <2s	75wt.% (bio-oil)	12wt.% char	13wt.%
Intermediate	~500°C. Moderate hot vapour residence time 5–30s	50wt.% in 2 phases	25wt.% char	25wt.%
Carbonisation (slow)	~400°C. Long hot vapour residence time hours → days	30wt.% in 2 phases	35wt.% char	35wt.%
Gasification (allothermal)	~750–900°C. Moderate hot vapour time 5s + in	3wt.%	1wt.% char	96wt.%
Torrefaction (slow)	~280°C, solids residence time ~10–60min	0wt.% unless vapours are condensed, then up to 15%	80wt.% solid	20wt.%

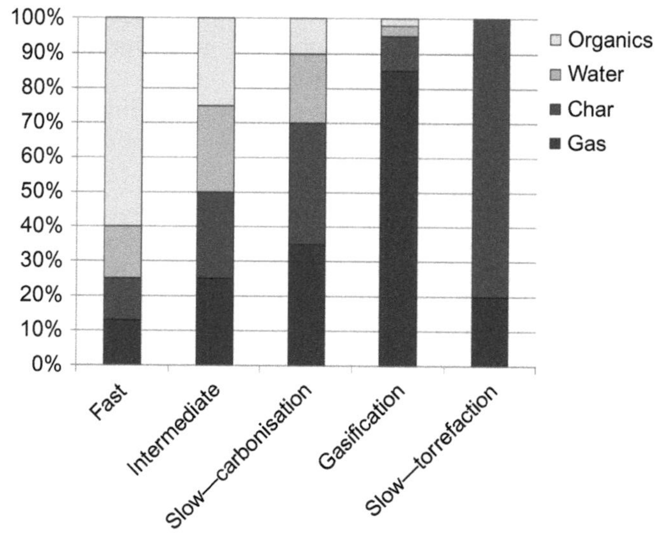

FIG. 8.4 Product spectrum from pyrolysis.

This is referred to as bio-oil and is the basis of the recent ASTM standard [39]. A high yield of liquid is obtained with most low ash biomass. Yields of up to 75 wt.% on dry biomass feed can be achieved. The essential features of a fast pyrolysis process for producing liquids are:

- Small particle sizes of typically less than 5 mm for high heating rates and rapid devolatilisation,
- Feed moisture content of less than 10 wt.% since all the feed water reports to the liquid phase along with water from the pyrolysis reactions,
- Very high heating rates and very high heat transfer rates at the biomass particle reaction interface. This usually requires a finely ground biomass feed of typically less than 3 mm as biomass generally has a low thermal conductivity. The rate of particle heating is usually the rate-limiting step. Preparation of the biomass (comminution, drying, and grinding) to this specification before pyrolysis needs to be considered in the overall process scheme.
- Carefully controlled pyrolysis reaction temperature of around 500°C to maximise the liquid yield for most biomass,
- Short hot vapour residence times of typically less than 2 s to minimise secondary reactions,
- Rapid removal of product char to minimise cracking of vapours,
- Rapid cooling of the pyrolysis vapours to give the bio-oil product.

As fast pyrolysis for liquids occurs in a few seconds or less, heat and mass transfer processes and phase transition phenomena, as well as chemical reaction kinetics, play important roles. The critical issue is to bring the reacting biomass particles to the optimum process temperature as quickly as possible and minimise both their exposure to lower temperatures that favours formation of charcoal and their exposure to higher temperatures that accelerates thermal cracking. One way this objective can be achieved is by using small particles, for example, in the fluidised bed processes. Another possibility is to transfer heat very fast only to the particle surface that contacts the heat source which is used in ablative pyrolysis. Several comprehensive reviews of fast pyrolysis for liquids production are available [38,40–45].

8.3.4.3 Systems Integration of Pyrolysis Technology

A conceptual fast pyrolysis process is depicted in Fig. 8.5 from biomass feed to collection of a liquid product. Each process step has several alternatives such as the reactor and liquid

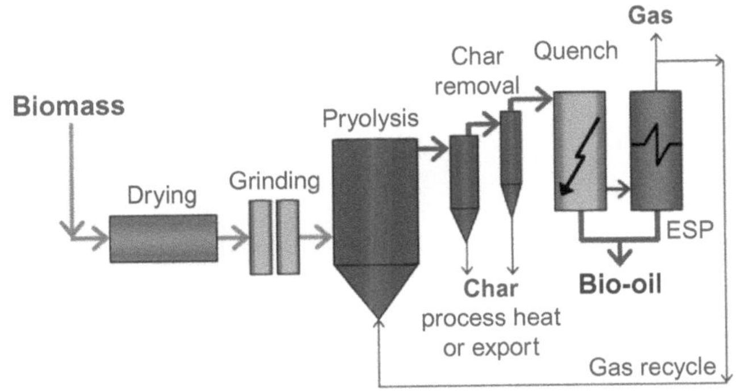

FIG. 8.5 Conceptual fast pyrolysis process.

collection, but the underlying principles are similar. At the heart of a fast pyrolysis process is the reactor. Although it probably represents only about 10%–15% of the total capital cost of an integrated system, most research and development has focused on developing and testing different reactor configurations on a variety of feedstocks, although increasing attention is now being paid to improvement of liquid collection systems and improvement of liquid quality. The rest of the fast pyrolysis process consists of biomass reception, storage and handling, biomass drying and grinding, product collection, storage and, when relevant, upgrading. A typical biomass feed specification is up to 10% moisture and so drying and grinding are key steps from an LCA mass-energy balance perspective.

8.3.4.4 Liquid Bio-Oil

Liquid bio-oil is formed by rapidly quenching and thus 'freezing' the intermediate products of flash degradation of hemicellulose, cellulose, and lignin. The liquid, thus, contains many reactive species, which contribute to its unusual attributes. Pyrolysis oil typically is a dark brown, free-flowing liquid and approximates to biomass in elemental composition (Table 8.3). It is composed of a very complex mixture of oxygenated hydrocarbons with an appreciable proportion of water from both the original moisture and reaction product. Some solid char may also be present. Typical organics yields from different feedstocks and their variation with temperature is shown in Fig. 8.6 [46]. Liquid yield depends on biomass type, temperature, hot vapour residence time, char separation, and biomass ash content, the last two having a catalytic effect on vapour cracking. It is important to note that maximum yield is not the same as maximum quality, and quality needs careful definition if it is to be optimised. From a LCA perspective, it is important to be confident that the process scheme being examined is able to deliver the product quality required, i.e. that the liquid fuel produced is

TABLE 8.3 Typical Properties of Fast Pyrolysis Bio-Oil

Physical Property	Typical Value
Moisture content	25%
pH	2.5
Specific gravity	1.20
Elemental analysis	
C	56%
H	6%
O	38%
N	0%–0.1%
HHV as produced	17 MJ/kg
Viscosity (40°C and 25% water)	40–100 MPa s
Solids (char)	0.1%
Vacuum distillation residue	Up to 50%

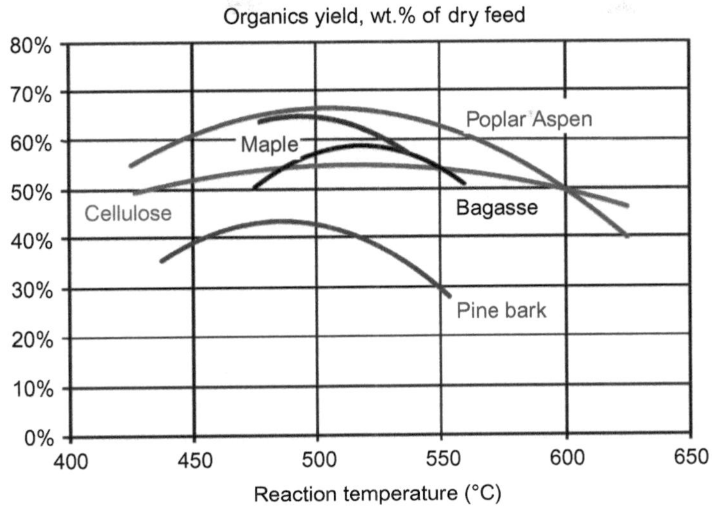

FIG. 8.6 Organics yield from different feedstocks [46].

appropriate for the application required. This is important for identifying the appropriate counterfactual in LCA, which is often critical in evaluating the GHG balance of different systems. Bio-oil quality and quality management and improvement have been reviewed by Bridgwater [47].

The liquid contains varying quantities of water, which forms a stable single-phase mixture, ranging from about 15 wt.% to an upper limit of about 30–50 wt.% water, depending on the feed material, how it was produced, and subsequently collected. As stated above, a typical feed material specification is a maximum of 10% moisture in the dried feed material, as both this feed moisture and the water of reaction from pyrolysis, typically about 12% based on dry feed, report to the liquid product. Pyrolysis liquids can tolerate the addition of some water, but there is a limit to the amount of water which can be added to the liquid before phase separation occurs; in other words, the liquid cannot be dissolved in water. Water addition reduces viscosity, which is useful; reduces heating value which means that more liquid is required to meet a given duty; and can improve stability. The effect of water is therefore complex and important. Bio-oil is miscible with polar solvents such as methanol, acetone, etc., but totally immiscible with petroleum-derived fuels. This is due to the high oxygen content of around 35–40 wt.%, which is similar to that of biomass, and provides the chemical explanation of many of the characteristics reported. Removal of this oxygen by upgrading requires complex catalytic processes which are described in the next sub-sections.

8.3.4.5 By-products (Char and Gas)

Char and gas are by-products, typically containing about 25% and 5% of the energy in the feed material, respectively. The pyrolysis process itself requires about 15% of the energy in the feed, and of the by-products, only the char has sufficient energy to provide this heat. The heat can be derived by burning char in an orthodox reaction system design such as a circulating fluid bed, which makes the process energy self-sufficient and results in a high temperature flue gas. As the ash in the biomass is substantially retained in the char, and thereby concentrated, combustion needs careful control. This is practiced commercially. There are many other ways of providing the heat for pyrolysis, but in-process combustion of char is currently preferred. The waste heat from char combustion and any heat from surplus gas or by-product gas can be used for feed drying and, in large installations, could be used for export or power generation.

An important principle of fast pyrolysis is that a well-designed and well-run process should not produce any emissions other than clean flue gas, i.e. CO_2 and water, although they will have to meet local emissions standards and requirements.

8.3.4.6 Bio-Oil Upgrading

Bio-oil can be upgraded in a number of ways—physically, chemically, and catalytically. This has been extensively reviewed [48–51] and only the more significant features and recent developments are reported here. A summary of the main methods for upgrading fast pyrolysis liquids and the products is shown in Fig. 8.7. Bio-oil can substitute fuel oil or diesel in many applications including boilers, furnaces, engines, and turbines for electricity generation.

From Fig. 8.7, it is clear that the different upgrading processes need to be taken into account in an overall assessment of the GHG balance of a bioenergy systems, since the energy requirement, losses, waste streams, and changes in fuel composition and conditions can all impact the overall life cycle performance.

8.3.4.7 Catalytic Upgrading of Bio-Oil to Bio-Fuels

Upgrading bio-oil to a conventional transport fuel such as diesel, gasoline, kerosene, methane, and LPG requires full deoxygenation and conventional refining, which can be accomplished either by integrated catalytic pyrolysis or by decoupled liquid phase hydrodeoxygenation as summarised in Fig. 8.5. There is also growing interest in partial upgrading to a product that is compatible with refinery streams in order to take advantage of the economy of scale and experience in a conventional refinery. This area has been well-reviewed (e.g. [47]).

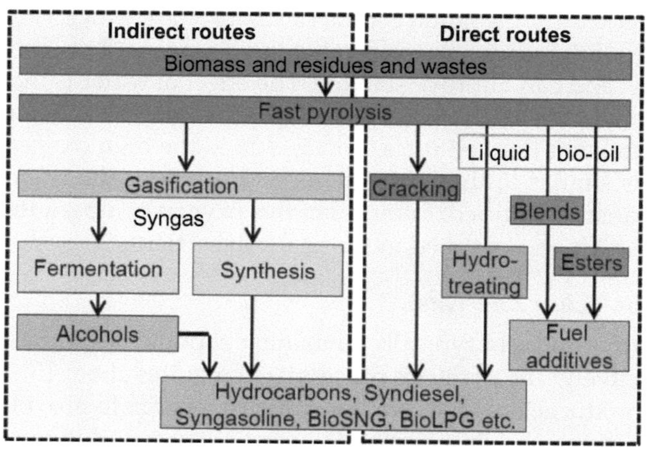

FIG. 8.7 Overview of fast pyrolysis upgrading methods

Integration into refineries by upgrading through cracking and/or hydrodeoxygenation has been reviewed by Huber and Corma [52]. The main methods are:

- Hydrodeoxygenation
- Catalytic vapour cracking—in situ or ex situ
- Gasification to syngas followed by synthesis to hydrocarbons or alcohols.

8.3.4.8 Hydrodeoxygenation

Hydrodeoxygenation rejects oxygen as water by catalytic reaction with hydrogen. This is a separate and distinct process to fast pyrolysis that can, therefore, be carried out remotely. The process is typically high pressure (up to 20 MPa) and moderate temperature (up to 400°C) and requires a hydrogen supply or source [53]. Most attention is now focussed on multiple step processes with increasingly severe conditions. The presence of phenols in the bio-oil makes complete deoxygenation very difficult, but full hydrodeoxygenation gives a naphtha-like product that requires orthodox refining to derive conventional transport fuels. This would be expected to take place in a conventional refinery to take advantage of know-how and existing processes. A projected typical yield of naphtha equivalent from biomass is about 25% by weight or 55% in energy terms excluding provision of hydrogen [50]. Inclusion of hydrogen production by gasification of biomass reduces the yields to around 15 wt.% or 33% in energy terms. The process can be depicted by the following conceptual reaction:

$$C_1H_{1.33}O_{0.43} + 0.77H_2 \rightarrow CH_2 + 0.43H_2O$$

A key aspect is production of hydrogen. Since the hydrogen requirement is significant, it should be renewable and sustainable. Few refineries have a hydrogen surplus, so this has to be provided. There are many ways of providing hydrogen such as gasification of biomass, shifting CO to H_2 followed by scrubbing CO_2; or bio-oil or the aqueous phase from a phase-separated product can be steam-reformed to hydrogen; or hydrogen can be generated locally by electrolysis of water. Supply of hydrogen from external sources is unlikely to be feasible due to the very high cost of storage and transport. The necessary purity of hydrogen is unknown, but some CO shifting may take place in the hydrodeoxygenation reactor removing the need for dedicated shift reactors.

8.3.4.9 Zeolite Cracking

Zeolite cracking rejects oxygen as CO_2, as summarised in the conceptual overall reaction below:

$$C_1H_{1.33}O_{0.43} + 0.26O_2 \rightarrow 0.65CH_{1.2} + 0.34CO_2 + 0.27H_2O$$

Cracking takes place at atmospheric pressure in a close coupled process to pyrolysis. There is no requirement for hydrogen or pressure. The projected yield is around 18 wt.% aromatics and the process was commercially developed by Kior, but this was not as successful as hoped and Kior has gone into liquidation. ZSM-5 has attracted most attention due to its shape selectivity to aromatics, with promoters such as Ga or Ni. A key disadvantage is that the catalyst rapidly cokes which requires frequent regeneration as in an Fluid catalytic cracking (FCC) unit.

Oxygen is thus removed as CO and CO_2 compared to H_2O in hydrodeoxygenation. Production of aromatics is also likely to be of significant interest to the chemicals sector.

8.3.4.10 *Gasification for Synfuels*

A recent concept that has attracted much interest is the decentralised production of bio-oil or bio-oil-char slurries for transportation to a central process plant for gasification and synthesis of hydrocarbon transport fuels by, for example, Fischer-Tropsch synthesis or alcohols. This is depicted above in Fig. 8.5. Although there is a small energy penalty from the lower pyrolysis energy efficiency, transportation energy, and additional bio-oil gasification stage, this is more than compensated by the economies of scale achievable on a commercial-sized gasification and transport fuel synthesis plant [54].

Although the concept of very large gasification plants of 5 GW or more has been promoted [55] based on importation of biomass on a massive scale to an integrated plant, based for example at Rotterdam, there are significant obstacles to be overcome. Decentralised fast pyrolysis plants of up to 100,000 t/year or 12 t/h are currently feasible and have been commercially realised. Bio-oil gasification in an entrained flow oxygen-blown pressurised gasifier is also feasible such as a Texaco or Shell system, with the added advantage that feeding a liquid at pressure is easier than solid biomass, offers lower costs, and the gas quality under such conditions is likely to be higher than from solid biomass. Finally, transport fuel synthesis at 50,000–200,000 t/day is also commercially realisable from the extensive gas-to-liquid plants currently operational around the world. Future Energy (now Siemens) has successfully conducted pressurised oxygen-blown gasification tests on both bio-oil and bio-oil/char slurries [56,57].

8.3.4.11 *Other Methods and Routes*

A wide variety of methods and catalysts including multiple stepped and combined processes have been investigated in recent years. It is important to emphasise the importance of maximising yield and minimising unwanted reactions especially minimising residues since these will have to be disposed of at a potential cost as well as lower efficiency.

8.3.5 Gasification

The gasification process thermo-chemically converts an organic feedstock (e.g. liquid or solid fuel) into its gaseous components which depend on the gasification temperature. Strictly speaking, synthesis gas or syngas is only $H_2 + CO$ which is primarily formed at higher gasification temperatures (above 1200°C). At lower gasification temperatures, product gas is formed comprising CO, H_2, CH_4, and CO_2 and may include tar compounds which adversely affect gasification performance and downstream end-uses. Tar and ash are potentially produced as by-products which may have added value and provide different accounting routes for GHG emissions during gasification. In a well-designed gasifier, char is not normally a by-product. In practice, however, a small amount of char may be produced which represents inefficient gasification. The gasification efficiency can be expressed as the % conversion of carbon. Syngas and producer gas (for convenience now referred to collectively as syngas) can be used to produce heat, power, and chemical products. Historically, both wood and coal have been gasified to produce convenient gaseous fuel (syngas) for many applications worldwide, but have not been commercially sustained due to low-cost fossil oil and natural gas availability.

Gasification produces a flexible energy source. There is potential to produce heat via combustion in a boiler, electricity via combustion in reciprocating engines or gas turbines, or converted to a liquid bio-fuel, or substitute natural gas (SNG) for injection into gas grids.

8.3.5.1 The Gasification Process

There are many different designs of modern gasifier, but, during the process of gasification, woody biomass undergoes three distinct transitions. Firstly, wood is dehydrated before temperatures exceed 200°C, at which point pyrolysis is initiated and char and vapours are produced. The presence of O_2 partially oxidises the char and vapours resulting in a gas containing CO, H_2, CO_2, and water as well as tars, ammonia, sulphur, and other impurities. A key parameter is the cold gas efficiency of the gasifier, which expresses the chemical energy in the outlet gas as a proportion of the chemical energy and sensible heat input to the gasifier. Cold gas efficiency will vary with gasifier design, scale, and conditions and is sensitive to the equivalence ratio (air flow rate/airflow for stoichiometric conditions ER). Changing the ER from 0.16 to 0.26 changed the cold gas efficiency from 57% to 74% [58].

A full review on gasifier design for biomass gasification is given by Sansaniwal et al. [59], but broadly speaking gasifiers can be classified in terms of their bed and gas flow direction, e.g. fixed bed (downdraft, updraft, or crossdraft), entrained flow, fluidised bed, or circulating bed. An oxidant is supplied to the combustion zone at typically 1/5 of the stoichiometric combustion ratio. Many gasifiers use catalysts (e.g. dolomite- or Ni-based) for catalytic steam reforming or partial oxidation of tars, or to enable catalytic upgrading to liquid bio-fuels [60]. Often the catalytic process is done downstream from the gasifier after sufficient gas cleaning and composition adjustments. Mass flows of catalysts, their potential recovery and impact on product output, and GHG emissions need consideration. Downdraft gasifiers are suitable for small scale operation, typically from 10 to 250 kWe; there are MW systems out there, fluidised bed systems typically operate between a few MWe to 100 MWe, and entrained flow from about 50 MWe to 1 GWe.

8.3.5.2 Gasification Systems

To allow informed comparison between LCA of different gasifier designs and plants, system boundaries should generally include the impact of feedstock production, feedstock pre-processing or pre-treatment, the gasification system (including gas clean-up), and end utilisation of the syngas. The mass-flow and end-use of by-products, such as ash and bio-char (see above), should be considered realising that their quality, quantity, and end-use will depend on the biomass feed characteristics, operational temperature profile, and residence time during the gasification process and gasifier design. The end-use of the syngas is largely determined by the plant's scale and end demand. Power generation can be achieved by co-firing, CHP or, for larger plants, Integrated Gasification Combined Cycle (IGCC). In IGCC, gas turbines are used due to their higher thermal efficiencies, but the inlet gas temperature and pressure have to be optimised and to match the turbine design (e.g. 400–500°C and 5–20 bar). Plant lifetime may be increased by ensuring complete tar cracking, tar removal, or careful temperature control to avoid tar condensation within the plant. In some circumstances, process efficiency can be increased by burning a tar-laden gas. End emissions from the plant should be minimised by gasification optimisation. Process inputs to the gasification plant, e.g. gas clean-up technology (e.g. scrubbing, cyclonic, electrostatic precipitation, plasma) will require additional energy, resulting in other GHG contributions that require accounting.

The sustainable available biomass feedstock is increasingly being dominated by waste-derived and residual materials including municipal solid waste (MSW) or refuse-derived fuel (RDF), a refined version of MSW. These are capable of reducing GHG emissions with less stress on land usage when waste products (e.g. agricultural and domestic) are used as the feedstock compared to energy crops. However, such feedstocks present technical challenges to gasification plants, e.g. RDF may contain 20% plastics. To combat potential process problems, researchers are investigating thermal plasma gasification to break chemical bonds of a wide range of feedstocks, from biomass to solid, hazardous wastes [61]. Plasma generation will likely have a significant electrical overhead, but may reduce the requirements for downstream cleaning, which itself can be done with plasmas. Given that power production is one potential output stream from gasification, the plasma can be run from a fraction (e.g. 5%) of the generated electricity.

Syngas can be directly combusted for electricity generation, but it is more common for it to be purified, as then almost any hydrocarbon compound may be synthesised from it, including premium fuels such as methanol, ethanol, methane, and transport fuels. The first stage in this synthesis is to adjust the proportions of H_2 and CO to the ratio required in the desired product. For example, methanol is CH_3OH, and therefore needs two H_2 molecules for each CO converted. Using the Fischer-Tropsch process, named after the chemists who developed it in the 1920s, syngas from biomass has been successfully converted to synthetic diesel. Additionally, the hydrogen which can be isolated from syngas has been used to power hydrogen fuel cells for the generation of electricity and to power electric vehicles [62]. However, these purification steps are costly and energy-intensive, and efficiency needs to be improved if they are to become widely used. In addition to the challenges which face purification of syngas, the implementation of gasification is limited because of the initial feedstock requirements. Moisture and ash content, homogeneity, particle size, bulk density, and energy content must all be tightly controlled if gasification is to be efficient and this has limited uptake of the technology. In 2010, only 0.5% of worldwide gasification relied on biomass, with the rest using coal, petroleum, natural gas, and petcoke as feedstocks [63]

8.3.5.3 Analysis of Gasification Systems

There are numerous review papers on gasification from a technology and performance point of view, but fewer examples of complete LCA. A critical review of gasification and LCA is given by Farzad et al. [64]. They consider conventional and emerging systems, co-gasification (coal and biomass), and different gasifying agents. Most of the work on LCA gasification focuses on GHG emissions with little analysis of environmental or social impact. Environmental drivers have, however, been shown to be dominated by biomass production, flue emissions, and electricity production with substantial improvement compared to other technologies in reduced emissions through H_2 production and CO_2 capture [65].

To complete LCA of gasification systems, the mass and energy flows within gasifiers need to be understood. Stoichiometric modelling of gasification [66] provides simple methods to identify impacts of biomass feedstock, oxidant, and operating conditions on syngas productivity. However, they lack the finesse of spatio-temporal models, but these require much longer computational times. Importantly, data from operating gasification plants can help to significantly improve and validate models by providing real data. A comprehensive review of biomass gasification is given by Sikarwar et al. [67] with consideration to the potential higher value products that can be made from gasification. Given that gasification can form the central technology behind biorefineries and chemical production, while enabling heat and power production, it is likely that gasification will continue to grow over the next few decades for both smaller and larger scale plants.

8.3.6 Combustion

Biomass has been burnt for heat since pre-historic times and it remains the most commonly used source of renewable energy in the world. A vast range of technologies are used from simple batch-operated designs in domestic homes to complex biomass-fired power stations with flue gas clean-up. This section describes the fundamental process involved in biomass combustion, the types of emissions evolved from combustion, and considerations around feedstock types. An overview is given of typical combustion systems at both domestic and large-scale power generation.

Biomass combustion is the simplest thermo-chemical conversion technology that takes place in the presence of air. Heat, power, or CHP are the main products of direct combustion of biomass; this process consists of consecutive heterogeneous and homogeneous reactions [11]. Biomass combustion essentially depends on the particle size and properties of the feedstock, temperature, and combustion atmosphere. High emissions of NO_x, CO_2, and particulate matter, in addition to ash handling, make combustion environmentally challenging.

8.3.6.1 The Biomass Combustion Process

Combustion can be split into four stages: (1) drying, (2) pyrolysis, (3) volatiles combustion, and (4) char combustion. These stages are shown in Fig. 8.8.

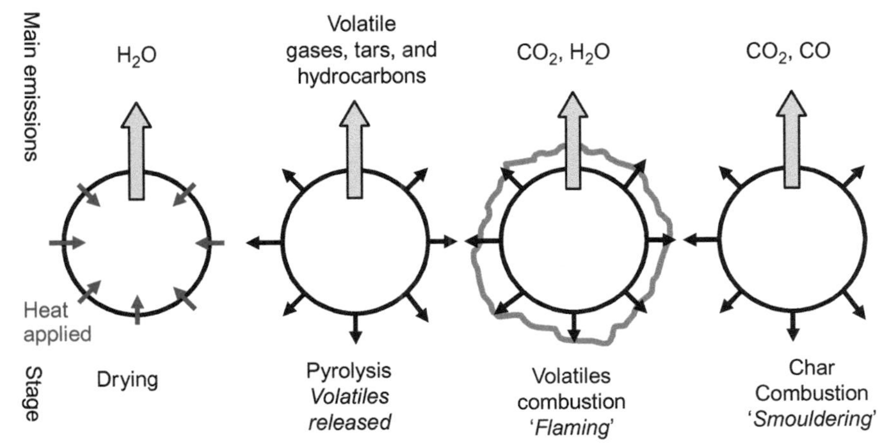

FIG. 8.8 Biomass combustion stages. *Modified from Brown RCE. Thermochemical processing of biomass: conversion into fuels, chemicals, and power. Wiley & Sons; 2011.*

8.3.6.2 Feedstock Quality

A wide range of biomass sources can be considered for combustion including wood chips, wood pellets, and wood sourced directly by the user including scrub and waste woods. Adherence to strict fuel quality parameters is required for large-scale systems. Fuel types for domestic use are often more varied and the quality is not consistent, leading to wide variation in emissions levels. There are many formulations of biomass-derived briquettes and increased interest in devolatilised fuels such as torrefied biomass and charcoal, which have lower particulate emissions. The best quality fuels contain high amounts of carbon and hydrogen and low amounts of other elements such as oxygen, nitrogen, sulphur, and trace elements. Biomass typically contains higher levels of oxygen than coal or other fossil fuels.

The fuel energy content (calorific value) is inversely proportional to the amount of oxygen present. A typical range of calorific values are shown in Table 8.4.

Nitrogen and sulphur species are associated with undesirable gaseous emissions. Trace elements such as potassium and sodium can lead to fouling within the combustion system, whereas chlorine leads to corrosion and silica causes excessive wear to milling equipment.

The first stage of combustion requires heat to dry the moisture; hence, it is desirable to have fuels with the minimal moisture content. Fresh woodchips can contain 50% moisture, whereas leaves can be over 90% water, which would not be easily combustible. Most stoves and boilers recommended biomass has less than 20% moisture content. Higher levels affect the combustion efficiency and increase the amount of smoke emitted.

8.3.6.3 Emissions From Combustion

The products from biomass combustion are essentially similar to emissions from fossil fuel combustion. Carbon dioxide (CO_2) and water are the most abundant emissions because they are the products of complete combustion. Carbon monoxide (CO), hydrocarbons (HC), and particulate matter are associated with incomplete combustion and are highest for the least efficient combustion systems. Methane and non-methane hydrocarbons (NMHC) are sometimes measured separately during flue gas analysis. Nitrogen oxides (NO_x) are also emitted and consist of nitric oxide (NO) and nitrogen dioxide (NO_2). NO_x emissions are formed according to three distinct mechanisms: Thermal NO_x, Prompt NO_x, and Fuel NO_x. Fuel NO_x is formed from the oxidation of nitrogen-containing species within the fuel. Prompt NO_x is formed from a reaction between the combustion air and the hydrocarbon radicals formed

TABLE 8.4 Typical Biomass Proximate Analysis and Calorific Values Based on Phylis Database [68] Dry Basis Values, Database Reference in Brackets

	Wood Pellet (3244)	Mixed Hardwood Chips (250)	Waste Wood (912)	Miscanthus (1743)	MSW (2920)
Ash wt.% (db)	1.10	0.93	3.23	3.41	16.82
Volatile wt.% (db)	83.00	83.41	77.57	86.50	72.60
Fixed carbon wt.% (db)	15.90	15.66	19.20	10.09	10.58
Gross calorific value MJ/kg	17.19	19.27	19.52	19.12	20.48

during combustion and is highest at fuel-rich conditions. Thermal NO_x is formed from reaction between oxygen and nitrogen in the combustion air and is highest at higher combustion temperatures (1000°C+). Using 100% biomass fuel results in lower flame temperatures compared to fossil fuels, so both thermal and prompt NO_x are relatively low. Another nitrogen species that can be present is nitrous oxide (N_2O), which is a GHG. Other species in flue gases can include a vast range of volatile species and partially combusted fuel fragments, including oxygenated species and polycyclic aromatic hydrocarbons. If the fuel contains sulphur species, the release of a proportional amount of SO_2 is likely. Some trace species can be used as markers to identify sources of combustion-generated emissions, examples are levoglucosan for biomass combustion and benzo(a)pyrene for diesel engine combustion.

8.3.6.4 Particulate Emissions

Particulate emissions from combustion sources consist of agglomerated chains of carbonaceous spherules, onto which other species are adsorbed. The adsorbed species include PAH, partially burned volatile fuel fragments, and ash. Small ash fragments can be present separately from the carbonaceous material. Visible smoke consists of both particulate matter (soot) and aerosols of volatile species.

The carbonaceous core of the particulate is termed either elemental carbon (EC) or black carbon (BC). EC is typically measured using thermal techniques, whereas BC is measured using optical methods. The adsorbed species are known as organic carbon (OC). Although the terms EC and BC are often used interchangeably, there can be differences if the OC fraction contains light absorbing material, such as tars. The proportion of BC/OC is dependent on the fuel and combustion conditions, but also on the sampling temperature. Samples taken at higher temperatures will contain less absorbed OC than those taken at ambient temperatures because the OC material is semi-volatile and will be released in the gaseous form rather than collected as particulate.

The particulate matter and smoke produced during combustion have recently been shown to have significant impact in terms of climate change. Recent work such as that by Bond et al. has demonstrated that the ratio of BC and OC is important [69]. The BC consists of the whole light absorbing fraction of the particles, so particles with a higher ratio of BC to OC absorb heat rather than reflecting it, affecting cloud albedo and decreasing the reflectivity of snow and ice. The results are complicated by the fact that the OC consists of PAH, which are known to be precursors to the formation of solid soot. Emissions are accompanied by NO_x, SO_x, and inorganic aerosols that can have an influence on cloud formation. Climate forcing from co-emitted species were estimated and used in the model by Bond et al. [69]. This study demonstrated the high levels of complexity and uncertainty involved in assessing net forcing impacts.

8.3.6.5 Combustion Efficiency

The efficiency of a combustion system is related to how completely the fuel is burnt. The maximum emissions of CO_2 for a given mass of fuel will correlate to the maximum extraction of energy from the fuel because CO_2 is the final product of complete combustion. The presence of CO, volatile fuel fragments, and soot indicate lower combustion efficiency. The most efficient combustion systems are those where there is sufficient flow of air for complete combustion, which is always more than the stoichiometric fuel air ratio. The excess air is typically between 5% and 50%, depending on the combustor design and fuel properties. The combustion efficiency of a domestic fireplace might only be 20%–30%, whereas around 75% is achievable for a larger furnace system.

8.3.6.6 Small-Scale Domestic Combustion for Heat

The size, design, and complexity of a domestic system will affect the amount of pollutants produced. Designs include overfed, underfed, and moving grate systems. Batch-fed designs are typically fireplaces or stoves for radiant room heat. Fuels for batch-fed domestic use are generally logs and briquettes, although waste wood pieces are also commonly used. Continuously fed systems use wood chips or wood pellets—the latter being less likely to block the fuel feed system but more expensive.

8.3.6.7 Medium-Scale Combustion for Heat and Power

There are many examples of the use of wood chips or waste-derived fuels at district scale for provision of heat and/or power using conventional combustion technology. The efficiency of the system is dependent on the scale of application and associated steam conditions. It should be noted that CHP systems are usually designed to operate to service a particular heat demand. Some facilities can be operated even when that heat demand is lower or absent, but usually this will have a significant impact on the efficiency of the plant. It is therefore important to take into account the heat demand in the assessment as this may affect the number of hours for which a facility is run. If frequent start-ups are then required, the airborne pollutants may be higher during the start-up phase than the routinely reported steady state figures, making it difficult to accurately assess the lifetime impact.

8.3.6.8 Utility Scale Co-firing for Electricity

Biomass co-firing with coal has been commonly deployed across Europe and North America to encourage the de-carbonisation of electricity production. It is a relatively low-cost technology for efficiently and cleanly converting lignocellulosic biomass to electricity [70]. During co-firing, the primary fuel (coal) is partially substituted by biomass in a high efficiency boiler. Depending on the boiler capacity and efficiency, the percentage of biomass co-firing varies between 5 and 20 by weight [32]. The substitution of biomass reduces both GHG emissions from coal combustion and also produces lower amounts of NO_x and SO_x. Co-firing takes the advantage of the power plant's economies of scale and saves fossil fuels. Biomass can be burned in fixed bed combustors with thermal outputs of 1 MW_{th} or larger.

The largest scale of biomass combustion is used for power generation using pulverised fuel in either converted boiler units or co-fired with coal. Energy is extracted from low moisture biomass residues and waste wood using incineration systems. There are a few cases of utility scale-dedicated biomass plant, e.g. Drax power station in the United Kingdom was a coal-fired plant that converted to 100% biomass firing on some units and similar-sized dedicated units have been proposed elsewhere. These benefit from the economic and efficiency advantages of larger scale plant, but need to be traded off against the disadvantage of having to collate very large quantities of biomass in one location with associated cost, GHG, and other burdens.

8.3.6.9 Efficiency Considerations in Combustion Systems

The efficiency of a boiler combustion process is assessed by how much energy from the fuel is transferred to the water for either heating or to produce electricity. Common sources of process inefficiency are heat loses via flue gases, convection and radiation to surroundings, and losses due to incomplete combustion of the fuel. A comparison of the combustion efficiencies for a pellet boiler and similarly sized oil and gas boilers was given by Monteleone et al. as shown in Table 8.5 [71]. It can be seen that the lowest efficiency is given by the biomass pellet boiler, which has a corresponding high level of PM_{10} particulate emissions. This indicates very poor levels of combustion, which could be improved by better air/fuel mixing, higher combustion temperatures, or a longer residence time for the fuel.

The actual overall system efficiency in a commercial power plant is lower than the combustion efficiency due to significant other losses, such as in the electrical conversion process and processing of the fuel. Table 8.6 shows a comparison of overall system efficiencies. It can be seen that the actual efficiency might be as low as 20%.

Unburned carbon remaining in the ash represents a loss of efficiency which adversely affects the GHG balance of the system. Most data available is for coal use rather than biomass, where the residual carbon and organic matter is often measured as 'Loss of Ignition' (LOI), although this can overestimate the actual unburned carbon content [75]. The LOI for coal in a pulverised coal boiler is typically only 2%–3%, whereas carbon in the fly ash can be as high as 10%–30% in a circulating fluidised bed [76]. The levels of carbon remaining in the ash are dependent on both the fuel's characteristics and the type of combustor used. The highest rates of burnout are achieved with finely milled particles of fuel, high levels of fuel air/mixing, and

TABLE 8.5 Comparison of Combustion Efficiency

	Pellet Boiler (15 kW)	Oil (15.5 kW)	Natural Gas (15 kW)
Efficiency/%	75.0	89.5	99.0
PM10/mg/Nm^3	47.6	60.0	0.3

Modified from Monteleone B, Chiesa M, Marzuoli R, Verma VK, Schwarz M, Carlon E, et al. Life cycle analysis of small scale pellet boilers characterized by high efficiency and low emissions. Appl Energy 2015;155:160–70. Available from: http://www.sciencedirect.com/science/article/pii/S0306261915007229.

TABLE 8.6 Comparison of Combustion Systems

System	Overall System Efficiency (%)	Rated Capacity (kWe)	References
Indirect-fired gas turbine	20	200	[73]
Pyrolysis/engine	28	300	[74]
Updraft gasifier/engine	28	2000	[74]
Fluidised bed boiler/steam turbine	32	2000	[72]

Modified from Abdul Malek ABM, Hasanuzzaman M, Abd Rahim N, Al Turki YA. Techno-economic analysis and environmental impact assessment of a 10 MW biomass-based power plant in Malaysia. J Clean Prod 2017;141:502–13.

high temperature combustion [77]. Biomass composition is extremely variable and some fuels have higher losses within the same combustor, for instance, a study measured an 18% loss attributed to unburned carbon when using rice husk compared to sawdust [78]. Use of low NO_x burners with reduced combustion temperatures can potentially increase the amount of unburned fuel remaining as carbon due to slower combustion and lower temperatures [79].

8.3.6.10 Bio-CCS

Bioenergy with carbon capture and storage (Bio-CCS) is a method of producing carbon negative power (i.e. removing carbon dioxide from the atmosphere via biomass conversion technologies and storage underground). CCS technology is currently at a demonstration phase, and research is mainly focused on reducing the costs of capture and storage so that it can be utilised in a new generation of large-scale, power stations with lower emissions. In the future, CCS could potentially be applied to several different energy plants, including those incorporating co-firing or co-gasification of sustainable biomass feedstocks (agricultural and wood wastes and energy crops), dedicated biomass energy plants, bio-fuel production facilities, or biorefineries. Eighty-five percent of IPCC scenarios (or 101 of 116 scenarios) consistent with the 2°C goal require global net negative emissions before 2100, typically through bio-CCS and afforestation [80]. A key technical challenge for bio-CCS is that biomass-fired power plants are usually smaller and more dispersed than coal plants, which may result in reduced economies of scale when compared to larger CCS plants. From an LCA perspective, while CCS can result in negative emissions, it is also important to take account of the additional emission costs associated with the carbon capture process. For example, gases captured will require compression and transportation typically through pipelines. This can increase the GHG burden of processing, so the energy and emission implications of this need to be fully assessed to calculate the net benefit of Bio-CCS.

8.3.7 Hydrothermal Processing of Biomass

Hydrothermal processing involves heating aqueous slurries of biomass or organic wastes at elevated pressures to produce an energy carrier with increased energy density. Feedstocks with high moisture and ash content are particularly suitable for hydrothermal processing and include feedstocks such as AD digestate, manures, sewage sludge, DDGS, food waste, municipal wastes, and aquatic biomass such as micro- and macroalgae [81]. Hydrothermal processing can be divided into three separate processes, depending on the severity of the operating conditions. Hydrothermal carbonisation (HTC) occurs at temperatures between 180°C and 250°C and pressure of 20–40 bar and produces a solid hydro-char with similar properties to a low rank coal [82]. This is the mildest of the three hydrothermal processing routes.

Hydrothermal liquefaction (HTL) occurs at temperatures between 250°C and 375°C and pressures up to 180 bar and produces a liquid bio-crude [83]. Bio-crude can be upgraded to the whole distillate range of petroleum-derived products by catalytic hydro-treatment [84]. Hydrothermal gasification (HTG) or supercritical water gasification (SCWG) occurs at temperatures above 375°C and higher pressures beyond 200 bar producing a syngas [85–87]. Hydrothermal processing routes can typically feed slurries up to 30 wt.% solids. The maximum solid loading is related to the ease in creating and pumping the slurry. High moisture feedstocks therefore still require dewatering, but do not require the same level of drying as is required for pyrolysis. In each case, the water is kept in the liquid state by operating at or above the saturation point which largely minimises the enthalpy change associated with the latent heat of vapourisation of water.

8.3.7.1 *Environmental Impact of Hydrothermal Processing*

The energy return on investment (EROI) for hydrothermal processing is typically between 1 and 3 and is favourable compared to all conventional bio-fuels, but not as high as petroleum-derived fuels [88]. Most reactor systems under development are continuous flow, allowing higher throughput, improved heat recovery, and recycling of process water reducing the water consumption. Continuous operation generally has a lower residence time than batch operation. As residence times decrease due to increasing flow rate, the reactor output increases and energy requirements decrease [89].

There has been limited LCA for HTC, although, generally, a positive energy balance has been observed from the limited studies reported [90]. The production of hydro-char from HTC of wood waste and subsequent use in replacement of coal for electricity generation indicates a net environmental saving in GWP; however, the magnitude is significantly influenced by the source of energy used in processing [91]. There is growing interest in processing nutrient-rich feedstock by HTC such as bio-solids and manure facilitating nutrient-recovery [92,93].

The majority of lifecycle assessments are based on HTL of microalgae. Comparisons of net energy ratio (NER) and GHG emissions between HTL and pyrolysis indicate significant benefits for HTL. Reported NER and GHG emissions for HTL of 1.23 and −11.4 g $CO_{2\text{-equiv.}}$ (MJ renewable diesel), respectively, are lower than pyrolysis with 2.27 and 210 g $CO_{2\text{-equiv.}}$ (MJ renewable diesel) largely due to drying requirements and combustion of co-products required for pyrolysis [94]. The energy consumption ratio (ECR) was similarly more favourable for HTL (ECR 0.44–0.63) compared to pyrolysis (ECR 0.92–1.24) [95]. In a similar manner to HTC, the availability of renewable electricity affects the overall environmental profile of HTL of microalgae [96]. Nutrient recycle is crucial in converting microalgae to bio-fuels and this is reflected in the NER and GHG emissions [97].

LCA comparisons vary, mainly due to different system boundaries; however, they indicate that HTL bio-crude with nutrient recycling is environmentally advantageous over petroleum [98]. The retention of N-containing heteroatoms in the bio-crude following HTL reduces the amounts of N recycle and has a significant impact on GHG emissions [99]. The GHG emissions for HTL are also impacted by location, as the source of nutrients is crucial and locating near, for example, a waste water treatment plant reduces the GHG emissions significantly [100]. Other feedstocks such as yeasts cultivated on waste water have been investigated and indicate comparable GHG emissions of 30 g $CO_{2\text{-equiv.}}$ (MJ of renewable diesel) [101]. Only limited LCA has been performed on the liquefaction of lignocellulosics.

A number of system analyses have been performed for hydrothermal gasification, e.g. Ro et al. compare the processing of manure by hydrothermal gasification and AD [102]. While catalytic hydrothermal gasification of wet manures is considerably more expensive than AD, advantages include destruction of pathogens and odorous compounds, production of relatively clean water, and the production of valuable by-products such as ammonia and phosphates for the fertiliser market. Additionally, the high rate of conversion of the organic matter into gas drastically decreases the land requirement for manure application and resulting costs in transportation and tipping fees [102].

8.4 PHYSIO-CHEMICAL CONVERSION

Physio-chemical conversion, also referred to as mechanical extraction, is a mechanical (physical) conversion process used to produce oil from the seeds of various biomass crops, such as oilseed rape (OSR) and linseed [3]. This process provides a liquid fuel which can undergo a further stage, known as esterification, which turns the oil to fatty acid methyl ester, more widely known as bio-diesel. The process produces not only oil, but also a residual solid or 'cake', which is suitable for animal fodder [1]. This technology is used on a wide scale in Europe using vegetable oils from crops, primarily OSR, but waste fats and oils are also used. The main use of bio-diesel is as a liquid transport fuel; most commonly blended with diesel derived from petroleum. Physio-chemical conversion is further considered in Chapter 10—Biodiesel from Argentinian Soy.

8.5 SUMMARY

There are a vast array of conversion technologies used to convert biomass from one form to another and release energy. When calculating the GHG balance of a bioenergy system, it is key to ensure that a complete system is actually being considered, i.e. that all required feedstock supply, pre-processing and upgrading steps are considered. It is equally important to ensure that data used to define the conversion steps in an LCA process is appropriate for the steps and feedstocks being considered. Wherever possible, datasets that correspond to equivalent feedstocks, reagents, and process conditions should be used. Mass-energy balances and ideally primary data from operating facilities should be included. Often, particularly when evaluating more advanced processes, this is not possible as appropriate datasets do not exist. Modelling can then be a useful tool whereby closed mass-energy balances can be generated by application of appropriate simulation tools, particularly where there is existing knowledge of a similar feedstock or similar set of process conditions. Performing a robust sensitivity analysis is important so the relative importance of uncertain parameters can be assessed. In a LCA of conversion processes there are often one or two primary variables that have the largest impact on GHG emissions (such as process energy or conversion efficiency), it is therefore often appropriate to focus on these rather than smaller (not material) parameters. Great care must be taken when transitioning between scales and there is no substitute for engineering knowledge here, but the following generic comments may provide some guidance:

- Combustion efficiency and control of airborne pollutants and unburned carbon are usually improved at larger scales.
- Theoretical conversion efficiency and process conditions are often different to those achievable in practice due to commercial constraints and feedstock variability.

- LCAs must make clear the system boundary so that the results can be used to evaluate different conversion technology options.
- The composition of syngas varies from one type of gasifier to another, as the ratio of key components reflects the equilibrium reached at the conditions pertaining inside the gasifier. This also reflects whether air or oxygen is used in the process; the former resulting in a syngas of significantly lower calorific value.
- Emissions from different conversion processes depend on key factors such as the feedstock composition, pre-processing, operating conditions, and emission abatement technology. These variables should be considered within LCAs as biomass conversion is inherently variable.
- The composition of fast pyrolysis liquid does not vary significantly with feedstock variations, but the ratio of char: liquid: gas is affected by the relevant process conditions.
- Gas engines and internal combustion engines are reasonably insensitive to variations in scale. However, steam turbines and steam cycles achieve very much higher efficiencies at higher steam conditions and so generally higher efficiencies are achieved at larger scales.
- Commercial gas turbines have been developed and optimised over many years to achieve high efficiencies. While in the long term it would be possible to optimise for a lower calorific value syngas, the most practical near-term option is to adapt existing gas turbines, generally resulting in efficiency impairment.
- Overall the conversion technology choice should be based on the feedstock available, its composition, and the required end use. LCA and techno-economic assessments can be utilised to optimise the economics and maximise GHG savings.

References

[1] McKendry P. Energy production from biomass (part 2): conversion technologies. Bioresour Technol 2002;83(1):47–54. Available from: http://linkinghub.elsevier.com/retrieve/pii/S0960852401001195 [Cited 6 March 2017].

[2] Adams PWR. An assessment of UK bioenergy production, resource availability, biomass gasification and life cycle impacts. Bath: University of Bath; 2011.

[3] FAO. Unified bioenergy terminology. Rome: FAO; 2004.

[4] Deublein D, Steinhauser A. Biogas from waste and renewable resources: an introduction. Weinheim: Wiley-VCH; 2008.

[5] Andersons Centre. A detailed economic assessment of anaerobic digestion technology and its suitability to UK farming and waste systems. 2nd ed. United Kingdom: The Andersons Centre, Melton Mowbray; 2010.

[6] Ecofys. Planning and installing bioenergy systems: a guide for installers, architects, and engineers. London: Deutsche Gesellschaft für Sonnenenergie. James and James; 2005.

[7] ADBA. The practical guide to AD. 1st ed. London: Anaerobic Digestion & Bioresources Association (ADBA); 2012.

[8] Larson ED. A review of life-cycle analysis studies on liquid biofuel systems for the transport sector. Energy Sustain Dev 2006;10(2):109–26. Available from: http://linkinghub.elsevier.com/retrieve/pii/S0973082608605360 [Cited 6 March 2017].

[9] von Blottnitz H, Curran MA. A review of assessments conducted on bio-ethanol as a transportation fuel from a net energy, greenhouse gas, and environmental life cycle perspective. J Clean Prod 2007;15(7):607–19. Available from: http://linkinghub.elsevier.com/retrieve/pii/S0959652606001016 [Cited 6 March 2017].

[10] Hertel TW, Golub AA, Jones AD, O'Hare M, Plevin RJ, Kammen DM. Effects of US maize ethanol on global land use and greenhouse gas emissions: estimating market-mediated responses. Bioscience 2010;60(3):223–31 Available from: http://www.bioone.org/doi/abs/10.1525/bio.2010.60.3.8.

[11] Patel M, Zhang X, Kumar A. Techno-economic and life cycle assessment on lignocellulosic biomass thermochemical conversion technologies: a review. Renew Sustain Energy Rev 2016;53:1486–99.
[12] Bridgwater AV. Thermal conversion of biomass and waste: the status. Birmingham: Bio-Energy Res Group, Aston University; 2001.
[13] Jenkins B, Baxter L, Miles T, Miles T. Combustion properties of biomass. Fuel Process Technol 1998;54(1–3): 17–46. Available from: http://linkinghub.elsevier.com/retrieve/pii/S0378382097000593 [Cited 7 March 2017].
[14] Williams RB, Jenkins BMND. Solid waste conversion: a review and database of current and emerging technologies—final report. California: California Integrated Waste Management Board; 2003.
[15] Zhang L, Xu C, Champagne P. Overview of recent advances in thermo-chemical conversion of biomass. Energy Convers Manag 2010;51(5):969–82. Available from: http://linkinghub.elsevier.com/retrieve/pii/S0196890409004889 [Cited 7 March 2017].
[16] Iribarren D, Peters JF, Dufour J. Life cycle assessment of transportation fuels from biomass pyrolysis. Fuel 2012;97:812–21.
[17] Czernik S, Bridgwater AV. Overview of applications of biomass fast pyrolysis oil. Energy Fuel 2004;18(2): 590–8. https://doi.org/10.1021/ef034067u.
[18] Isahak WNRW, Hisham MWM, Yarmo MA, Yun HT. A review on bio-oil production from biomass by using pyrolysis method. Renew Sustain Energy Rev 2012;16(8):5910–23. Available from: http://www.sciencedirect.com.ezproxy1.bath.ac.uk/science?_ob=ArticleListURL&_method=list&_ArticleListID=-1158972228&_sort=r&_st=13&view=c&md5=64754a904d4db8d3928b38a9fcf67bb9&searchtype=a [Cited 7 March 2017].
[19] Nachenius RW, Ronsse F, Venderbosch RH, Prins W. Biomass pyrolysis; 2013 pp. 75–139. Available from: http://linkinghub.elsevier.com/retrieve/pii/B978012386505200002X [Cited 7 March 2017].
[20] Williams PT, Besler S. The influence of temperature and heating rate on the slow pyrolysis of biomass. Renew Energy 1996;7(3):233–50. Available from: http://linkinghub.elsevier.com/retrieve/pii/0960148196000067 [Cited 7 March 2017].
[21] Kebelmann K, Hornung A, Karsten U, Griffiths G. Intermediate pyrolysis and product identification by TGA and Py-GC/MS of green microalgae and their extracted protein and lipid components. Biomass Bioenergy 2013;49:38–48. Available from: http://linkinghub.elsevier.com/retrieve/pii/S0961953412005041 [Cited 7 March 2017].
[22] Horne PA, Williams PT. Influence of temperature on the products from the flash pyrolysis of biomass. Fuel 1996;75(9):1051–9. Available from: http://linkinghub.elsevier.com/retrieve/pii/0016236196000816 [Cited 7 March 2017].
[23] Maggi R, Delmon B. Comparison between "slow" and "flash" pyrolysis oils from biomass. Fuel 1994;73(5): 671–7. Available from: http://linkinghub.elsevier.com/retrieve/pii/0016236194900078 [Cited 7 March 2017].
[24] Garcìa-Pèrez M, Chaala A, Roy C. Vacuum pyrolysis of sugarcane bagasse. J Anal Appl Pyrolysis 2002;65(2): 111–36. Available from: http://linkinghub.elsevier.com/retrieve/pii/S016523700100184X [Cited 7 March 2017].
[25] Garcìa-Pérez M, Chaala A, Pakdel H, Kretschmer D, Roy C. Vacuum pyrolysis of softwood and hardwood biomass. J Anal Appl Pyrolysis 2007;78(1):104–16. Available from: http://linkinghub.elsevier.com/retrieve/pii/S0165237006000702 [Cited 7 March 2017].
[26] de Jongh WA, Carrier M, Knoetze JH. Vacuum pyrolysis of intruder plant biomasses. J Anal Appl Pyrolysis 2011;92(1):184–93. Available from: http://linkinghub.elsevier.com/retrieve/pii/S0165237011001082 [Cited 7 March 2017].
[27] Peacocke GVC, Bridgwater AV. Ablative plate pyrolysis of biomass for liquids. Biomass Bioenergy 1994; 7(1–6):147–54. Available from: http://linkinghub.elsevier.com/retrieve/pii/096195349400054W [Cited 7 March 2017].
[28] Alauddin ZABZ, Lahijani P, Mohammadi M, Mohamed AR. Gasification of lignocellulosic biomass in fluidized beds for renewable energy development: a review. Renew Sustain Energy Rev 2010;14(9):2852–62.
[29] Parthasarathy P, Narayanan KS. Hydrogen production from steam gasification of biomass: influence of process parameters on hydrogen yield—a review. Renew Energy 2014;66:570–9. Available from: http://linkinghub.elsevier.com/retrieve/pii/S0960148113007027 [Cited 7 March 2017].
[30] Zhou J, Chen Q, Zhao H, Cao X, Mei Q, Luo Z, et al. Biomass–oxygen gasification in a high-temperature entrained-flow gasifier. Biotechnol Adv 2009;27(5):606–11. Available from: http://linkinghub.elsevier.com/retrieve/pii/S0734975009000664 [Cited 7 March 2017].
[31] Arce EM, Saavedra Á, Míguez LJ, Granada E, Cacabelos A. Biomass fuel and combustion conditions selection in a fixed bed combustor. Energies 2013;6:5973–89.

[32] Sebastián F, Royo J, Gómez M. Cofiring versus biomass-fired power plants: GHG (Greenhouse Gases) emissions savings comparison by means of LCA (Life Cycle Assessment) methodology. Energy 2011;36(4):2029–37.

[33] Zhang B, von Keitz M, Valentas K. Thermochemical liquefaction of high-diversity grassland perennials. J Anal Appl Pyrolysis 2009;84(1):18–24. Available from: http://linkinghub.elsevier.com/retrieve/pii/S016523700800123X; [Cited 7 March 2017].

[34] Kumar M, Gupta RC, Sharma T. Effects of carbonisation conditions on the yield and chemical composition of Acacia and Eucalyptus wood chars. Biomass Bioenergy 1992;3(6):411–7. Available from: http://linkinghub.elsevier.com/retrieve/pii/096195349290037Q; [Cited 7 March 2017].

[35] Prins MJ, Ptasinski KJ, Janssen FJJG. From coal to biomass gasification: comparison of thermodynamic efficiency. Energy 2007;32(7):1248–59. Available from: http://www.sciencedirect.com/science/article/pii/S0360544206002106 [Cited 7 March 2017].

[36] McKendry P. Energy production from biomass (part 1): overview of biomass. Bioresour Technol 2002;83(1): 37–46. Available from: http://linkinghub.elsevier.com/retrieve/pii/S0960852401001183 [Cited 6 March 2017].

[37] Jahirul IM, Rasul GM, Chowdhury AA, Ashwath N. Biofuels production through biomass pyrolysis—a technological review. Energies 2012;5(12):4952–5001.

[38] Bridgwater AV. Review of fast pyrolysis of biomass and product upgrading. Biomass Bioenergy 2012;38:68–94.

[39] ASTM D7544-12, Standard specification for pyrolysis liquid biofuel. West Conshohocken, PA: ASTM International; 2012.

[40] Mohan D, Pittman CU, Steele P. Pyrolysis of wood/biomass for bio-oil: a critical review. Energy Fuel 2006;20:848–89.

[41] Kersten SRA, Wang X, Prins W, van Swaaij WPM. Biomass pyrolysis in a fluidized bed reactor. Part 1: literature review and model simulations. Ind Eng Chem Res 2005;44:8773–85.

[42] Bridgwater AV. Renewable fuels and chemicals by thermal processing of biomass. Chem Eng J 2003;91:87–102.

[43] Bridgwater AV, Czernik S. The status of biomass fast pyrolysis. In: Bridgwater AV, editor. Fast pyrolysis of biomass a handbook. vol. 2. Newbury: CPL Press; 2002. p. 1–22.

[44] Bridgwater AV. Fast pyrolysis of biomass. In: Bridgwater AV, Hofbauer H, van Loo S, editors. Thermal biomass conversion. Newbury: CPL Press; 2009.

[45] Venderbosch RH, Prins W. Fast pyrolysis technology development. Biofuels Bioprod Biorefin 2010;4(2): 178–208. https://doi.org/10.1002/bbb.205.

[46] Toft AJ. Ph.D. thesis, Aston University; 1998.

[47] Bridgwater AV. Upgrading biomass fast pyrolysis liquids. In: Brown RC, editor. Thermochemical processing of biomass: conversion into fuels, chemicals and power. Wiley Series in Renewable Resources. Oxford, United Kingdom: Wiley-Blackwell; 2011. ISBN: 978-0-470-72111-7. [chapter 6].

[48] Maggi R, Elliott D. Upgrading overview. In: Bridgwater AV, Boocock D, editors. Developments in thermochemical biomass conversion. Amsterdam, Netherlands: Blackie Academic and Professional; 1997. p. 575–88.

[49] Bridgwater AV. Production of high-grade fuels and chemicals from catalytic pyrolysis of biomass. Catal Today 1966;29:285–95.

[50] Bridgwater AV. Catalysis in thermal biomass conversion. Appl Catal A Gen 1994;116:5–47.

[51] Qi Z, Jie C, Wang Tiejun XY. Review of biomass pyrolysis oil properties and upgrading research. Energy Convers Manag 2007;48:87–92.

[52] Huber GW, Corma A. Synergies between bio- and oil refineries for the production of fuels from biomass. Angew Chem Int Ed 2007;46(38):7184–201.

[53] Elliott DC. In: Klass D, editor. Energy from biomass and wastes X. Chicago, USA: IGT, Institute of Gas Technology; 1983. p. 765–82.

[54] Bridgwater AV. Technical and economic assessment of thermal processes for biofuelsReport to NNFCC. Shetland: COPE Ltd; 2009. Available from: www.copeltd.co.uk.

[55] Calis HPA, Haan JP, Boerrigter H, van der Drift A, Peppink G, van den Broek R, et al. Preliminary techno-economic analysis of large-scale synthesis gas manufacturing from imported biomass. In: Bridgwater AV, editor. Proceedings of pyrolysis and gasification of biomass and waste, Strasbourg, September 2002. Strasbourg: CPL Scientific Press; 2003.

[56] Morehead H. Siemens gasification and IGCC update. Washington, DC: Gasification Technologies; 2008.

[57] Volkmann D. Update on technology and projects. In: Gasification technologies conference, Washington DC, October; 2004.

[58] Kumar A, Jones D, Hanna M. Thermochemical biomass gasification: a review of the current status of the technology. Energies 2009;2:556–80.

[59] Sansaniwal SK, Pal K, Rosen MA, Tyagi SK. Recent advances in the development of biomass gasification technology: a comprehensive review. Renew Sustain Energy Rev 2017;72:363–84.

[60] Bulushev D, Ross J. Catalysis for conversion of biomass to fuels via pyrolysis and gasification: a review. Catal Today 2011;171:1–13.

[61] Sanlisoy A, Carpinlioglu M. A review on plasma gasification for solid waste disposal. Int J Hydrogen Energy 2017;42:1361–5.

[62] E4Tech. Review of technologies for gasification of biomass and wastes: NNFCC project 09/008 final report. London: E4tech; 2009.

[63] NETL. Syngas composition. Albany, OR: National Energy Technology Laboratory; 2013.

[64] Farzad S, Mandegari M, Görgens J. A critical review on biomass gasification, co-gasification, and their environmental assessments. Biofuel Res J 2016;12:483–95.

[65] Susmozas A, Iribarren D, Zapp P, Linβen J, Dufour J. Life-cycle performance of hydrogen production via indirect biomass gasification with CO_2 capture. Int J Hydrogen Energy 2016;41(42):19484–91.

[66] La Villetta M, Costa M, Massarotti N. Modelling approaches to biomass gasification: a review with emphasis on the stoichiometric method. Renew Sustain Energy Rev 2017;74:71–88.

[67] Sikarwar V, Zhao M, Clough P, Yao J, Zhong X, Memon M, et al. An overview of advances in biomass gasification. Energy Environ Sci 2016;9(10):2927–3304.

[68] Netherlands ERC of the Phyllis 2 database for biomass and waste. Available form: https://www.ecn.nl/phyllis2/.

[69] Bond TC, Doherty SJ, Fahey DW, Forster P, Berntsen T, et al. Bounding the role of black carbon in the climate system, a scientific assessment. J Geophys Res Atmos 2013;118:5380–552.

[70] Agbor E, Zhang X, Kumar A. A review of biomass co-firing in North America. Renew Sustain Energy Rev 2014;40:930–43. Available from: http://linkinghub.elsevier.com/retrieve/pii/S1364032114006558 [Cited 7 March 2017].

[71] Monteleone B, Chiesa M, Marzuoli R, Verma VK, Schwarz M, Carlon E, et al. Life cycle analysis of small scale pellet boilers characterized by high efficiency and low emissions. Appl Energy 2015;155:160–70. Available from: http://www.sciencedirect.com/science/article/pii/S0306261915007229.

[72] Xuan Do T, Lim Y-i, Heejung Y. Techno-economic analysis of bio oil production process from palm empty fruit bunches. Energy Convers Manag 2014;80:525–34.

[73] Sahoo BB, Niranjan Sahoo UKS. Effect of H2:CO ratio in syngas on the performance of a dual fuel diesel engine operation. Appl Therm Eng 2012;49:139–46.

[74] Abdul Malek ABM, Hasanuzzaman M, Abd Rahim N, Al Turki YA. Techno-economic analysis and environmental impact assessment of a 10 MW biomass-based power plant in Malaysia. J Clean Prod 2017;141:502–13.

[75] Fan M, Brown RC. Comparison of the loss-on-ignition and thermogravimetric analysis techniques in measuring unburned carbon in coal fly ash. Energy Fuel 2001;15:1414–7.

[76] Duan L, Liu D, Xiaoping Chen CZ. Fly ash recirculation by bottom feeding on a circulating fluidized bed boiler co-burning coal sludge and coal. Appl Energy 2012;95:295–9.

[77] Bartoňová L. Unburned carbon from coal combustion ash: an overview. Fuel Process Technol 2015;134:136–58.

[78] Permchart W, Kouprianov VI. Emission performance and combustion efficiency of a conical fluidized-bed combustor firing various biomass fuels. Bioresour Technol 2004;92(1):83–91.

[79] Backreedy RI, Jones JM, Pourkashanian M, Williams A. Burn-out of pulverised coal and biomass chars. Fuel 2003;82:2097–105.

[80] Clarke L, et al. Assessing transformation pathways. In: Edenhofer O, et al., editors. Climate change 2014: mitigation of climate change. Contribution of Working Group III to the fifth assessment report of the intergovernmental panel on climate change. Cambridge: Cambridge University Press; 2014. p. 413–510. Available from: https://www.ipcc.ch/pdf/assessment-report/ar5/wg3/ipcc_wg3_ar5_chapter6.pdf.

[81] Biller P, Ross AB. Production of biofuels via hydrothermal conversion. Handbook of biofuels production. New York: Woodhead Publishing; 2016. pp. 509–47 [chapter 17].

[82] Titirici M-M, Funke A, Kruse A. Hydrothermal carbonization of biomass. Recent advances in thermo-chemical conversion of biomass. New York: Elsevier; 2015. pp. 325–52 [chapter 12].

[83] Elliott DC, Biller P, Ross AB, Schmidt AJ, Jones SB. Hydrothermal liquefaction of biomass: developments from batch to continuous process. Bioresour Technol 2015;178:147–56.

[84] Biller P, Sharma BK, Kunwar B, Ross AB. Hydroprocessing of bio-crude from continuous hydrothermal liquefaction of microalgae. Fuel 2015;159:197–205.
[85] Elliott DC. Catalytic hydrothermal gasification of biomass. Biofuels Bioprod Biorefin 2008;2(3):254–65.
[86] Yakaboylu O, et al. Supercritical water gasification of biomass: a literature and technology overview. Energies 2015;8(2):859.
[87] Waldner MH, Vogel F. Renewable production of methane from woody biomass by catalytic hydrothermal gasification. Ind Eng Chem Res 2005;44(13):4543–51.
[88] Zhu Y, Biddy MJ, Jones SB, Elliott DC, Schmidt AJ. Techno-economic analysis of liquid fuel production from woody biomass via hydrothermal liquefaction (HTL) and upgrading. Appl Energy 2014;129:384–94.
[89] Jazrawi C, et al. Pilot plant testing of continuous hydrothermal liquefaction of microalgae. Algal Res 2013;2(3):268–77.
[90] Li L, et al. Hydrothermal carbonisation of food waste and associated packaging materials for energy source generation. Waste Manag 2013;33:2478–92.
[91] Berge ND, et al. Assessing the environmental impact of energy production from hydrochar generated via hydrothermal carbonisation of food wastes. Waste Manag 2015;43:203–17.
[92] Ekpo U, et al. A comparison of product yields and inorganic content in process streams following thermal hydrolysis and hydrothermal processing of microalgae, manure and digestate. Bioresour Technol 2016;200:951–60.
[93] Heilmann SM, et al. Phosphorus reclamation through hydrothermal carbonization of animal manures. Environ Sci Technol 2014;48:10323–9.
[94] Bennion EP, Ginosar DM, Moses J, Agblevor F, Quinn JC. Lifecycle assessment of microalgae to biofuel: comparison of thermochemical processing pathways. Appl Energy 2015;154:1062–71.
[95] Vardon DR, Rajagopalan K, Strathmann TJ. Thermochemical conversion of raw and defatted algal biomass via hydrothermal liquefaction and slow pyrolysis. Bioresour Technol 2012;109:178–87.
[96] Patel B, et al. Environmental profile of algal hydrothermal liquefaction—a country specific case study. Algal Res 2016;16:127–40.
[97] Quinn JC, et al. Microalgae to biofuels lifecycle assessment—multiple pathway evaluation. Algal Res 2014;4:116–22.
[98] Patel B, et al. A review on hydrothermal pre-treatment technologies and environmental profiles of algal biomass processing. Bioresour Technol 2016;199:288–99.
[99] Frank EF, et al. Life cycle comparison of hydrothermal and lipid extraction pathways to renewable diesel from algae. Mitig Adapt Strateg Glob Chang 2013;18:137–58.
[100] Fortier MP, et al. Life cycle assessment of biojet fuel from hydrothermal liquefaction of microalgae. Appl Energy 2014;122:73–82.
[101] Summers HM, et al. Techno-economic feasibility and life cycle assessment of diary effluent to renewable diesel via hydrothermal liquefaction. Bioresour Technol 2015;196:431–40.
[102] Ro KS, et al. Catalytic wet gasification of municipal and animal wastes. Ind Eng Chem Res 2007;46(26):8839–45.

Further Reading

[1] Brown RCE. Thermochemical processing of biomass: conversion into fuels, chemicals, and power. Chichester, United Kingdom: Wiley & Sons; 2011.

CHAPTER

9

GHG Emissions From Biomethane Gas-to-Grid Injection via Anaerobic Digestion

Paul Adams

University of Bath, Bath, United Kingdom

9.1 INTRODUCTION

9.1.1 Biomethane

Biomethane is a renewable fuel that can be produced from a wide range of different feedstocks, technologies, and different scales for conversion. It is a versatile fuel that can be used as a substitute to natural gas (NG), which is increasingly an important resource for meeting the demand of heat, electricity, and fuel in many countries [1]. Natural gas is a significant all-round energy carrier with an already well-developed infrastructure in many countries such as pipelines (gas grids), filling stations, and new infrastructure such as road transport via heavy duty vehicles or marine transport via tanker in the form of compressed natural gas (CNG) or liquefied natural gas (LNG) [2]. In locations with existing gas pipelines, NG is easy to transport and has lower greenhouse gas (GHG) emissions in comparison to coal and oil [3].

Biomethane is defined as methane produced from biomass (ISO 16559:2014), with very similar properties to NG [4]. There are two main pathways to producing substitute NG:

- via thermo-chemical conversion (e.g., gasification and methanation), the methane-rich product gas is normally referred to as bio-based synthetic natural gas (bio-SNG) (see Chapter 8); or
- via bio-chemical conversion when it is produced by biological processes, including anaerobic digestion, landfills, and waste water treatment; the initial product is raw biogas which must be cleaned and upgraded to reach a high methane content (referred to as biomethane) suitable for grid injection.

Bio-SNG and biomethane from upgraded biogas are essentially chemically identical and must meet the same technical specification to be injected into natural gas pipelines [2,5].

Greenhouse Gas Balances of Bioenergy Systems
https://doi.org/10.1016/B978-0-08-101036-5.00009-4

Biomethane, in principle, can be used for exactly the same applications as natural gas, if the final composition is in line with the different NG qualities on the market [2,6,7]; therefore, it can be used as a substitute for liquid transport fuels, to produce combined heat and power (CHP), heat alone, and serve as feedstock for the chemical sector. In contrast to liquid biofuels such as biodiesel and bioethanol, biomethane and natural gas are fully interchangeable from an end-user perspective [2].

The focus of this chapter is on the GHG emissions that can arise from the biomethane production chain from feedstock supply through to gas grid injection and end-use. Biomethane is now commonly implemented at a commercial scale in several countries including the United Kingdom and Germany [8]. In contrast, bio-SNG is at an earlier stage of development and so is not the main focus of this chapter [9,10]. This chapter therefore assesses GHG emissions from anaerobic digestion (via bio-chemical conversion) and subsequent biogas upgrading, using farm-derived feedstocks including manures, slurries, residues, and energy crops (as silage).

9.1.2 Farm-Based Anaerobic Digestion

Farm production of beef and pork produces very significant quantities of cattle and pig slurry, which require management, treatment, and disposal. These usually contain valuable nutrients which are important to reuse for sustainable farm management. Therefore, it makes sense to spread these on land to recover and reuse the nutrient value and avoid mineral fertilizer application. However, the traditional methods of broadcast or even injection spreading can result in significant releases of GHG emissions [11,12]. Anaerobic digestion of the slurry is attractive to farmers because it deals with the waste management in a way that provides wider benefits to the farming system, since digestate is available as a source of nutrients and organic matter for use on farm. Previous work has shown that this farm integration is often a more important stakeholder driver for technology uptake than energy production [13]. In addition, desires to accelerate progress in renewable energy deployment at minimal cost have led to implementation of solutions that make the best use of existing infrastructure and so there have been increases in injection of upgraded biogas into large-scale gas grids, similar to co-firing of biomass in existing coal-fired power plants. It is, therefore, important to verify that the use of farm-derived feedstocks used to produce biogas which is upgraded to biomethane for the gas grid is actually delivering low carbon energy.

9.2 AN OVERVIEW OF GHG EMISSIONS FROM BIOMETHANE PRODUCTION

To ensure that GHG emissions are reduced by biomethane production, it is important that the GHG balance of production and delivery is favorable. This is a complex subject as emissions from biomethane supply chains cannot be completely measured directly and therefore require sophisticated models to calculate GHG emissions [14,15]. All bioenergy systems have different emissions profiles depending on the biomass source used, conversion process, and end-use [16]. Therefore, each biomass-to-energy production pathway needs to be considered individually and the GHG emissions calculated alongside specified sustainability criteria [17]. Biomethane and other bioenergy options can then be compared to the conventional fossil fuel alternatives to assess performance in comparison to the nonrenewable energy sources they may displace.

The generation of biomethane is based on a multifaceted process in various stages (see Chapter 8, Section 8.2.1). Many factors influence the process from the cultivation of biomass to the application of the fed-in biomethane. At each stage of the supply (or 'value') chain, GHG emissions arise from agricultural processes, energy consumption, fugitive emissions (e.g., methane leakage), and other minor sources. Carbon sequestration also occurs during the cultivation of biomass, and carbon dioxide can be captured in the upgrading process. The extent of net GHG emissions for any given biomethane value chain will depend upon a number of factors including biomass source, agricultural practice, transportation, AD facility operational practices, fugitive emission management, technology choices, etc. Fig. 9.1 portrays the main process stages in the biomethane value chain with potential GHG emissions sources, main system inputs, energy, and carbon flows [14].

Biogas is regarded as a climate-neutral fuel since the carbon in the biogas is fixed from atmospheric carbon dioxide (CO_2). The CO_2 released when combusting biogas is therefore regarded as being biogenic (with a CO_2 equivalent value of zero) rather than fossil (which has a CO_2 equivalent value of one). Furthermore, any consumption of fossil fuels replaced by biogas will lower the total CO_2 emissions. Biogas consists mainly of methane (CH_4), and since methane in itself is a strong greenhouse gas (28–34 times compared to CO_2) [18], it is important to consider losses that might occur in the biogas production chain, and subsequently it is imperative to minimize these emissions [19].

Fig. 9.1 shows that emissions do arise at each stage of the biomethane production process. The following sections present an overview of these potential GHG emissions of the whole chain (so-called 'well to wheel' or 'cradle to grave') of biogas and biomethane production. Key potential sources of GHG emissions are presented from the different stages of the value chain (known as 'unit processes') in the context of the relative importance of the GHG emission sources and possible ways to mitigate these.

9.3 FEEDSTOCK SUPPLY

Biogas can, in principle, be generated from any organic compound, meaning the biomass feedstocks available to produce biogas are diverse. Biomass feedstock (known as 'substrates' when used in AD) can come from farms in the form of plants or energy crops. Another source is the by-products of industrial or agricultural processes in the form of animal slurries and manures, damaged crops, or waste materials such as sewage sludge and food waste. GHG emissions arising from biomass production therefore depend primarily on the biomass source, its location, and the cultivation or collection method used to produce or obtain the biomass [20].

9.3.1 Farm Manures, Slurries, and Residues

Definitions of waste and residue vary depending on the policy and regulatory context. For example, waste streams are assumed to have a zero economic value and are intended to be discarded by the holder [21]. Biomass sustainability criteria separates wastes from residues, with wastes defined as above, but residues having potential uses in alternative markets [22]. Biomass wastes are assumed to not have the environmental impacts from their original production attributed, and therefore, have a zero GHG burden [23]. The diversion of waste from landfill can also alleviate some of the environmental pressures of methane emissions from

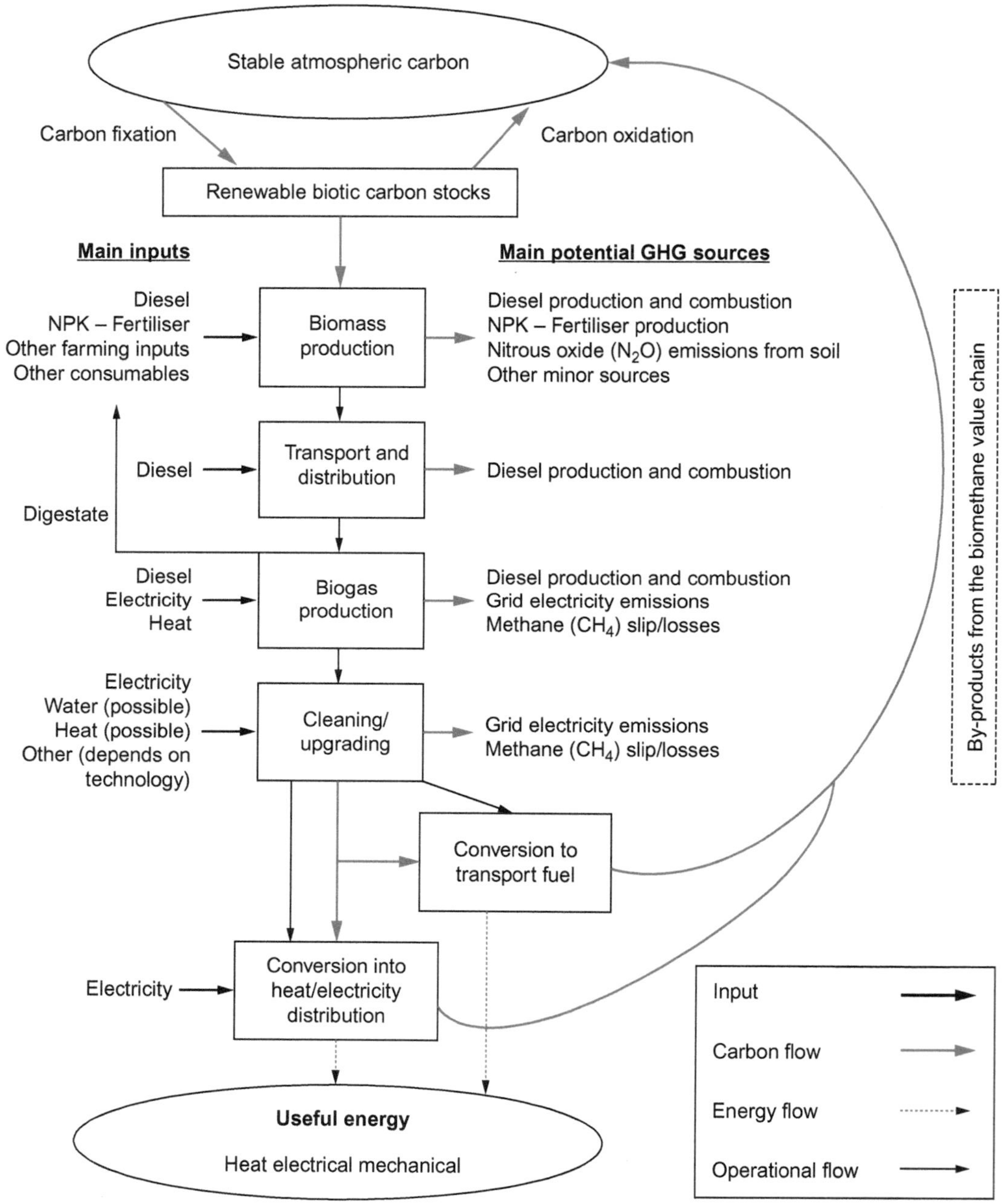

FIG. 9.1 Flow chart for the calculation of GHG emissions for biomethane production [14].

the anaerobic decomposition of biomass in landfill [24]. Advantages of biogas production by agricultural biomass and organic wastes are emphasized in the Renewable Energy Directive [25]: *'The use of agricultural material such as manure, slurry and other animal and organic waste for biogas production has, in the view of the high greenhouse gas emission saving potential, significant environmental advantages in terms of heat and power production and its use as biofuel (as biogas).'*

Examples of waste and residues suitable for AD include organic household waste, organic commercial and industrial waste, animal by-products, organic farm wastes, and sewage sludge. Suitability of different biogas production technologies is dependent on the input feedstock characteristics [26]. Farm-based AD facilities treating agricultural residues may require supplementary feedstocks in the summer months when livestock are outdoors and manures and slurries are less available. Farms are often unlikely to consider municipal sources of organic waste due to the more complex planning and licensing required. Therefore, nonfarm-derived wastes and residues are more commonly utilized by industrial AD facilities which require waste separation and sometimes pretreatment prior to being input to the AD process. Farm-based AD has therefore developed using crop silage in addition to wastes and residues.

Wastes and residues are considered to have no GHG emissions associated with their production; therefore, the only real GHG implications arise from the acquisition, transportation, separation, pretreatment, and storage. This is a clear advantage over purposely grown energy crops as it is commonly found that cultivation is the largest source of GHG emissions (see Section 9.3.2). Nonetheless, farm wastes and residues usually have lower methane yields [27], incur variations in their regional and seasonal availability, and often have additional legislative requirements (e.g. environmental permitting).

A key consideration for the GHG balance of wastes and residues used to produce biogas and biomethane is the counterfactual or alternative management or use of the feedstock. This is important as the relative GHG savings can vary substantially depending on the counterfactual (see Section 9.8).

9.3.2 Energy Crop Silage

The term 'energy crop' refers to biomass feedstocks which are cultivated especially for the purpose of energy production. For a farm system, this can involve sophisticated land management and crop husbandry regimes designed to maximize the biomass available from a given area of land. Purposely grown energy crops are an important feedstock for biogas plants, both mixed with farm and other organic waste and on their own. Along with offering a high level of energy generation, they can make a significant contribution to the economic and environmental sustainability of farming [5]. Nonetheless, the role of purposely grown crops in increasing the sustainability of agriculture in conjunction with use as a feedstock can be a delicate balance [11,13].

Best practice measures focus on using locally sourced crop-based feedstocks which do not impact upon food production by fitting into the normal crop rotation and working as a break crop [28]. This can provide additional financial benefits for farmers as well as promoting biodiversity and boosting soil conditions on marginal lands previously unsuitable for food production. In addition to this, biogas systems are very flexible; feedstocks often include plants that are not used directly for human consumption including grass. Inappropriate siting of feedstocks leading to changes in carbon stocks or land-use change (LUC) is likely to cause significant increases in GHG emissions; therefore careful crop management is crucial [28,29]. A possible consequence of LUC is the displacement of food production where agricultural expansion can contribute to large GHG emissions and biodiversity loss through land transformation, or 'indirect land use change' (ILUC), which can dominate life cycle GHG burdens [30,31].

The main inputs to a crop cultivation system which produce GHG emissions include the use of fertilizers, pesticides, lime, herbicides, fuels combusted in farm machinery use, electricity, and irrigation (where used). The extent of GHG emissions will depend primarily on the agronomic practice, i.e., how much and what type of fertilizers and other agro-chemicals are applied to the crop, as well as application method [32]. Nitrogen-based fertilizer is the largest source of GHG emissions from crop cultivation due to the upstream use of natural gas in manufacturing ammonium-based fertilizer and through soil N_2O emissions from fertilizer application (see Chapter 16).

Maize is especially well-adapted for use in biogas facilities, with cereals (such as rye), sugar beet, and grass silage also commonly used [28,33–35]. Crops most suitable for use in biogas generation vary from region to region, and therefore the GHG emissions will vary depending on the local climate, soil quality (e.g., N_2O emissions from fertilizer use), water availability, nutrient supply, and so on [36]. From a GHG mitigation and ecosystem protection perspective, energy crops must be chosen against the local conditions. It is imperative to ensure that local land use changes associated with biomass cultivation do not create negative ecological effects [16].

Most food crops are not yet subject to GHG reporting; however, stringent monitoring of crops for bioenergy ensures that GHG savings are achieved [14]. The cultivation process for any crop establishment can result in changes to soil carbon and nitrate leaching [35,37–39]. Good soil management and appropriate application of digestate can help to improve and enhance the organic matter content of soil under cultivation, with benefits for soil structure and reduced incidence of run-off [28]. However, cultivation and harvesting are usually accountable for the largest source of emissions in the crop-based AD life cycle. Fig. 9.2 provides

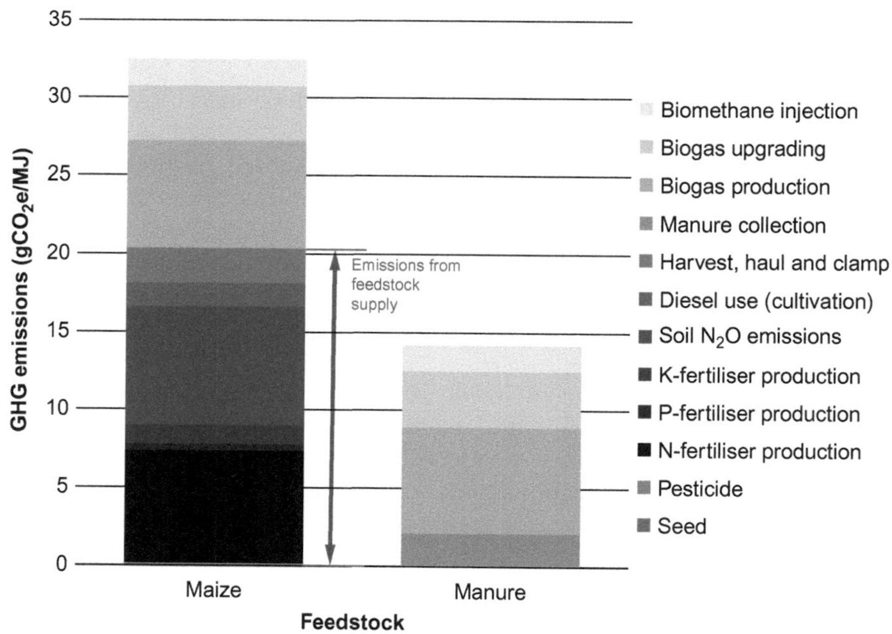

FIG. 9.2 Typical GHG emissions from biomethane production using maize or manure. *Adapted from Adams PWR, Mezzullo WG, McManus MC. Biomass sustainability criteria: greenhouse gas accounting issues for biogas and biomethane facilities. Energy Policy 2015;87:95–109.*

some typical results for manure and maize to illustrate the relative importance of feedstock supply in the life cycle of biomethane. The data used follows the RED methodology using an attributional LCA approach as described in previous work [14].

9.3.3 Transportation and Distribution of Biomass

Transport and distribution GHG emissions are relatively minor in local arable systems [40,41], although this is sensitive to mode (road, sea, or rail) and distance. GHG emissions from transport arise primarily from the combustion of diesel and other liquid fuels. Total contribution is usually <10% of the total supply chain GHG emissions [5], as seen in Fig. 9.2. Due to the local nature of biomethane systems, road transport (lorry, truck, or tractor/trailer) is most commonly utilized with distances generally minimized depending on the size of the biogas production plant. Long distance delivery of feedstock is often impractical when the economic, energetic, and emission barriers are considered. As biomass is only harvested during certain short periods of the year, these barriers require a well-planned biomass logistics chain [27]. Transportation distances should therefore be minimized as far as possible in order to reduce the economic and energy cost, alongside curtailing GHG emissions associated with logistics.

To calculate GHG emissions from transport, it is common to work out the mass (weight) of the feedstock and the distance travelled; this gives a metric ordinarily used as ton-km (tkm). GHG emissions can then be worked out using standard emission factors for transportation. In addition to CO_2, it is also important to consider tail pipe emissions of N_2O and CH_4; these are calculated by multiplying the distance transported by emission factors for non-CO_2 GHG emissions, such as those provided by DEFRA and HEBFA [42,43].

9.4 BIOGAS PRODUCTION

Biogas is generated through the fermentation of feedstock where microorganisms facilitate the process of material conversion. Raw biogas consists of 45%–70% methane (CH_4); the second largest component of the gas is carbon dioxide (CO_2), making up 25%–50% of the total [27]. The flammable and thus relevant component of biogas is methane, which varies primarily with the feedstock used. While methane is the essential component of biogas, it is also an area of potentially significant emissions, which could result in worsening GHG savings of biogas production and use [5].

Methane emissions from biogas plants (so-called 'methane slip') occur if unburned biogas escapes into the atmosphere (e.g., diffusion through foils, leaks, high-pressure cut-off valves, engines failures, etc.) or if fermentation residue which has not completely decomposed leaves the closed system of the biogas plant, i.e., fermentation residue tank not gas-tight covered [5]. GHG emissions need to be considered in terms of both economic and environmental aspects. Clearly, any methane slip has a direct impact on the biogas yield as well as a GHG penalty, resulting in reduced economic returns and more difficulty in adhering to sustainability standards [14].

Several aspects around the biomethane facility influence the level of methane loss. These include the type of feedstock or mixture of feedstocks on site, the practice in storing the feedstock and feeding it to the digesters, digester and gas storage design and construction, the type of digestion process, digestate storage practice, and the processes involved in end-use of biogas (i.e., upgrade and injection into the grid, or combustion in CHP/boiler).

Other sources of GHG emissions which may arise from the AD facility operation may include indirect emissions such as from electricity consumed from the grid, or direct emissions such as diesel used for handling feedstocks and manures [14,44].

9.4.1 Feedstock Storage

Generally, stored silage does not emit significant amounts of methane, which is primarily due to the low apparent pH. Fugitive emissions primarily occur when there are delays in delivery or processing. These emissions occur prior to biogas production. Methane loss can also arise from the pumping of slurry and delivery of feedstock to feeders. Liebetrau et al. found that the maximum methane measured was 0.008 g/kWh, which equals 0.004% of the methane converted [45]. Similarly, the Baltic Biogas Bus project found the range of emissions from feedstock storage and feeding for existing plants is around 0%–0.1% [46]. In terms of the overall GHG balance of biomethane, the methane emissions from silage storage are therefore considered negligibly low.

9.4.2 Feeding System

Several scenarios exist for possible emissions from digester feeding systems. Fugitive methane emissions are likely to occur when the fresh substrate is mixed with digestate in an open tank prior to the feeding procedure. For instance, Liebetrau et al. measured one AD facility which showed methane emissions of 0.48 g/kWh (or 0.27% of the converted biogas). In comparison, the fugitive methane emissions measured in three input systems showed that feeding silage indirectly amounted on average to 0.055 g/kWh [45]. A tank for mixing feedstocks is required to ensure feedstock composition is homogeneous. Therefore, with regard to fugitive methane emissions, the direct intake of solid material is the preferred technology.

Depending on the technology employed, process energy may be required such as electricity to power the feeding system or diesel to power tractors or feeding equipment. GHG emissions from process energy requirements are likely to be minimal when considered against the whole system, but should still be considered in a LCA.

9.4.3 Digester

Digesters are usually fitted with either a concrete or foil digester cover and it is expected that no methane losses should happen from the digester. Head space on the digester performs as gas collection and gas storage space. Liebetrau et al. demonstrated that in most digesters no detectable emissions were measured and digesters with a concrete roof showed no emissions [45]. The survey revealed that emissions can occur where the foil connects to the digester with one case reporting fugitive methane emissions of 0.29 g/kWh (or 0.17% of the converted biogas). However, poor maintenance and management of digestion facilities can result in methane emissions of up to 6% [47], with Liebetrau et al. also measuring one facility with 5.46% of converted biogas that was released into the atmosphere [45]. Biogas release during maintenance of the chamber is not usually accounted for in LCA studies as it is not normal operation; however, this should be considered within the system boundary if the data is available.

Some methane slip will occur from gas safety release valves, which are designed to open when the container experiences too much internal pressure. If the valves are not calibrated correctly to match normal chamber pressure, they can become significant fugitive emission sources releasing at lower pressures than the chamber can handle. Uncalibrated valves can also cause damage to chambers, leading to leakage at seams, or can cause punctures to the container in the case of sealed membrane digesters.

Requirements exist for biogas plants so that methane can no longer escape into the atmosphere. This is achieved through technical measures, such as double membrane roofs on the fermentation tanks, gas measurement and analysis devices, a covered end tank for fermentation residue, and fermentation residue distribution on fields using drag hoses, as well as the frequent checking and measuring of the methane slip [5,45,47]. Other best practice options include optimizing plant design and implementing methane slip detection systems to assess possible leaks from the fermenter, piping, joints, and valves. Methane leak detection should be carried out on a regular basis to ensure any plant deterioration does not increase methane emissions from the AD plant [47].

9.4.4 Gas Storage

Gas holders and storage systems are designed to store gas and prevent leakage. The design of the AD plant can vary and gas storage units could be integrated with the digester itself or separate from it, leading to different methane loss into the atmosphere. Poorly maintained facilities can cause leaks and have the potential to suffer from large methane emissions. Gas storage is nonetheless expected to be a minor source of emissions with existing plants around 0%–0.2% of biogas production [46].

9.4.5 Digestate Storage Tanks

Digestate is produced throughout the year and must therefore be stored until the growing season, which is the only appropriate time for its application as a fertilizer, since its application is not allowed in winter. Digestate is stored in open or closed tanks, with closed storage additional biogas released is mostly recovered. Like manure, if liquid digestate is stored in open tanks, ammonia and methane gases (residual biogas) are given off and are therefore a potential source of GHG emissions [48]. From the economic point of view, the operator is interested in attaining as low as possible remaining gas potential in the digestate. An evaluation of biogas plants in Germany (Biogas-Messprogramm) revealed a gas potential of 1.5%–3.5% remaining after digestion [49]. However, Liebetrau et al. found that in summer temperatures up to 10% of the total biogas production is generated in the digestate store. The determination of emissions occurring in reality is far from being simple and depends on numerous variables such as the available substrate, temperature, mixing activities, or weather conditions [45]. Moreover, most modern biogas plants have secondary digestion, so the residual gas in digestate is much lower.

TABLE 9.1 Open Digestate Storage—Examples of Potential Impact on GHG Emissions [14]

Source	Maize	Manure	Reference
Joint Research Centre (JRC)	$0.44gCH_4/MJ_{biogas}$ $11.0gCO_2e/MJ_{biogas}$	$2.0gCH_4/MJ_{biogas}$ $50.0gCO_2e/MJ_{biogas}$	[48]
Liebetrau et al.—summer (mean)	$22.5gCO_2e/MJ_{biogas}$	–	[45]
Boulamanti et al.	$15.1gCO_2e/MJ_{biogas}$	$45.9gCO_2e/MJ_{biogas}$	[36]
Buratti et al.	$0.81gCH_4/MJ_{biomethane}$ $20.3gCO_2e/MJ_{biomethane}$	–	[50]
Biograce	$15.2gCO_2e/MJ_{biogas}$	$78.2gCO_2e/MJ_{biogas}$	[3]

Table 9.1 provides some examples of the impact open storage can have on GHG emissions and demonstrates many crop and waste biogas facilities could fail sustainability criteria if included within the system boundary. Open storage of digestate can result in GHG emissions of above 20 gCO_2 eq. per MJ of biomethane [50]. These emissions are avoidable through covering the digestate storage, assuming no leakage of biogas. Best practice measures focus on the use of a protective layer over the liquid digestate to reduce emissions. Some European countries with a developed biogas sector (e.g., Germany, Denmark, and Austria) now have financial incentives to establish covered digestate stores, with the main objective of reducing emissions [5].

9.4.6 Digestate Use

AD digestate can replace artificial fertilizers to realize several environmental benefits. As the nutrient profile and fertilizer value of digestate is dependent on the feedstock composition, crops for AD generally have greater dry matter content and more favorable nutrient composition than manure slurries, resulting in a more concentrated and valuable digestate [51]. It is important to follow best practice when applying digestate and organic manures to crops for AD. Using a diverse range of feedstocks will provide a greater application window for digestate as well as manures, for both predrilling and growing crops. The IEA Bioenergy Task 37 working group has developed guidelines on the application of digestate as a fertilizer, focusing on key environmental considerations such as managing odor release, biosecurity, plant pathogen reduction, and reduction of weed seeds [52]. Such practices will result in lower gaseous emissions into the atmosphere as well as less diffuse pollution from surface run-off and leaching through the use of artificial fertilizers [5].

Digestate as an organic fertilizer with nitrogen content will result in N_2O emissions from soil. Calculations of nitrous oxide are important due to its high Global Warming Potential (GWP). As with inorganic fertilizer, both direct and indirect N_2O emissions are released when N from digestate is applied to land, which can contribute to over 50% of the GHG emissions from feedstock production [3, 18, 41].

9.5 BIOGAS CLEANING AND UPGRADING

As methane is the primary product desired in biogas, the amount contained in the biogas should be increased to as large a percentage as possible. The upgrading of biogas' methane content is achieved through the optimization of facility systems for particular feedstock, and through the implementation of measurement and regulation systems [27]. Biogas upgrading and the production of biomethane is a state-of-the-art process of gas separation [2]. A number of different technologies to fulfil the task of producing a biomethane stream of sufficient quality to act as a vehicle fuel or to be injected into the natural gas grid are already commercially available and have proven to be technically and economically feasible [2,27]. Nevertheless, a small percentage of methane is lost during the upgrade stage, the amount of loss being dependent on the technology type, gas quality, maintenance, operation, and local factors [53, 54].

Research is still in progress to optimize and further develop technologies to minimize methane slip. A range of different technologies currently exist, each with its own specific advantages and disadvantages [2, 53, 54]. It is apparent that no technology is the optimal solution to each and every biogas-upgrading situation [5]. The right choice of the economically optimal technology is strongly dependent on the quality and quantity of the raw biogas to be upgraded, the desired biomethane quality, and the final utilization of this gas [54]. An overview of methane loss from different upgrade technologies is provided in Table 9.2:

Methane slip from biomethane upgrade plants is typically around 1%–2% of the total methane production, although this can be up to 10% and as low as <0.1% [54]. Technology choice and economics are key drivers for emissions from biogas upgrading with technologies like cryogenic separation requiring a large amount of energy. The cost of the technology can influence the efficiency of recovery and amount of loss. Therefore, selecting a type of upgrade technology that generates the lowest amount of methane slip is crucial, with the energy requirements also an important consideration for the LCA. Posttreatment techniques can be implemented to deal with methane slip; these include:

- Regenerative thermal oxidizer
- Recuperative thermal oxidizer
- Biological de-methanation.

In a study performed by Starr et al., GHG impacts were evaluated for three biogas- upgrading technologies. An in-depth evaluation was conducted on high-pressure water scrubbing

TABLE 9.2 Parameters of Different Upgrade Technologies Including Methane Losses [27, 54]

Criteria	PSA	PWS	CS	Membrane	Genosorb	MEA	DEA
Adsorption process	Physical	Physical	Cryogenic	Physical	Physical	Chemical	Chemical
Precleaning necessary?	Yes	No	No	Yes	No	Yes	Yes
Methane loss	2%–4%	1%–2%	<1%	0.5%–1%	1%–4%	<0.1%	<0.1%
Required power consumption [kWh/Nm^3]	0.25	<0.25	0.2–0.5	0.15–0.25	0.25–0.33	<0.15	<0.15

PSA, pressure swing adsorption; PWS, pressurised water scrubbing; CS, cryogenic separation; MEA, monoethanol amine washing; DEA, diethanol amine washing.

(HPWS), as well as alkaline with regeneration (AwR) and bottom ash upgrading (BABIU), which additionally offer carbon storage [53]. AwR and BABIU are two novel technologies that utilize waste from municipal solid waste incinerators—namely bottom ash (BA) and air pollution control residues (APC)—and are able to store CO_2 from biogas through accelerated carbonation processes. These were compared to high-pressure water scrubbing (HPWS) which is a widely used technology in Europe. Starr et al. found that AwR and BABIU have a particularly low GHG impact as a result of the immediate storage of the CO_2. It was also found that cryogenic separation and pressure swing adsorption (PSA) were the most energy-intensive, and all upgrading technologies rely on electricity. For most of the technologies, the electricity use had the highest impact, but it can be reduced by increasing the supply from renewables. The impact of the climate change category can be minimized if the removed CO_2 is stored or used in industry. Sun et al. provide an in-depth review of energy demands and methane losses from a range of biogas cleaning and upgrading technologies [54].

9.6 BIOMETHANE GRID INJECTION (FEED-IN)

Feeding upgraded biogas as biomethane into the natural gas grid is an efficient energy solution, even if the sites in which the gas is to be applied are far away from the sites at which it is produced [27]. Gas feed-in is facilitated via a compressor, a device raising the pressure level of the biomethane to that of the gas in the closed pressurized lines of the grid. Given European regulations, new gas producers have the opportunity to feed gas into the conventional gas grid. For purposes of injection, however, the gas must meet the quality specifications of the relevant legal provisions and may only deviate within the range of these quality standards [6, 27]. Such standards are realized using technologies for reconditioning gas. Fugitive emissions of methane are the only potential source of direct GHG emissions, but can be minimized by ensuring pipes, joints, and valves are not leaking [47].

Since a nonnegligible quantity of energy is necessary for gas compression, the energy balance and the economic feasibility of the compression and feed-in process must be reviewed on a case-by-case basis. Depending on the amount required and the source of electricity, indirect GHG emissions associated with gas compression could be significant.

9.7 END-USE ACCOUNTING

There is some debate in the literature about how GHG emissions in use should be accounted for [16]. Clearly, for conventional fossil fuels, the emissions arising from use are important and contribute significantly to Global Warming Potential (GWP). In contrast, the combustion of biomethane will result in predominantly biogenic CO_2 and therefore contributes to more stable atmospheric carbon. It is therefore useful to make reference to the most relevant legislation here, and the impact this has on the goal and scope definition of GHG accounting. As stated by the Renewable Energy Directive 2009/28/EC, Annex V, C-13 'emissions from the fuel in use shall be taken to be zero for biofuels and bioliquids' [25]. Hence the combustion of biomethane should not be taken into consideration, although there is some debate in the scientific community concerning this scenario. Nonetheless, regardless of the methodology applied, biogenic CO_2 is considered to have a GWP of zero; hence, the end-use of biomethane will not have a material impact on the overall GHG balance of the biomethane value chain. There is however some debate in the literature about methane slip when biomethane is used in vehicles, due to uncombusted methane in non-stationary engines.

9.8 COUNTERFACTUALS

A final consideration in the life cycle GHG emissions assessment of biomethane production is the counterfactual use of the feedstock, i.e., what would be the alternative use of the feedstock if it wasn't used in AD and what are the emissions associated with this other management option. This should be considered only where following a consequential LCA (cLCA) approach.

9.8.1 Manure and Slurry

For a plant utilizing agricultural wastes as feedstock, the counterfactual could be considered as the expected amount of methane which would be emitted into the atmosphere if the feedstock was not used in the AD plant. The counterfactual will vary depending on the type of agricultural feedstock used (e.g., manure, slurry, crop residues, etc.). For manure and slurry this will depend on what management system is assumed. Several manure and slurry management systems are given in the IPCC Guidelines (2006) [55]. These include daily spread on land, drying and burning as fuel, and storage for spreading on land at a later stage.

The most common practice in many European countries for manure is to collect and leave it as piles on farm or possibly in the field or off-site. Slurry is generally collected and transported for storage in tanks, ponds, or lagoons on-site in the vicinity of the animal housing, for spreading on land at a later date. The storage could be in open or closed tanks and may last for several months or even a year. With storage in closed lagoons, biogas can be collected and either utilized or flared releasing CO_2 to the atmosphere which is likely to be an uneconomical option for many farmers.

A counterfactual commonly considered is that manure is stored in open piles or stacks, potentially for several months. Manure can be stacked due to the presence of sufficient animal bedding material and then due to loss of moisture due to evaporation. The counterfactual for animal slurry could therefore be that storage occurs for a period of several months in an open tank with minimal addition of water outside the animal housing, or as excreted.

IPCC Guidelines (2006) can be used to quantify methane release into the atmosphere for these counterfactuals. The assumptions for potential manure and slurry counterfactuals and estimated reference methane emissions are given in Table 9.3.

Methane released is dependent on the composition of the manure or slurry and methane emissions are related to the organic matter. Defining a reference slurry composition is a difficult task as it depends on type, age, and diet of the animal [56].

The total methane emissions, whether released from feedstock storage, the digester area, the digestate storage, or from the biogas combustion area, will not exceed the maximum methane emissions which would have been released if the feedstock was not used in the AD plant. There is no doubt that if biogas is not produced, the management of manure and slurry on land will lead to higher methane emissions compared to leakage from the AD plant (from digesters, digestate storage, etc.). Piles of manure are under more aerobic conditions and so produce much lower methane emissions than storage of liquid slurry in tanks, lagoons, or ponds which operate under anaerobic conditions.

TABLE 9.3 Methane Leakage

Feedstock	Counterfactual Description	Counterfactual Assumption	Methane Leakage, kg/tonne of feedstock	Reference
Manure	The collection and storage of manure is typically for a period of several months in unconfined piles or stacks. Manure is able to be stacked due to the presence of a sufficient amount of bedding material or loss of moisture by evaporation.	Manure is commonly mixed with animal bedding. Counterfactual assumes dry solids content of 25% in fresh material and 80% organic matter content in the dry matter. This will vary depending on manure type.	0.9	Calculation based on IPCC guidance [56]
Animal slurry	Waste is collected and transported in a liquid form to tanks or silos for storage near animal housing before it is spread on land. Liquid is stored for up to six months depending on season.	Assumes dry solids content of 8% and 4.5% respectively in fresh material for cattle and pig and 80% organic matter content in the dry matter.	1.8	Calculation based on IPCC guidance [56]
Energy crops	Energy crops do not exist as they are grown for the specific purpose of using in the AD plant	Crops are not grown and so do not contribute to feedings animals	none	
MSW	Collecting and sending biodegradable MSW to landfill where 25% of the methane is vented to the atmosphere and rest either utilised or flared. The collection rate will depend on the landfill so is likely to vary.	25% of methane produced is vented to atmosphere. The typical methane production is 100 kg/tonne MSW, but this does depend on the type of waste.	25	Estimated and calculated based on typical published values
Sewage sludge and industrial waste	Waste collected using a flush system and transported to lagoons for storage.	AD plant exists anyway as treatment is a requirement by legislation.	None	

9.8.2 Other (Nonfarm) Waste

The counterfactual for waste feedstocks depends on the existing processing route which could include recycling, composting, digestion, landfill, or incineration. Recycling is usually preferred due to its position on the waste hierarchy, although the energy and carbon balance should be considered first. Composting is an aerobic process with a large fraction of biodegradable fraction converted to CO_2. Methane can be formed in anaerobic segments of compost, but is then largely oxidized in the aerobic segments with estimated methane released into the atmosphere between 1% and 3% of the initial carbon content in the material [57].

For landfill waste, methane emissions will depend on the technology employed and operation at the landfill site. Total emissions for a landfill can be modelled using waste decomposition rates, the type of waste deposited, and the age of the waste. Most landfills will capture a sizeable portion of the landfill gas which is either flared or used for energy production with the remainder vented. The gas capture technology and age of the landfill will effect the emission profile. Using a typical biogas yield from landfill, it is therefore possible to estimate an average methane leakage rate counterfactual based on the amount vented. Further description of this approach is included in the Clean Development Mechanism methodology for landfill gas [58].

Sewage sludge and industrial wastes require treatment before they can leave a site. AD is commonly used in the sewage and water treatment sectors, the counterfactual is therefore AD itself and no avoided emissions need to be taken into account.

9.8.3 Energy Crop Silage

If crops grown such as maize, grass, and rye silage were not used in AD plants to produce biogas, it could be considered they would not been grown at all, hence the counterfactual would be zero emissions from land use and cultivation. In reality, a different crop would likely have been grown, so emissions would arise from alternative land use. The convention is therefore that alternative land use emissions are not considered as a counterfactual regardless of whether the previous crop was more or less intensively grown. Where previous land use is accounted for, this is direct land use change (dLUC) and would only be considered if following a cLCA approach.

As many AD sites use a mix of feedstocks, the counterfactual considered could be based on a weighted average of the proportions used. For example, where maize silage and manures are both used, the counterfactual should calculate the emissions from the relative proportions of each feedstock, measured in CO_2e/t of feedstock.

9.9 SUMMARY

This chapter has described the main possible sources of GHG emissions associated with the biomethane production value chain. It shows that significant variations are possible which arise from both the plant design and operation, feedstock and technology choice, uncertainty in measurements, and the GHG accounting methodology [14]. By following best practice, it is possible to achieve GHG savings compared to the fossil fuel alternative of 60%–80% using crops, with even greater reductions possible when utilizing wastes and residues. Negative emissions are also possible if counterfactuals such as waste management and CO_2 capture are accounted for [25,48]. GHG emissions vary depending on the process stage in the biomethane production; key sources include biomass feedstock cultivation (where purposely grown energy crops are used), fugitive methane losses imported grid electricity, and biogas upgrading (methane slip).

The use of fertilizers in feedstock cultivation can produce nitrous oxide emissions from soil and can cause around 40%–50% of the emissions associated with cultivation. Using waste as a feedstock offers a zero or negative emission feedstock, but with seasonal variability and often much lower methane yields. Opportunities exist for utilizing digestate obtained as a co-product from the anaerobic digestion process as an alternative nutrient-rich organic fertilizer.

Methane slip in biogas production and biomethane upgrading can cause significant emissions if it is not appropriately managed. However, following best practice and voluntary schemes (such as in Sweden) shows that methane emissions can be minimized. It is clearly in the plant operator's interest to mitigate any potential methane losses in order to maximize economic returns.

Overall, the literature shows that biogas production has an important potential for the production of biomethane as both a transport fuel and for heat and electricity, both in terms of energy inputs and GHG emissions. Biomethane is currently the most important renewable option for gas supplies and is fully compatible with natural gas. With the construction of advanced plants and continuous improvements in crop cultivation and plant operation, reductions of over 80% compared with natural gas appear feasible [59].

It has been found that the most effective ways of achieving low GHG emissions and high process performance in terms of energy demand and costs are:

- Use of residue substrates such as manure or biowaste can avoid up to 97% of the emissions [59]
- Follow best practice guidance when using crop feedstocks in AD [28]
- Utilization of digestate (fermentation residues) as fertilizer [52]
- Appropriate digestate storage [50]
- Minimization of silage losses by appropriate silage storage [59]
- Use of highly efficient upgrading technologies requiring little electricity with low methane losses in the off-gas [27, 53, 54]
- Use of gas-tight equipment and monitoring of diffuse methane losses [5, 45].
- Regular monitoring of methane loss [47]

References

[1] IEA. World energy outlook 2016. Paris: IEA; 2016.
[2] IEA. Biomethane—status and factors affecting market development and trade. DBFZ, Germany: IEA; 2014. ISBN 978-1-910154-10-6.
[3] Biograce. Harmonised calculations of biofuel greenhouse gas emissions in Europe, Intelligent Energy Europe (IEE); 2013. Available from: http://www.biograce.net/home.
[4] ISO. ISO 16559:2014 preview solid biofuels—terminology, definitions and descriptions. International Standards Organisation; 2014. Available from: https://www.iso.org/standard/57111.html.
[5] Green Gas Grids. Greening your biomethane production chain—a best practice guide for reducing greenhouse gas emissions; 2013.
[6] Billig E, Thrän D, Pu P, Yu C. The standardisation, production and utilisation of biomethane in Europe and China - a comprehensive analysis. Int J Oil Gas Coal Technol 2017;14(1/2):110. Available from: http://www.inderscience.com/link.php?id%20=%2081105.
[7] Green Gas Grids. Standards for biomethane as vehicle fuel and for injection into the natural gas grid. 2013.
[8] Horschig T, Adams PWR, Röder M, Thornley P, Thrän D. Reasonable potential for GHG savings by anaerobic biomethane in Germany and UK derived from economic and ecological analyses. Appl Energy 2016;184:840–52. Available from: http://linkinghub.elsevier.com/retrieve/pii/S0306261916310480.
[9] DBFZ. Focus on biomethane. Leipzig: Deutsches BiomasseForschungsZentrum gemeinnützige GmbH (DBFZ); 2012. Available from: www.energetische-biomassenutzung.de.
[10] National Grid. The future of gas—supply of renewable gas. 2016. Available from: http://www2.nationalgrid.com/UK/Industry-information/Future-of-Energy/Gas/.

[11] Styles D, Dominguez EM, Chadwick D. Environmental balance of the UK biogas sector: An evaluation by consequential life cycle assessment. Sci Total Environ 2016;560:241–53. Available from: http://www.sciencedirect.com.ezproxy1.bath.ac.uk/science/article/pii/S0048969716306714.

[12] Welfle A, Gilbert P, Thornley P, Stephenson A. Generating low-carbon heat from biomass: life cycle assessment of bioenergy scenarios. J Clean Prod 2017;149:448–60. Available from: http://www.sciencedirect.com.ezproxy1.bath.ac.uk/science/article/pii/S0959652617302391.

[13] Röder M. More than food or fuel. Stakeholder perceptions of anaerobic digestion and land use; a case study from the United Kingdom. Energy Policy 2016;97:73–81. Available from: http://www.sciencedirect.com.ezproxy1.bath.ac.uk/science/article/pii/S030142151630355X.

[14] Adams PWR, Mezzullo WG, McManus MC. Biomass sustainability criteria: greenhouse gas accounting issues for biogas and biomethane facilities. Energy Policy 2015;87:95–109.

[15] Aylott M, Higson A, Evans G, Mortimer N. Measuring the energy and greenhouse gas balances of biofuels and bio-based chemicals using LCA (No. 12-023). York, UK: National Non Foods Crops Centre; 2012.

[16] Adams P, Bows A, Gilbert P, Hammond J, Howard D, McNamara N, Thornley P, Whittaker C, Whitaker J. Understanding the greenhouse gas balances of bioenergy systems. Supergen Bioenergy Hub, Tyndall Centre for Climate Change Research, University of Manchester, UK 2013. Available from: http://opus.bath.ac.uk/47285/1/Adams_2013_GHG_balances.pdf.

[17] Society TR. Sustainable biofuels: prospects and challenges. London: The Royal Society; 2008.

[18] IPCC. Climate change 2013: the physical science basis. Contribution of working group I to the fifth assessment report of the intergovernmental panel on climate change (IPCC). Cambridge; 2013.

[19] Energiforsk. Measurement of methane emissions from biogas production. Report 2015:158; 2015.

[20] Adams PWR, McManus MC. Characterisation and variability of greenhouse gas emissions from biomethane production via anaerobic digestion of maize. J Clean Prod 2017 (in press).

[21] BSI. Guide to PAS 2050: how to assess the carbon footprint of goods and services. London, UK: British Standards Institute (BSI); 2008.

[22] OFGEM. Non-domestic renewable heat incentive (RHI sustainability self-reporting guidance). 2016. Available from: https://www.ofgem.gov.uk/publications-and-updates/non-domestic-rhi-main-guidance.

[23] Wiloso EI, Heijungs R, de Snoo GR. LCA of second generation bioethanol: a review and some issues to be resolved for good LCA practice. Renew Sustain Energy Rev 2012;16(7):5295–308. Available from: http://linkinghub.elsevier.com/retrieve/pii/S1364032112003103.

[24] Cherubini F. GHG balances of bioenergy systems—overview of key steps in the production chain and methodological concerns. Renew Energy 2010;35(7):1565–73. Available from: http://linkinghub.elsevier.com/retrieve/pii/S0960148109005230.

[25] Commission E. Directive 2009/28/EC of The European Parliament and of The Council of 23 April 2009 on the promotion of the use of energy from renewable sources. Official Journal of the European Union 2009, 16-62, Brussels; 2009.

[26] Arnold K. Greenhouse gas balance of bio-methane—which substrates are suitable? Energy Sci Technol 2011;1:67–75.

[27] Biogaspartner. Biogas grid injection in Germany and Europe—market, technology and players, biogaspartner—a joint initiative, article no. 5010. Berlin: Deutsche Energie-Agentur GmbH (dena); 2011.

[28] ADBA. Voluntary guidelines on best practice for the use of crop feedstocks in Anaerobic Digestion. London: the Anaerobic Digestion and Bioresources Association (ADBA); 2013.

[29] Palmer RC, Smith RP. Soil structural degradation in SW England and its impact on surface-water runoff generation. Soil Use Manag 2013;29(4):567–75. Available from: https://doi.wiley.com/10.1111/sum.12068.

[30] Hamelin L, Jørgensen U, Petersen BM, Olesen JE, Wenzel H. Modelling the carbon and nitrogen balances of direct land use changes from energy crops in Denmark: a consequential life cycle inventory. GCB Bioenergy 2012;4(6):889–907. Available from: https://doi.wiley.com/10.1111/j.1757-1707.2012.01174.x.

[31] Warner E, Inman D, Kunstman B, Bush B, Vimmerstedt L, Peterson S, et al. Modeling biofuel expansion effects on land use change dynamics. Environ Res Lett 2013;8(1):15003. Available from: http://stacks.iop.org/1748-9326/8/i=1/a=015003?key=crossref.a71a3c5f97279dad62bd4fb061d9b783.

[32] Ferchaud F, Vitte G, Machet J-M, Beaudoin N, Catterou M, Mary B. The fate of cumulative applications of 15N-labelled fertiliser in perennial and annual bioenergy crops. Agric Ecosyst Environ 2016;223:76–86. Available from: http://www.sciencedirect.com.ezproxy1.bath.ac.uk/science/article/pii/S0167880916301074.

[33] Andersons. A detailed economic assessment of anaerobic digestion technology and its suitability to UK farming and waste systems; 2010.

[34] KWS. Biogas in practice; 2012.

[35] Gissén C, Prade T, Kreuger E, Nges IA, Rosenqvist H, Svensson S-E, et al. Comparing energy crops for biogas production—yields, energy input and costs in cultivation using digestate and mineral fertilisation. Biomass Bioenerg 2014;64. Available from: http://www.sciencedirect.com/science/article/pii/S0961953414001901.

[36] Boulamanti AK, Donida Maglio S, Giuntoli J, Agostini A. Influence of different practices on biogas sustainability. Biomass Bioenerg 2013;53:149–61. Available from: http://linkinghub.elsevier.com/retrieve/pii/S0961953413000949.

[37] Gan Y, Liang C, Chai Q, Lemke RL, Campbell CA, Zentner RP. Improving farming practices reduces the carbon footprint of spring wheat production. Nat Commun 2014;5:5012. Available from: https://doi.org/10.1038/ncomms6012.

[38] Felten D, Fröba N, Fries J, Emmerling C. Energy balances and greenhouse gas-mitigation potentials of bioenergy cropping systems (Miscanthus, rapeseed, and maize) based on farming conditions in Western Germany. Renew Energy 2013;55:160–74. Available from: http://www.sciencedirect.com/science/article/pii/S0960148112007598.

[39] Kochsiek AE, Knops JMH. Maize cellulosic biofuels: soil carbon loss can be a hidden cost of residue removal. GCB Bioenergy 2012;4:229–33.

[40] AEA. BEAT2 (Biomass Environmental Assessment Tool) v2.1. Oxford, UK: AEA Technology & North Energy; 2010.

[41] E4Tech. Solid and gaseous carbon calculator (B2C2). London; 2016. Available from: http://www.e4tech.com/.

[42] DEFRA. Greenhouse gas reporting—conversion factors 2016; 2016.

[43] HEBFA. The handbook of emission factors for road transport (HBEFA) v3.3; 2017.

[44] Ricardo. Biomethane GHG Calculator Tool v4.1: User Guide; 2016.

[45] Liebetrau J, Clemens J, Cuhls C, Hafermann C, Friehe J, Weiland P, et al. Methane emissions from biogas-producing facilities within the agricultural sector. Eng Life Sci 2010;10(6):595–9. Available from: https://doi.wiley.com/10.1002/elsc.201000070.

[46] Jonerholm K, Lundborg H SEA. Methane losses in the biogas system. Stockholm; 2012.

[47] Petersson. Biogas sustainability, The Swedish Voluntary System for control of methane emissions, IEA Bioenergy Task 37, SP Technical Research Institute of Sweden; 2012.

[48] JRC. Solid and gaseous bioenergy pathways: input values and GHG emissions. Report EUR 26696 EN, Ispra, Italy: Joint Research Centre (JRC); 2014.

[49] Gemmeke B, Rieger C, Weiland P. Biogasmessprogramm, 2009. II– 61 Biogasanlagen im Vergleich, 1st Edn, Fachagentur f.ur Nachwachsende Rohstoffe e. V. (Ed), Gulzow. 2009.

[50] Buratti C, Barbanera M, Fantozzi F. Assessment of GHG emissions of biomethane from energy cereal crops in Umbria, Italy. Appl Energy 2013;108:128–36. Available from: http://linkinghub.elsevier.com/retrieve/pii/S0306261913001979.

[51] Al M, et al. Fertilization of arable crops with digestate of agricultural biogas plants. Landwirtschaftliches Technologienzentrum Augeustenbuerg in Karlsruhe; 2008.

[52] Lukehurst C, Frost P, Al Seadi T. Utilisation of digestate from biogas plants as biofertiliser. IEA Bioenergy Task 37 2010. Available from: http://www.iea-biogas.net/_download/publi-task37/Digestate_Brochure_Revised_12-2010.pdf.

[53] Starr K, Gabarrell X, Villalba G, Talens L, Lombardi L. Life cycle assessment of biogas upgrading technologies. Waste Manag 2012;32(5):991–9.

[54] Sun Q, Li H, Yan J, Liu L, Yu Z, Yu X. Selection of appropriate biogas upgrading technology—a review of biogas cleaning, upgrading and utilisation. Renew Sust Energ Rev 2015;51:521–32.

[55] IPCC Guidelines for Inventories. Emissions from Livestock and manure management; 2006 [Chapter 10]. Available from: http://www.ipcc-nggip.iges.or.jp/public/2006gl/pdf/4_Volume4/V4_10_Ch10_Livestock.pdf.

[56] Danish Ministry for the Environment. Life Cycle Assessment of Slurry Management Technologies. Available from: http://www2.mst.dk/udgiv/publications/2009/978-87-92548-20-7/pdf/978-87-92548-21-4.pdf.

[57] Brown S, Kruger C, Subler S. Greenhouse gas balance for composting operations. J Environ Qual 2008;37:1396.

[58] UNFCC. Flaring or use of landfill gas v18.0, 4 May 2017. Available from: https://cdm.unfccc.int/UserManagement/FileStorage/0X2IE6B1PJDLKMWN89AZGTFUHR3VYS.

[59] Adelt M, Wolf D, Vogel A. LCA of biomethane. J Nat Gas Sci Eng 2011;3:646–50.

CHAPTER

10

Biodiesel from Argentinean Soy

Patricia Thornley

University of Manchester, Manchester, United Kingdom

10.1 INTRODUCTION

Energy and climate change policy incentives often encourage liquid biofuels as a means to reduce carbon emissions and increase renewable in the transport sector. These liquid fuels have higher energy density than raw biomass and a global infrastructure already exists by which petroleum and related products are traded and transported around the world from point of production to point of use/demand. Consequently, it is relatively simple for liquid fuel users to seek low carbon alternatives that are produced in a geographically distant region. The result is a trend to decarbonize by importing liquid biofuels for transport and chemical production that have been grown and manufactured overseas. Essentially, liquid biofuel vectors are amenable to long distance transport in a way that heat and electricity are not and so particular care needs to be exercised in ensuring that they have been produced in a way that achieves the greenhouse gas benefits claimed.

First-generation liquid biofuel production has focused on taking the most amenable plant-based feedstock (liquid oils) and converting them to liquid fuels that meet transport standards and specifications. This has been very successful in proving the feasibility of delivering liquid transport fuels from plants. However, there are three major drawbacks of this approach:

1. The yield of the usable proportion of the biomass product (in this case the oil) is often not very high and so the land areas required can be large [1]. This means that it is important to take into account the impact of the biofuel production on soil carbon when carrying out a greenhouse gas balance assessment. The principle should be that the state of the stock reserve (soil) should remain unchanged by biofuel cultivation, so that there is no effective transfer of carbon from the ecosphere to the atmosphere via reduction of the terrestrial carbon stock during the lifetime of biofuel crop growth. If this can be achieved, then a scope of analysis for the biofuel system can be defined which does not have any negative external impacts. If this cannot be achieved, then the biofuel analysis should take into account the environmental impacts associated with any definable changes in the associated soil system. For the purposes of greenhouse gas balances, the key parameter is soil carbon content, but the principle applies equally to other parameters, e.g., if a biofuel

Greenhouse Gas Balances of Bioenergy Systems
https://doi.org/10.1016/B978-0-08-101036-5.00010-0

crop results in a long-term reduction of soil nutrients such as phosphorus or potassium, an equivalent restoration could be taken into account in the scope of system if the goal of the study includes maintaining soil health for future crop cultivation.
2. In addition to low oil yields, there is often (as is the case with the soy case study here) a second product from the crop and so the attribution of emissions between the different products can very significantly impact the results, depending on the methodological approach used.
3. The major global oil-producing crops have been established to provide food and chemical grade oils. There is a long history of agronomic development to achieve the relevant quality and the result is a relatively intensive agronomic regime, often utilising agrochemicals to enhance crop qualities which may not be as relevant for biofuel production as they are for fuel production. While fertilisers have significant embodied energy/GHG emissions and also result in significant soil emissions, this is not generally the case with herbicides, pesticides, etc., which tend to have lower GHG impacts. However, they can have very significant other environmental impacts which should be taken into account in a comprehensive system life cycle assessment.

In addition to these issues, it is worth noting that there is little incentive to minimise GHG emissions associated with food production in current markets. So established practises for crop production have been developed to incorporate agronomic practises (including agrochemical application) that enhance characteristics valued in the food market. These may not always be as relevant or valuable in the fuel market and so there may be scope to optimise future crop production with lower GHG intensity; however, early market exploitation will reduce risk and maximise economy by relying at least initially on established practises.

In this chapter, production of biodiesel from Argentinean soy will be examined in order to explore the three key aspects of GHG life cycle assessment discussed above. The case study and results used have previously been published in a journal paper and the interested reader can find a more complete description of the system/assumption/results there [2]. The main focus of this chapter is the methodological issues raised by such an assessment. It should, however, be noted at this point that the argument presented in point (1) above is a key driver for advanced third-generation biofuel production, which can access the whole crop rather than just the oil fraction and thus attain improved environmental performance [1].

10.2 BIODIESEL PRODUCTION FROM ARGENTINEAN SOY

10.2.1 Goal

As stated in Chapter 4, it is always important to start an LCA assessment by considering the goal. In this case, we are evaluating the production of biodiesel for overseas (European) markets from Argentinean soy. Therefore, the goal is to compare the life cycle greenhouse gas emissions of mineral diesel production with the life cycle greenhouse gas emissions of soy oil production, transesterification, and transport to Europe (United Kingdom). In particular, this work was interested in comparing the environmental impacts of different scales of production (since there is an increasing trend towards very large-scale, industrial production rather than smallholder scale with accompanying shifts in land ownership) [3].

10.2.2 Scope Definition

Given the goal of comparing two different production systems, the scope of assessment was chosen to be fully consistent with these. A key factor being explored is the land-use associated with different production regimes. Therefore, it is essential that the scope of system includes emissions from soil during the period. If land clearance is required in order to plant the crop, the emission associated with this land clearance should be included if they have been driven by the process being considered: in this case soy farming for biodiesel production. However, soy is commonly planted as part of a rotation where it will often have been preceded by a more agrochemical intensive crop. In these circumstances, it would be appropriate to attribute emissions associated with soy establishment to the soy/biodiesel production system, but not emissions associated with the land-use change, since this is a temporary change, which will be reverted the following year. By contrast for the large-scale, industrial soy production, the soy crop would be planted for continuous years.

All agronomic steps associated with soy production must then be included, as outlined in Table 10.1.

Traditionally, land would have been ploughed between rotations, but the development of genetically modified soy strains which are herbicide-tolerant facilitates 'no-till' planting, with attendant savings in cost and soil disturbance (which preserves soil carbon and reduces GHG emissions). Therefore, this option (B) was compared to smaller scale plantation on previous agricultural land within a rotational context where no land-use change is required (A). The continuous regime (A) requires fertiliser input in the long term, but this is not the case within the rotationary context.

This raises an important point in terms of the timescale chosen for analysis. LCA was originally developed as a technique to examine the cradle to grave impact of products manufactured in an industrial context. Therefore, it was important to consider the full 'life cycle' through manufacture, use, and disposal. The assessment would take into account material and energy flows at each stage in the life cycle. The unwritten assumption in such a calculation is that the production process is the main focal point and other interfacing systems remain unchanged and so do not need to be considered.

Care is needed when transposing this approach to land-based agricultural production. The equivalent 'unwritten assumption' here is that the land/soil base remains unchanged. This is important from a sustainability perspective. If biomass production changes land/soil

TABLE 10.1 Agronomic Regime Modelled for Soy Biodiesel Assessment [2]

Option A	Option B
Production on previous agricultural land within a rotational context (no land-use change)	Production on previous scrub land with inclusion of land-use change GHG emissions
No-till planting	No-till planting
Two applications of postemergence herbicide	Two applications of postemergence herbicide
No fertiliser or soil conditioner application	Single annual application of a compound fertiliser Application of lime once every 5 years
Combining of crop at harvest	Combining of crop at harvest

characteristics in a way that impacts fertility, it could reduce the capacity of the land to produce crops in future and so negative impact needs to be captured somehow in the LCA assessment. The most practical way of doing this is to ensure that the system is defined in such a way as to include steps to ensure that the soil is returned at the end of the assessment in the same state it was at the start. In this case, artificial fertilisers are not routinely used in the soy production system and so option A does not include them. However, there is evidence that such production may result in a gradual long-term decline in soil's nutrient content. Therefore, it is necessary to include a compensation for this in the form of application of nutrients in order to make sure this is a 'fair comparison' and that the bioenergy production is not depleting nutrient stocks. This means that we incorporate in the life cycle assessment steps that may not actually routinely happen in the supply chain.

In addition, the timescale of assessment is important. When dealing with continuous industrial/manufacturing processes, the material flows are fairly constant and so it does not make a huge difference if life cycle assessment is carried out over a 1h, 1 year, or 30-year period. However, in bioenergy systems, there are sometimes changes that occur over a longer timescale and so the assessment period should be defined in a way that is appropriate to the system being studied. Therefore, if a crop is established and maintained over a 25-year period, it makes sense for the assessment to cover that period. If there are steps that occur only every few years, the considered time period must be long enough to include a proportionate quantity of these.

This presents obvious challenges when considering systems that change over significantly longer timescales and some of these temporal issues are discussed further in Chapter 5 which deals with carbon sequestration and the carbon cycle. A key example is forestry systems, where carbon sequestration takes place over many years and the forest has an important ecosystem function associated with its carbon stock. In principle, any LCA of forestry assessments should therefore ensure that the forest standing stock is the same at the start- and end-point of the assessment. If not, an adjustment must be made to account for the difference in carbon stock, which is discussed further in Chapter 14.

10.2.3 Key Results

The results for the greenhouse gas balance are shown in Fig. 10.1 and Table 10.2, where these are compared to default values used by the UK's Renewable Fuels Agency, to give some indication of the level of difference between the assumed default values and specific values assigned to the supply chain calculated.

10.3 ACCOUNTING FOR INTERFACES WITH OTHER SYSTEMS

The most striking feature of Fig. 10.1 is the large difference in calculated GHG value between option A and option B. This is driven by whether or not the land-use change emissions are included. Therefore, it is critically important to consider at the outset whether a bioenergy LCA should include land-use change and/or indirect land-use change. Both can have a major impact on results. This decision must be made with knowledge of the production context, including land-use competition, alternative uses, and product demand. It is key to understand

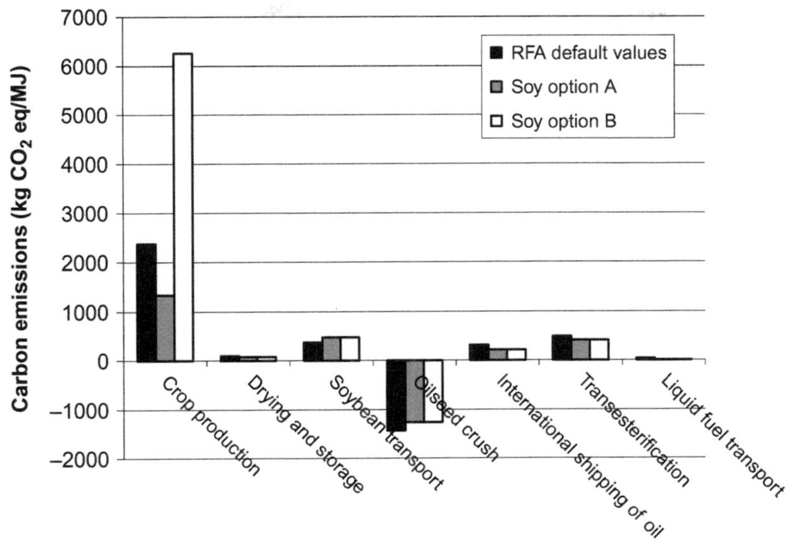

FIG. 10.1 Carbon emissions from soy biodiesel supply chain.

TABLE 10.2 Greenhouse Gas Savings and Intensity for Biodiesel Production from Argentinean Soy Compared to Mineral Diesel

	RFA Default	Option A	Option B
Carbon savings (%)	39.8	65.9	−66.5
Greenhouse gas intensity (kg CO_2/MJ)	0.052	0.029	0.144

if the bioenergy production is driving/causing the land-use change. Often, it can be difficult to establish this fully and sometimes a probabilistic approach may be adopted where it may be appropriate to consider that a certain amount of bioenergy expansion could occur with no land-use change, but beyond that some would be required and so land-use change should be accounted for, on average, some of the supply chain.

A second striking feature of Fig. 10.1 is the significant positive credit attributed for soy meal. This is a second product of the soy production systems and the accounting procedures for dealing with this are discussed in more detail in Section 10.4. However, the significant feature here is that if there is a by-product or co-product that is carbon-intensive, it can have a very significant impact on the carbon balance of the bioenergy system, since it is assumed that it has displaced or avoided production of that carbon-intensive product and therefore the bioenergy system is 'credited' with that GHG benefit. It is important to realise that this only holds true and achieves an actual real GHG reduction if there is a market demand for the by-product and production of it actually displaces other sources of production. If there is no real market demand, the by-product may become a waste product that requires disposal, in which case there is no logic in applying a carbon credit, however energy- and GHG-intensive the material may be to product. Similarly, if the by-product is used in a way that constitutes a new/additional demand rather than displacing a preexisting demand that would have been fulfilled in a more carbon-intensive way, then it has not actually achieved a GHG reduction and should not be ascribed a carbon credit.

10.4 CONSEQUENTIAL AND ATTRIBUTIONAL LIFE CYCLE ASSESSMENT

As discussed in Section 10.3, it is key to decide whether the LCA question being asked is one where we are looking at the direct consequences only of a particular supply chain or looking more widely at the consequences of implementing bioenergy. If (as is often the case with policy evaluations) we are considering the wider consequences of implementation, then it is appropriate to consider the consequences of growing soy in terms of the land-use change incurred. This is termed *consequential LCA* where we look at the impact of the bioenergy system and compare it to what would have happened (including what land-use would have taken place) if we had not implemented the bioenergy system, i.e., the LCA is evaluating the consequences of implementing bioenergy.

The term used for what would have taken place otherwise is the 'counterfactual.' With consequential LCA, the role of the *counterfactual* can be particularly important. Work carried out for the UK Department of Energy and Climate Change in 2015 highlighted this [4]. It showed (Fig. 10.2) that, for a given bioenergy system (shown along the *x*-axis), a wide range of GHG variations was possible depending on the assumptions made about the counterfactual system. In this work, the focus was anaerobic digestion of farm wastes and so the counterfactual assumptions focused on how these wastes would have been treated if they had not been used to produce biogas via anaerobic digestion.

Fig. 10.2 shows that using slurry for bioenergy generation instead of the counterfactual agricultural processes of storing in lagoons, or using waste for bioenergy instead of the counterfactual of disposing of food wastes within landfill result in large potential GHG savings. However, where bioenergy solutions are associated with mitigation of relatively low GHG intensity processes (such as closed tank slurry storage), the GHG savings associated with using the resource for bioenergy generation are much diminished. With biodiesel production from soy, the counterfactual for the biodiesel would be most obviously fossil fuel-derived diesel. However, there is also a meal co-product and the pertinent question then becomes what is the appropriate counterfactual for this. As discussed above, the carbon intensity of the soy meal credit has a very significant impact on the net greenhouse gas balance and needs to be chosen very carefully. This is related to the issues of goal and scope in Chapter 4 and the use of consequential or attributional LCA in this chapter. A fairly narrow consequential LCA might assume that the soy meal is displacing another carbon-intensive high protein animal feed, but a more nuanced approach might instead assume that a strong focus on carbon reductions is also incentivizing lower carbon intensity animal feed or indeed a lower consumption of animal feed due to global shifts away from a meat diet, so that the elasticity of the link between meal supply and demand is affected and soy meal becomes a low or zero value waste rather than a co-product with a market to be serviced. This may seem like an extreme example, but it is illustrative of the implicit assumptions about the future that are embedded within calculation methodologies and the need to think very carefully about the most appropriate choices.

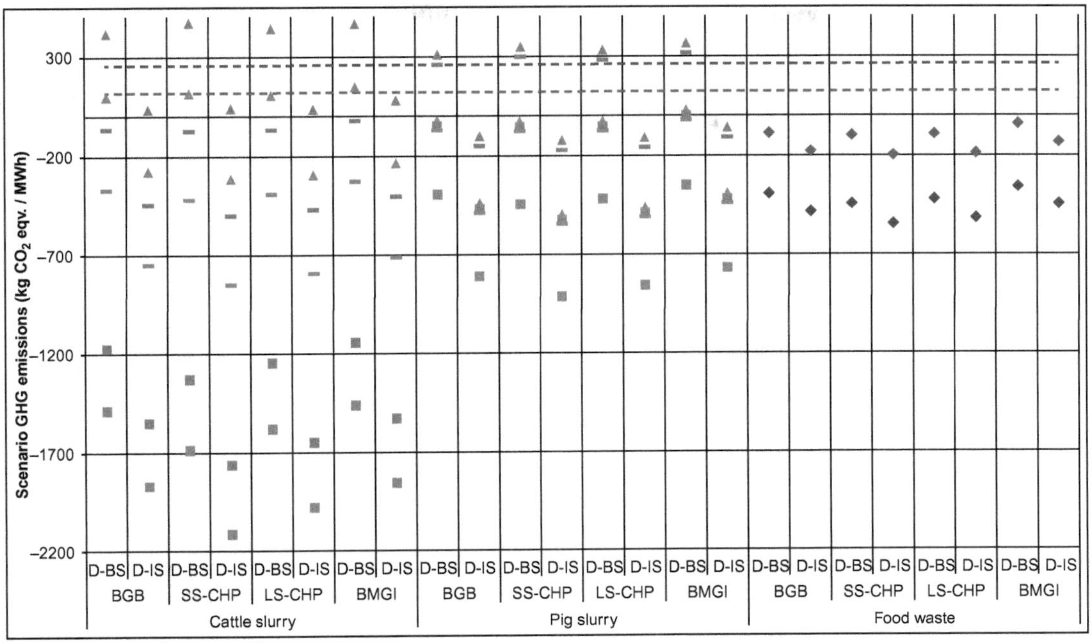

Counterfactual scenarios characteristics: explanations of the results markers plotted within the figures	
▲	Wastes are stored in **closed tank** systems; and **injection** spread into soils
-	Wastes are stored in **open tank** systems; and **injection** spread into soils
■	Wastes are stored in a **lagoon**; and **injection** spread into soils
▲	Wastes are stored in **closed tank** systems; and **broadcast** spread on land
-	Wastes are stored in **open tank** systems; and **broadcast** spread on land
■	Wastes are stored in a **lagoon;** and **broadcast** spread on land
▲	Wastes are stored in **closed tank** systems; and **injection** spread into soils. Food wastes are **composted**
-	Wastes are stored in **open tank** systems; and **injection** spread into soils. Food wastes are **composted**
■	Wastes are stored in a **lagoon**; and **injection** spread into soils. Food wastes are **composted**
▲	Wastes are stored in **closed tank** systems; and **injection** spread into soils. Food wastes are sent to **landfill**
-	Wastes are stored in **open tank** systems; and **injection** spread into soils. Food wastes are sent to **landfill**
■	Wastes are stored in a **lagoon;** and **injection** spread into soils. Food wastes are sent to **landfill**
▲	Wastes are stored in **closed tank** systems; and **broadcast** spread on land. Food wastes are **composted**
-	Wastes are stored in **open tank** systems; and **broadcast** spread on land. Food wastes are **composted**
■	Wastes are stored in a **lagoon**; and **broadcast** spread on land. Food wastes are **composted**
▲	Wastes are stored in **closed tank** systems; and **broadcast** spread on land. Food wastes are sent to **landfill**
-	Wastes are stored in **open tank** systems; and **broadcast** spread on land. Food wastes are sent to **landfill**
■	Wastes are stored in a **lagoon**; and **broadcast** spread on land. Food wastes are sent to **landfill**
♦	Food wastes are **composted**
♦	Food wastes are sent to **landfill**

FIG. 10.2 GHG intensity of UK Agricultural & Food Waste of Scenarios—Generating heat bioenergy following the anaerobic digestion of sole feedstocks: Cattle slurry, pig slurry, and food wastes.

Analysis by the European Environment Agency [5] advised that counterfactuals should be included in GHG balances for bioenergy systems, but the choice of counterfactual can be subjective and contentious. Sometimes it is difficult to 'know' what the real counterfactual would have been, and even harder to quantity any consequential LCA emissions that may/may not have resulted from that counterfactual. Therefore, it is important to be aware that inclusion of counterfactuals often increases the uncertainties of any outputs.

The alternative approach is *attributional LCA,* which is commonly used for regulatory or company accounting purposes. This takes a narrower perspective and looks only at a particular, well-defined supply chain and what environmental impacts can justifiably be directly attributed to that supply chain. This narrower view will often exclude land-use change (unless there is very explicit evidence that land has been cleared or its use changed explicitly because of the bioenergy system) and will always exclude indirect land-use change since its system boundary is more finite, focused on the bioenergy system only. This is appropriate for a supply chain manager, who will have little control over what happens in the wider region as a result of his planting a particular energy crop and so needs to focus on managing the environmental impacts over which they have control and influence. However, such calculations should not be used to make wider judgements and recommendations about the impact of bioenergy systems or management practises more generally. Therefore, the EU RED uses aLCA for soy as a mechanism to apportion and attribute emissions to suppliers as part of a regulatory mechanism, but this is of limited utility when trying to assess the global climate impact of a change, for which a consequential approach is required.

10.5 ALLOCATION PROCEDURES

Soy oil is only one product of soy bean production. The second product is soy meal, which is used as high protein animal feed. While economic market conditions do vary, the revenue a farmer receives for the oil and meal can be roughly equivalent. This presents a challenge from an LCA perspective. LCA aims to allocate the environmental burdens of a production system to a unit of that product. The unit chosen is the *functional unit* and, for bioenergy systems, could be 1 t of a certain biomass or 1 MJ of energy or 1 L of biofuel. With some biomass material, it is obvious what the main product of a system is. For example, farmers grow wheat because of the value of the wheat grain. While some may also use or even receive revenue for the wheat straw, the value of this is much less than the wheat grain and so is not the driver for production.

The ideal way to deal with this problem of multiple products is by 'system expansion' in which we would look at the secondary product, e.g., soy meal, and would expand our system to consider what would have provided the meal if it hadn't been soy and substitute the environmental impact of that product. For example, de Menna (2015) [6] gives an explicit comparison of the difference that consequential and attributional LCA approaches can make to results for an LCA of peach nectar with and without bioenergy integration in the production process. System expansion is ideologically sound, but can result in extensive secondary calculations (which may themselves require system expansion) and only works if the analyst is quite confident that there is a standard alternative source of the product with a fairly well-defined production system that can be analysed.

The alternative is to accept that we need to deal with two products and assign one as the primary product and one as a 'by-product.' It is then possible to look at the total environmental impact and split it in some way. The challenge is in deciding an appropriate mechanism or formula for apportioning the impacts. It is intuitively obvious that if we are looking at wheat production, the grain would be the primary product and we would apportion impacts between grain and straw in some ratio. Practitioners commonly resort to apportioning by economic value (often used when one product is clearly the main product which has driven production and the other is a waste or by-product), energy or calorific value (often used where there are multiple energy products, e.g., where gasification or pyrolysis yield a solid, liquid, and gaseous fuel), or mass (often used where products may have multiple uses or values). However, there are obvious drawbacks with this approach, one of the most significant of which is that the environmental burden assigned to a particular energy product can change just because the market price has changed. This can be misleading when trying to draw relative comparisons between different products or fuels (e.g., choosing the lowest carbon intensity biofuel), since the real environmental impact of the production has not changed—we have simply changed how we split the assignment.

With soy, this distinction is difficult since the meal and grain values can be roughly equal and so it can be very difficult to determine which is the primary product and which the by-product. Also, both products are traded in the global food market and so there can be significant external variability in the prices, resulting in apparent changes in environmental performance that are entirely due to markets and prices!

10.6 ENVIRONMENTAL TRADE-OFFS BETWEEN GHG'S AND OTHER ENVIRONMENTAL IMPACTS

Adequately addressing the research challenge outlined at the end of Section 4.3 is extremely time-consuming. The framework and method require appropriate consideration and a vast number of data inputs are required. Many LCA practitioners resort to specialist software linked to databases, which have a huge advantage in providing ready access to very detailed information on the environmental impacts of different systems. For example, it is reasonable to expect bioenergy researchers to have knowledge in and take an interest in the yields and conversion efficiencies for the plants systems they are evaluating and to check that these are consistent and relevant. However, it is unlikely that bioenergy researchers will be well-informed, for example, about the heavy metal releases to soil associated with slow erosion of the tyres on trucks as they make land journeys transporting biomass.

In such situations, specialist software and databases are invaluable in allowing a comprehensive and consistent approach. However, there is always a danger in using data sets with which the user has limited familiarity. It, therefore, behoves researchers to ensure that they consider which aspects of a bioenergy system are likely to have a significant impact on end-results and to ensure they are using appropriate, up-to-date figures for those inputs, which should also be considered for sensitivity analysis once results are computed.

With an area as diverse as bioenergy, it is therefore useful to start with a high level indication of which factors are likely to be significant for different types of bioenergy system. Table 4.1 in Chapter 4 has shown the factors that are likely to dominate the greenhouse gas balance calculations for different types of bioenergy system, disaggregated by feedstock. This can be used as a guide to focus attention on ensuring appropriate figures for the most critical factors in an LCA assessment.

10.7 SUMMARY

KEY POINTS

- Biodiesel from soy has been examined and shown to be capable of delivering very significant greenhouse gas savings.
- GHG emission from land-use change can impact significantly the overall carbon balance. Also, significant credit is assumed from displacement of soy meal production in the overall soy system. Emissions from agrochemical use are less significant than those from land-use change and occur in a longer time frame than annual production. Despite the positive GHG balance, there are other significant environmental impacts that need to be taken into account in an overall environmental LCA.
- It is important to consider not only the energy and material flows from and to the system under study, but also any changes in the characteristics of stock reserves associated with the system.
- Where bioenergy systems interface closely with other (food, land, waste, etc.) systems, careful consideration needs to be given to the scope of system being studied. It may be appropriate to expand the system scope to ensure all impacts are covered.
- The time period chosen for analysis in an LCA should be chosen appropriate to the system being studied to ensure that all appropriate impacts are taken into account. Where there is potential for impacts on soil characteristics or other stocks, this will likely require a longer analysis time frame than the production timescale.

References

[1] Thornley Patricia P. Biofuels review: Report for Government Office for Science. University of Manchester; 2012.
[2] Thornley P, Gilbert P. Biofuels: balancing risks and rewards. Interface Focus 2013;3(1):20120040.
[3] Tomei J, Upham P. Argentinean soy-based biodiesel: an introduction to production and impacts. Energ Policy 2009;37(10):3890–8.
[4] Welfle A, Gilbert P, Thornley. T. Performance of UK Biomass Resources for Heat Bioenergy Pathways. University of Manchester and DECC; Forthcoming.
[5] EEA. Opinion of the EEA Scientific Committee on Greenhouse Gas Accounting in Relation to Bioenergy. Copenhagen; 2011.
[6] De Menna F, Vittuari M, Molari G. Impact evaluation of integrated food-bioenergy systems: a comparative LCA of peach nectar. Biomass Bioenerg 2015;73:48–61.

Further Reading

[1] DIRECTIVE 2009/30/EC—amendment to Directive 98/70/EC on environmental quality standards for fuel (Fuel Quality Directive). European Commission.

CHAPTER

11

Combustion of Energy Crops for District Heating

Paul Gilbert

University of Manchester, Manchester, United Kingdom

11.1 INTRODUCTION

As outlined during the 2015 United Nations Framework Climate Change Conference, COP21, climate change is a pivotal global issue that needs to be tackled over the coming decades. Here, the international community was able to acknowledge and quantifiably define it: to hold *'the increase in … temperature to well below 2°C … and to pursue efforts to limit the temperature increase to 1.5°C.'* Focusing further on Europe and the United Kingdom in particular, under the Kyoto Protocol [1], European Governments had at the time greenhouse gas (GHG) emission and renewable energy targets and the European Commission's Renewable Energy Directive [2,3]. At a United Kingdom level, the 2008 Climate Change Act outlines a series of budgets and targets out to 2050 that limit greenhouse gas emissions to 80% below 1990 levels, and a benchmark target to reduce carbon emissions by 35% below 1990 levels by 2020 [4]. In similar keeping to the majority of nations, the United Kingdom's energy sector is a substantial contributor to greenhouse gas emissions. This comprises energy consumption for power, heat, and transport demand for a range of subsectors, resulting in primary energy demand in 2014 being 201.4 Mtoe and emissions being 422.0 $MtCO_2$ [5]. Of this, ~46% of final energy consumed is to provide heat [6]. However, heat is regarded as a challenging sector to decarbonize with high uncertainty relating to electricity grid decarbonization, deployment of heat pumps, role of heat networks [6], and issues relating to the low carbon provision of high-grade heat for industry. When exploring alternative energy sources for the provision of heat, bioenergy is regarded as having an increasingly important role [4], due to its flexible nature and ability to provide gaseous fuels that could be integrated into existing heat networks and utilised remotely. Given available technology and practices, this is especially true for low-grade heating demand, for example, domestic space heating and hot water.

Greenhouse Gas Balances of Bioenergy Systems
https://doi.org/10.1016/B978-0-08-101036-5.00011-2

When considering feasible bioenergy feedstock and pretreatment processes for district heating, it is important to consider what feedstocks are available indigenously and, of these feedstocks, what are also suitable for upgrading via thermal conversion [7]. This reduces the need to import feedstocks that are in demand among competing nations and regions. Using the United Kingdom as an example, energy crops are a promising bioenergy feedstock, as they exhibit low nutrient requirements and have limited implications associated with pest or disease [8]. This results in low demands for soil cultivations, fertilisers, and agrochemicals, providing subsequent benefits from a sustainability perspective. Depending on the land use implications, they are able to demonstrate greenhouse gas savings compared to their fossil fuel alternative [9]; however, there are still associated levels of greenhouse gas emissions from the use of intensive agronomic practises and agricultural inputs, such as fertilisers. With regard to pretreatment processes, drying and pelletization is the preferred option, as it is able to provide a standardised fuel to a boiler at a high bulk density—there are also benefits for transportation. Nevertheless, pelletization has increased heat and power demand compared to conventional chipping. Given the anticipated increase in the uptake of bioenergy to meet future heat, power, and transport demand globally [10,11], the wider environmental impacts need to be understood fully and recommendations made for policy and regulation [12]. This holds not just for energy crop cultivation for district heating, but for all forms of bioenergy use.

The research aim of this chapter is twofold and is achieved using life cycle assessment (LCA) methodology. Firstly, it investigates the environmental life cycle impact of delivering district heating from energy crop combustion when compared to natural gas combustion; and secondly, examining more closely the production of the feedstocks, it investigates the sensitivity to the main agricultural input, fertilisers, and to what extent this has an influence on the associated environmental impacts. Additionally, the chapter evaluates the implication for the use of different life cycle impact allocation procedures on the findings. The objectives for this research are as follows:

- Outline the appropriate LCA method and accompanying goal and scope to achieve the aim;
- Gather and assemble the inventory data appropriate for the given system boundaries;
- Complete an impact assessment to explore and quantify Global Warming Potential (GWP) and wider environmental impacts;
- Discuss implications of the impact assessment in light of decarbonizing the UK heat sector.

Fulfilling this research aim and subsequent objectives should highlight the significance for considering environmental impacts beyond GWP, in order to advise industry and policy accordingly, particularly for the use of alternative fertiliser regimes such as using sewage sludge. These recommendations would go beyond the series of nonlegally binding sustainability criteria developed by the EU [13]; the guidelines within this criteria cover only the levels of greenhouse gas savings that should be achieved through the use of bioenergy from a given biomass resource in comparison to a fossil fuel-derived alternative.

11.2 METHODOLOGY

An LCA approach is the most frequently applied methodology for evaluating the environmental impacts and performance of bioenergy systems. The primary purpose of the LCA approach is to provide inventory data and to assess the full impact of a given bioenergy system. This can include quantifying the mass, energy, and emission flows associated with the processes used to grow and harvest, pretreat, distribute, convert, and utilise the bioenergy feedstock [14]. Each potential variation within a bioenergy system effectively equates to assessing a slightly different 'LCA question' and hence the goal and scope of the study is very important.

11.2.1 Goal and Scope

The goal of this study relates back to the research aim and is to investigate the GWP impacts associated with delivering district heating from energy crop combustion when compared to natural gas combustion (Goal 1). It then explores feedstock production in more detail to investigate the wider environmental impacts for large-scale energy crop cultivation under United Kingdom conditions, using different fertiliser options for crop nutrition (Goal 2). In particular, it examines acidification and eutrophication potential (AP and EP, respectively). Conventional inorganic fertilisers and sewage sludge are examined, as well as no fertiliser control. Direct emission allocation from sewage sludge application is also examined, as it can be argued that the emissions associated with this 'waste' product could be allocated to either the farmer or the wastewater treatment companies [15]. As fertiliser application results in substantial direct emissions from the cultivation of bioenergy, sensitivity analysis also assessed the impact of N_2O leaching on total greenhouse gas emissions.

There are two functional units within this study. In Goal 1, the functional unit is 1 MJ heat delivered to the end-user. In Goal 2, the functional unit is 1 MJ biomass delivered to the end-user, as this allows for comparison between 1 MJ natural gas, also delivered to the end-user (including infrastructure and construction life cycle implications) [16]. This functional unit is appropriate to assess the implications of varying the fertiliser regime, as it places emphasis on the growing of the feedstock, rather than the energy delivered.

11.3 FEEDSTOCK SELECTION AND SEWAGE SLUDGE OVERVIEW

The United Kingdom has limited natural resources of biomass feedstock. The largest potential lies with agricultural residues, food waste, and energy crops [7]. To increase the indigenous supply level, as well as to promote associated socio-economic benefits [17], there has been a substantial focus on energy crops. In this study, short rotation coppice (SRC) is selected as the case study energy crop, as it has demonstrated significant potential in the United Kingdom to date. It should be noted that during the period that this study was completed, energy crops were in their infancy and so agronomic requirements, including fertiliser application rates, were not fully understood. More recent research suggests that fertiliser and pesticide application rates for energy crops are lower than what is presented here. Nonetheless, the results are still of relevance when other crops are considered, in particular crops associated with the production of biofuels, such as rapeseed that still have notable fertiliser demand.

A key feature of this study is that the use of sewage sludge as a fertiliser is assessed alongside conventional inorganic fertilisers. Approximately 1.35 million tons of sewage sludge (dry solids) is produced annually in the United Kingdom, of which 88% is either recycled to land as an agricultural fertiliser, used for the recovery of contaminated sites, or used as a fuel [18]. Under the Waste Management Licensing Regulations [19], sewage sludge is recognised to restore, reclaim, or improve the land to which it is applied and is therefore exempt from requiring a waste management licence [20]. Using it as a fertiliser and soil conditioner is a beneficial disposal route when compared to landfill and incineration [21], as sewage sludge contains valuable macronutrients such as nitrogen (N) and phosphorous (P) [22], micronutrients [23], and acts as a soil conditioner. However, in practice [24–26] and based on the LCA literature [21,27], the environmental impacts of disposing of sewage sludge to land are mixed, and accordingly, there are therefore mixed opinions on this disposal route. Further insight into the use of sewage sludge as a fertiliser is covered elsewhere [28].

11.4 SYSTEM DEFINITION AND INVENTORY

An overview of the bioenergy system is outlined in Fig. 11.1. The biomass cultivation scheme and inventory is based on previously published work [9,28,29] and Ecoinvent v2.1 [30]. The specific assumptions and quantities relating to agricultural practises and agricultural inputs are presented elsewhere [28]; however, the application rate of the fertiliser options is presented in Table 11.1, with an accompanying description. The natural gas reference case data is based on Ecoinvent v2.1 and includes the supply and combustion of the gas in a district heating system [16].

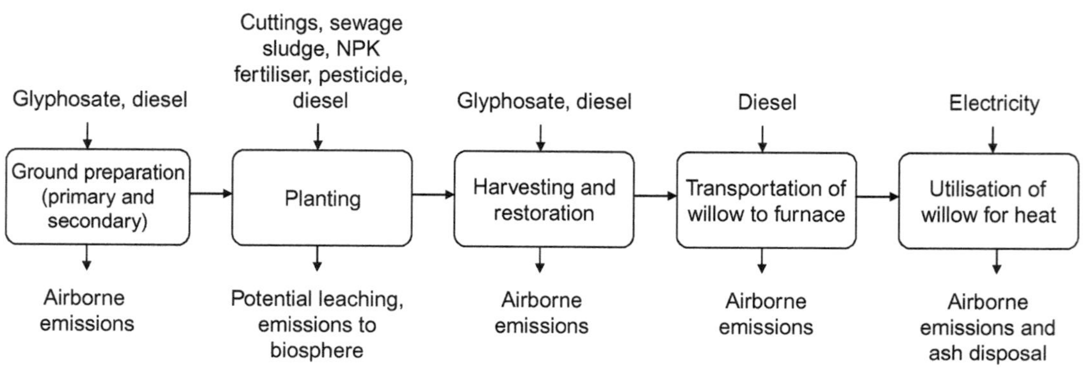

FIG. 11.1 Schematic of the woodchip district heating system.

TABLE 11.1 Fertiliser Application Rates for Scenarios

Scenario	Fertiliser	N	P	K
1	No fertiliser	–	–	–
2	Inorganic fertiliser	90 kg/ha/3 year AN[a]	24.2 kg/ha/3 year TSP[b]	60 kg/ha/3 year MoP[c]
3	Digested cake @ 40 t/ha/3 year	300 kg (15%, 10%, 5% available in years 1, 2, 3)	360 kg (50% available)	60 kg/ha/3 year MoP[c]

[a] *Ammonium nitrate.*
[b] *Triple superphosphate.*
[c] *Muriate of potash.*

11.4.1 SRC Willow Ground Preparation, Planting, and Harvesting

The SRC willow is cultivated on previously arable cropland. At the start of ground preparation, a broad spectrum herbicide, such as glyphosate, is applied to kill perennial weeds. SRC willow is established from lengths of willow cane, planted at a density of 15,000 cuttings/ha. The planting is assumed to take place during late winter. After the first year of growth, the crop is cut back to encourage coppicing. No allowance has been made for greenhouse gas emissions associated with the decomposition of biomass material left lying in the field. Once established, the SRC willow requires very little agronomic management or inputs and it is assumed that the SRC willow is grown for three years after cut back before the first harvest. The crop is maintained for seven full rotations (21 years productivity plus one year establishment and one year restoration). Typical yields are 30 odt/ha per 3 year rotation, at 50% moisture content, equating to an average annual yield of 10 odt/ha, as received [31]. Harvesting the SRC is completed using a self-propelled forage harvester with a specialist SRC header to produce 35 mm wood chips. Following harvesting, the woodchip is dried naturally for a period of 30 days to reduce the moisture content from 50% to 30% ([32] from [29]). fertiliser application rates are outlined in Table 11.1. Losses during drying, storage, and transit are assumed to be 5% (3% for field drying of chips and 1% assumed per transfer operation). After 21 years of productivity, the land is returned to its original condition by applying further herbicide and subsoiling and mulching of the roots to prevent further growth.

11.4.2 Upgrading of Woodchip to Heat (Goal 1)

The dried woodchips are transported 60 km to a district heating plant where they are stored in a 100 m^3 storage area that provides 1 week fuel in winter and 3–4 weeks fuel in summer. The woodchip is automatically fed into a 300 kW boiler, is ignited, and burnt along a moving grate. Ash falls through the grate where it is automatically transported to a collection point for disposal. The high-temperature flue gas passes through a heat exchanger and a cyclone separator ensures low dust emissions in the flue gas. The heat generated is used for domestic space heating and hot water supply. The efficiency of the boiler is 76% and ~0.1 kg of fresh biomass (30% moisture content) is required to produce 1 MJ of heat. The biomass flow rate is ~60 kg/h and the flue-gas flow rate is ~1500 Nm^3/h. 0.005 kWh/MJ electricity is required for the operation of the plant. The heater is assumed to operate for 200 days per year and for 21 years.

11.4.3 Fertiliser Application Scenarios (Goal 2)

To assess the impact of varying fertiliser rates, three scenarios are investigated: Scenario 1 assumes no fertiliser application; Scenario 2 provides all expected nutrient requirements from inorganic fertiliser with a known composition and; Scenario 3 uses digested sewage sludge cake to provide an equivalent amount of nitrogen as Scenario 2, based on the proportion of N in the sewage sludge that is 'available' for crop growth. These are outlined in Table 11.1.

11.4.4 Emission Allocation for Sewage Sludge Applications

Typically when dealing with a product and co-product (for example between wheat and straw), sharing the burden of the energy inputs and emissions to air, soil, and water is made on the basis of an allocation between the products. This could be on the basis of mass ratios, economic value ratios, or energy ratios. The choice of allocation procedure is somewhat subjective and must be based on an assessment of the context of the individual application. In this study, an allocation is to be made for the waste disposal function for the sewage sludge and a choice must be made of how to apportion the environmental impacts associated with the application of the sludge.

Sewage sludge application provides both the benefits of nutrients to the crop and the penalties in terms of emissions to air. There is the release of N_2O, a significant contributor to GWP; volatilization of NH_3 contributing to AP; and leaching of phosphates contributing to an increase in EP. Arguments can be made for the emission penalties to be allocated to either the wastewater treatment system or the bioenergy system. A sensitivity analysis is completed to consider the impacts of different allocation options for sewage sludge on the SRC willow system. The emission allocation is as follows: 0% attributed to grower (SRC willow Option 3a) and 100% attributed to grower (SRC willow Option 3c). An additional option is assessed (SRC willow Option 3b), based on Defra guidelines, which indicate that 30% of the nitrogen in the sludge is typically available as a crop nutrient over 3 years (15% in the first year, 10% in the second year, and 5% in the third year) [22]. This final allocation option reflects the value of the nitrogen in the sludge to the farmer in terms of potential crop yield benefits.

11.4.5 Emission factor for N_2O

Over a 100-year period, N_2O has 298 times more impact on GWP compared to CO_2. There is a direct release of N_2O emissions from the application of nitrogen-based fertilisers on the soil. Consequently, the significance of soil N_2O emissions on the overall environmental impact is assessed through a sensitivity analysis. The emission factor set by the Intergovernmental Panel on Climate Change (IPCC) for the release of N_2O from N fertiliser application, and used in this study, is 1% of applied N [33]. This factor could, however, vary due to: environmental factors, for example, climate, soil organic carbon content, drainage, and soil pH; and N management, for example, fertiliser type, application rate, and crop type [33]. As a result, the N_2O emission factor is varied from 0% to 3% to comply with the IPCC upper uncertainty range [33]. Therefore, the sensitivity analysis tests the N_2O sensitivity for the use of inorganic fertiliser and sewage sludge with 30% emission allocation. The results are presented in Table 11.3.

11.5 RESULTS FOR IMPACT ASSESSMENT AND SENSITIVITY ANALYSIS

The results presented are split into the two main goals of the study. This is followed by the results for the sensitivity analysis. The results are then discussed in Section 11.6.

11.5.1 Results for the GWP Impacts Associated With Delivering District Heating from Energy Crop Combustion (Goal 1)

Table 11.2 provides a summary of the main findings related to the GWP impacts of the district heating system combusting energy crops. Besides presenting total greenhouse gas emissions and savings compared to the natural gas system, two additional metrics are added. This style of reporting builds on a study by Thornley et al. [9] that presents LCA findings of typical bioenergy systems using a range of metrics that set about to address different policy- and industry-relevant questions. The two additional metrics are greenhouse gas savings per oven-dried ton (odt) of biomass utilised and greenhouse gas savings per unit of land occupied.

11.5.2 Results for the Wider Environmental Impacts for Energy Crop Cultivation (inc Sensitivity to Sewage Sludge Allocation) (Goal 2)

The impact assessment results for GWP, AP, and EP are presented in Fig. 11.2 for the energy crop cultivation. As noted, the functional unit is 1 MJ of biomass/natural gas delivered to the end-user. GWP has a negative saving for the SRC willow scenarios due to uptake of CO_2 during cultivation.

TABLE 11.2 Summary of Key Greenhouse Gas Metrics for Studied System

GWP Metric	Value	Unit
Absolute greenhouse gas emissions per unit of heat generated (SRC willow)	0.004	kg CO_2-eq/MJ heat
Absolute greenhouse gas emissions per unit of heat generated (natural gas)	0.067	kg CO_2-eq/MJ heat
Relative greenhouse gas savings per unit of heat generated	0.06	kg CO_2-eq/MJ heat
Relative greenhouse gas savings per unit of heat generated	93.7	%
Greenhouse gas savings achieved per odt of biomass utilised	1203.8	kg CO_2-eq/odt biomass
Greenhouse gas savings per unit of land occupied	572.8	kg CO_2-eq/ha.a

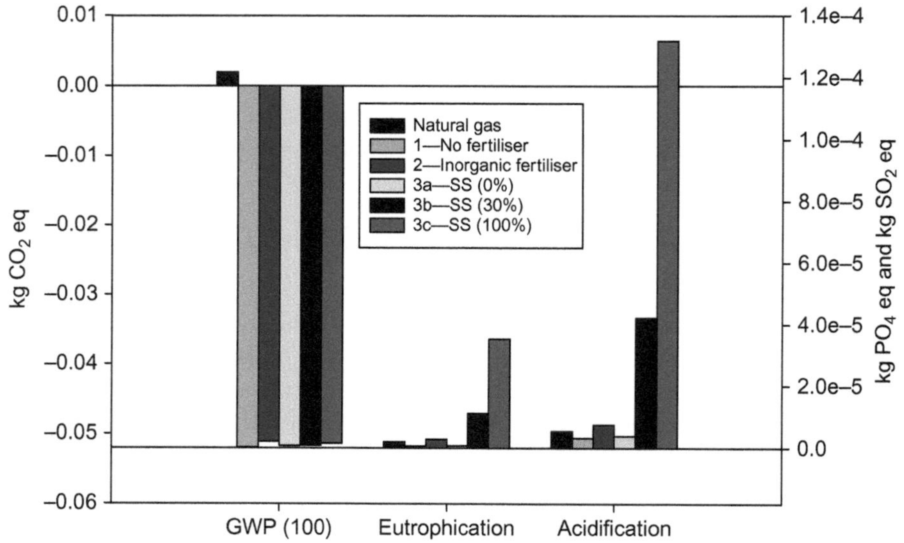

FIG. 11.2 Impact assessment for 1 MJ SRC willow, Scenarios 1–3c, with 1 MJ natural gas comparison. Global warming potential (GWP) is measured in kg CO_2 eq using the left hand axis, and eutrophication potential in kg PO_4 eq and acidification potential in kg SO_2 eq using the right hand axis. *From Gilbert P, Thornley P, Riche AB. The influence of organic and inorganic fertiliser application rates on UK biomass crop sustainability, Biomass Bioenergy 2011; 35:1170–1181.*

TABLE 11.3 Sensitivity Analysis for N_2O Emission Factor (kg CO_2 eq/MJ SRC Willow)

N_2O Emissions (%)	SRC Willow, Inorganic Fertiliser	SRC Willow, Sewage Sludge (30% Emission Allocation)
0	−0.0513	−0.0518
1	−0.0511	−0.0516
3	−0.0508	−0.0514

11.5.3 Results for Emission Factor for N_2O

The sensitivity analysis related to the N_2O emission factor is outlined in Section 11.4. The results are presented in Table 11.3.

From conducting the sensitivity analysis, the maximum % change observed in GWP compared to the original 1% emission factor is 1.2%. Therefore, it is concluded that the impact here is minimal when put into context with the larger emission savings of the SRC willow case when compared to the natural gas case.

11.6 DISCUSSION

11.6.1 Greenhouse Gas Implications for the Systems Studied (Goal 1 and 2)

The greenhouse gas burden for the energy crop district heating system is low (Table 11.2) when compared to other bioenergy systems presented elsewhere by this author [9], as the supply chain emissions are small. There is limited preprocessing (chipping instead of the more energy-intensive pelletization); these chips are dried naturally and the transportation is minimal as the crops are indigenous. For similar reasons, the greenhouse gas savings achieved per odt of biomass utilised is high. The implications for the size of the savings, however, would be different if land use change is taken into account. In this study, it is assumed that the energy crops are grown on previously arable land and so emissions relating to land use change are not considered. Increasing the fertiliser rates would impact yield, yet there would also be a trade-off with the added emissions to consider and understand further (both N_2O from direct application to soil and fertiliser production).

The findings in Fig. 11.2 also demonstrate that energy crop production provides a substantial GWP saving compared to natural gas production and delivery to the end-user. Focusing further on the role of fertilisers, there is marginal variation in GWP between the 3 Scenarios. Compared to Scenario 1 (no fertiliser), the application of inorganic fertiliser increases GWP by 2%. This is primarily due to the production of ammonium nitrate. The application of sewage sludge increased GWP by a lesser extent when compared against Scenario 2 (inorganic fertiliser). This is because the sewage sludge is less energy-intensive to produce.

11.6.2 Importance of Wider Environmental Impacts (Goal 2)

The two environmental impacts explored in this study are EP and AP. The EP in Scenario 1 (no fertiliser) is 231% lower than for the natural gas reference case; the AP is approximately equal in Scenario 1 and the natural gas reference case. When exploring the impact of fertiliser application, the EP and AP increase substantially for Scenario 2 and 3; fertiliser has a significant impact on these impact categories compared to the impact on GWP.

Applying inorganic fertiliser (Scenario 2) increased total EP by 383%. This increase is mainly due to leaching of nitrates and phosphates from soil. The EP was higher still with the sewage sludge case using the 30% emission allocation (Scenario 3b) where EP increases from 5.78×10^{-7} kg PO_4 eq to 1.11×10^{-5} kg PO_4 eq.

The increase in AP is due to the volatilization of NH_3 and the production of nitric acid, as a part of ammonium nitrate production in the case of inorganic fertiliser application (Scenario 2). In the case of sewage sludge (Scenario 3c), higher emissions result from a higher proportion of the inorganic content volatilizing shortly after spreading onto the land. Here, the AP increases by 460%, compared to inorganic fertiliser application.

11.6.3 Importance of Allocation of Sewage Sludge Emissions (Goal 2)

The emission allocation for sewage sludge emissions is outlined in Section 11.4 and consisted of 0% allocation (Scenario 3a), 30% allocation (Scenario 3b), and 100% allocation (Scenario 3c). Large environmental impacts are observed when 100% of the sewage sludge-related emissions are allocated to the grower. For 0% allocation, the EP and AP are lower than that of applying inorganic fertiliser (Scenario 2). Additionally, the GWP is −0.0514 kg $CO2_{eq}$ for Scenario 3c and −0.0516 kg CO_{2eq} for Scenario 3b, compared to the natural gas reference case. While both options offer substantial greenhouse gas savings, it is clear that guidance on which allocation factor to be used is just as relevant for GWP calculations. Consequently, the allocation factor is critical in assessing the actual environmental benefits or costs incurred for bioenergy systems—in particular if their supply chain includes the application of agricultural inputs such as fertilisers.

11.7 CONCLUSIONS

The GWP and wider environmental impacts of combusting SRC willow for use in a district heating system are assessed using a life cycle assessment approach and the findings are compared to a natural gas reference case. Examining the full life cycle for both systems (Goal 1) demonstrates that the energy crop system can deliver greenhouse gas savings of up to 94%. The high savings primarily come about due to the limited transportation over the feedstock life cycle and from the minimal preprocessing involved prior to combustion. Examining the production and delivery of the SRC willow in comparison to the natural gas reference case (Goal 2) provides an opportunity to understand further the impact of different fertiliser regimes and the impact of different allocation schemes for the sewage sludge emissions. The results show that the EP and AP of unfertilised SRC willow (Scenario 1) are marginally less than for the natural gas reference case. Applying inorganic fertiliser (Scenario 2) increases greenhouse gas emissions due to the manufacturing of the fertiliser. The emissions to air and soil from the fertiliser application increase the EP and AP impacts, making them comparable with those of the natural gas reference case. Applying sewage sludge as a fertiliser (Scenario 3) results in a substantial increase in EP and AP impacts compared to the natural gas reference case; however, greenhouse gas emissions increased to a lesser extent compared to application of inorganic fertiliser. Varying the assumption for the N_2O emissions factor had only a marginal impact on the results, but the emission allocation split between wastewater treatment (Scenario 3a) and crop grower (Scenario 3c) is very significant in affecting the EP and AP attributed to the grower. The results indicate the need to consider wider environmental criteria above and beyond Greenhouse gas savings when evaluating bioenergy systems as impacts such as EP and AP can be considerably higher than the fossil-based counterfactual. The results also demonstrate the impact emission allocation can have on the performance of a system and so allocation assumptions need to be outlined clearly when assessments are made.

References

[1] United Nations. Kyoto Protocol to the United Nations Framework Convention on Climate Change; 1998. Available from: http://unfccc.int/resource/docs/convkp/kpeng.pdf.
[2] European Commission. Renewable energy roadmap: renewable energies in the 21st century; 2006. Available from: http://ec.europa.eu/energy/energy_policy/doc/03_renewable_energy_roadmap_en.pdf.
[3] European Commission. 20 by 2020, Europe's climate change opportunity; 2008.
[4] DECC. The UK renewable energy strategy. 2009Available from: www.official-documents.gov.uk/document/cm76/7686/7686.pdf.
[5] DECC. Energy: Chapter 1, Digest of United Kingdom Energy Statistics (DUKES); 2015.
[6] Chaudry M, et al. Uncertainties in decarbonising heat in the UK. Energy Policy 2015;87:623–40.
[7] Welfle A, Gilbert P, Thornley P. Securing a bioenergy future without imports. Energ Policy 2014;68:1–14.
[8] Riche AB, et al. Grass and woody biomass species—agronomic requirements learnt from 15 years field experimentation. Glasgow: International Plant Protection Congress; 2007.

[9] Thornley P, et al. Maximizing the greenhouse gas reductions from biomass: the role of life cycle assessment. Biomass Bioenergy 2015;81:35–43.
[10] Eurelectric. Biomass 2020: opportunities, challenges & solutions; 2011. Available from: http://www.eurelectric.org/media/26720/resap_biomass_2020_8-11-11_prefinal-2011-113-0004-01-e.pdf.
[11] Hewitt J. Flows of biomass to and from the EU: an analysis of data & trends; 2011. Available from: http://www.fern.org/sites/fern.org/files/Biomassimports%20to%20the%20EU%20final_0.pdf.
[12] Junginger M, et al. Opportunities and barriers for international bioenergy trade. Energ Policy 2011;39(4):2028–42.
[13] European Commission. Report on sustainability requirements for the use of solid and gaseous biomass sources in electricity, heating and cooling, 2010: Brussels.
[14] Brander M, et al. Consequential and attributional approaches to LCA: a guide to policy makers with specific reference to greenhouse gas LCA of biofuels. Technical Paper TP-090403-A, 2009, Ecometrica.
[15] Doka G. Life cycle inventories of waste treatment services, Ecoinvent report No 13, 2007, Swiss Centre for Life Cycle Inventories: Dubendorf.
[16] Faist-Emmenegger M, Heck T, Jungbluth N, Teil V. Erdgas. Ecoinvent Report No. 6-V, 2007, Swiss Centre for Life Cycle Inventories: Villigen.
[17] Defra. Biomass Task Force; 2005.
[18] Water UK Report—Towards Sustainability 2005–2006. Water UK: London; 2006.
[19] The Waste Management Licensing Regulations 1994, Statutory Instrument 1994 No. 1056, 1994, HM Government.
[20] Enviros. The beneficial use of sewage sludge in land reclamation; 2004. Available from: http://www.water.org.uk/home/policy/positions/sewage-sludge.
[21] Lundin M, et al. Environmental and economic assessment of sewage sludge handling options. Resour Conserv Recy 2004;41(4):255–78.
[22] Defra. Fertiliser recommendations for agricultural & horticultural crops (RB209). 7th ed. Norwich: Defra; 2003.
[23] EPA. Process design manual. Land application of sewage sludge and domestic septage. Washington, DC: United States Environmental Protection Agency; 2005.
[24] Barraoui D, Labrecque M, Blais J-F. Decontamination of sludge by the METIX-AC process. Part II: effects on maize growth and bioaccumulation of metals. Bioresour Technol 2008;99(5):1450–64.
[25] Moffat AJ, Armstrong AT, Ockleston J. The optimization of sewage sludge and effluent disposal on energy crops of short rotation hybrid poplar. Biomass Bioenergy 2001;20(3):161–9.
[26] Wang X, et al. Studies on land application of sewage sludge and its limiting factors. J Hazard Mater 2008;160(2-3):554–8.
[27] Suh Y-J, Rousseaux P. An LCA of alternative wastewater sludge treatment scenarios. Resour Conserv Recy 2002;35(3):191–200.
[28] Gilbert P, Thornley P, Riche AB. The influence of organic and inorganic fertiliser application rates on UK biomass crop sustainability. Biomass Bioenergy 2011;35(3):1170–81.
[29] Thornley P, et al. Integrated assessment of bioelectricity technology options. Energ Policy 2009;37(3):890–903.
[30] Kagi T, Nemecek T. Life cycle inventories of agricultural production systems, ecoinvent. Report No. 15, 2007, Swiss Centre for Life Cycle Inventories: Zurich and Debendorf.
[31] SAC. In: Beaton C, editor. The farm management handbook 2008/09, Vol. 29. Edinburgh, Scotland: The Scottish Agricultural College; 2008.
[32] Gigler JK, Meerdink G, Hendrix EMT. Willow supply strategies to energy plants. Biomass Bioenergy 1999;17(3):185–98.
[33] IPCC. Volume 4: Agriculture, forestry and other land use. 2006 IPCC Guidelines for National Greenhouse Gas Inventories; 2006.

CHAPTER

12

Production of Wood Pellets from Waste Wood

Carly Whittaker

Rothamsted Research, Harpenden, United Kingdom

12.1 INTRODUCTION

This book has examined various global biomass feedstocks. This chapter describes a case study where wood pellets are manufactured from waste wood. In the United Kingdom, the total arisings of waste wood total around 4 million t [1]. This is spread between the construction, demolition, commercial, industrial, and domestic sectors and encompasses a wide range of materials in terms of quality and composition [2]. Any supply chain utilising wood waste must be equipped to handle the generally unpredictable quality of the stream [3]. Most countries have a waste wood classification system to help handlers predict its quality and potential contaminants [4–6]. Taking the United Kingdom as an example, the classifications tend to be split as follows:

- Nontreated
  - *Grade A*: Clean waste wood. Includes off-cuts of clean timber and untreated woody packaging. It can contain a small amount of plastic, nails, paints, and surface coatings.
  - *Grade B*: Mixed waste wood. Can contain a blend of 60% Grade A wood and up to 40% of lightly painted wood, and materials such as chipboard.
- Treated woods
  - *Grade C*: Contains a higher content of wood treated with coatings such as plastic, melamine, or paints. Also includes woods impregnated with heavy metal preservatives.
  - *Grade D*: Treated with hazardous substances such as arsenic, chromium, and creosote.

The opportunities for reuse, recycling, or combusting different waste wood grades are dependent on feedstock quality and the specification required by the end user [2]. A clear distinction between Grades A and B and C and D waste is that the latter two are usually treated with heavy metals or halogenated compounds to provide protection against fungal decay or insect damage. Due to the presence of potentially harmful substances, emissions

Greenhouse Gas Balances of Bioenergy Systems
https://doi.org/10.1016/B978-0-08-101036-5.00012-4

from the combustion of this wood are controlled under the European Waste Incineration Directive (WID, [7] for plants sited in Europe and similar legislation is likely to apply elsewhere). This requires additional plant provisions and monitoring equipment, which often add significant costs. This is not the case for Grade A or B wood, hence it is highly desirable for a range of recycling or bioenergy uses [4]. According to the waste hierarchy principles, waste should be reused before recycling, then recycled before being incinerated or landfilled. Therefore, there is a high reuse potential for Grade A wood waste, though the chipboard industry routinely uses 70%–90% of it [8,9]. It can also be used for animal bedding, horticultural mulches, or for bioenergy purposes. It is less easy to recycle Grade B wood mainly due to the presence of resin binders, and also this means it is not suitable for pellet manufacture.

This study examines the GHG emission balance from the production of pellets from Grade A waste wood. The production system is based in the United Kingdom and data was derived directly from a pellet manufacturer. The following section describes the system boundaries in more detail and suggests the various options for the counterfactual systems with which to compare this.

12.2 LIFE CYCLE ASSESSMENT: SYSTEM BOUNDARIES AND COUNTERFACTUALS

12.2.1 The Pellet Supply Chain

The goal of this study is to examine the GHG balance of producing wood pellets from waste wood, which are commonly used to substitute for coal in large power stations (Section 12.2.2). The study takes an attributional approach to LCA (ALCA), so it examines the system flows under direct or indirect control of the operator [10,11]. In this case, the operator is the wood pellet producer, and the embodied emissions from producing the biomass fall outside of the system boundaries. The functional unit will be 1 t of wood pellets produced and stored at the factory for final distribution. The final unit of measurement will be the GHG emissions (in kg CO_2 eq.) per ton of wood pellets.

The systems boundaries include GHG emissions from:

- Diesel fuel consumed by processing machinery,
- Electricity consumption by processing machinery,

The GHG savings are calculated by comparing the emissions from coal combustion, but the biogenic carbon from combustion of wood pellets is excluded.

The waste wood arrives at the site in skips where they are tipped and stored in their respective graded piles. These 9 t skips are sourced from various distances, though it is assumed that the majority is sourced within 60 km of the site. While visiting the site, a visual inspection of the piles indicated that there was a small amount of contamination of plastic and metal. A default content of 10% waste was assumed. Also, the waste is likely to be contaminated with lower quality grades of wood waste, including some painted wood and particle boards. The latter can contain nitrogen-based urea formaldehyde binders that, during combustion, can lead to higher nitrous oxide emissions. It is estimated that 1% of the nitrogen in urea formaldehyde is converted to nitrous oxide during combustion [12]. The concentration of nitrogen in particle board, or chipboard, is around 3% [13,14]; therefore, it is excluded from this study. It must be acknowledged that some uncertainty exists over composition of this material, as well as the emissions from combustion under different conditions. This would be a more important issue if Grade B or C wood was being used.

Waste wood pelleting involves the following stages: reception of raw material, shredding, screening, drying, hammermilling, pelleting, and cooling. Magnets are positioned at the inlet and outlet of almost every stage in order to remove metals. The wood is shredded, as this method is best suited to handle the irregular shapes received. This is performed using a tub-grinder; a large and powerful diesel-powered machine, though at larger scales could be powered from auxiliary power from the plant. The shredder required 41 diesel/t processed. The chips are then passed through an air knife screen which literally lifts plastics away from the woody stream. The chips then enter a drum-drier which uses ~0.5 t of wood to dry 9 t of wood from 20% to 13%–15% MC. The dried biomass then enters the hammermill where it is reduced to particles about 5 mm long before entering a mixing chamber where steam and additives can be added. The steam is used to condition the biomass and increase the temperature to around 120°C for pelleting A additives and may be used to improve pellet quality, depending on feedstock composition, and would be limited to just 2% of the mass of the pellets [15], but was not being used in this case. In the pellet mill, the biomass is forced through a circular die that is perforated with holes up to 10 mm in diameter. The pelleting plant had two mills, with a total throughput of 9 t/h. On leaving the mill, the pellets are cooled by blowing cooled (0–25°C) air over them until they reach within 5°C of ambient temperatures. This is necessary to ensure that the lignin re-solidifies and forms durable pellets [16]. The final pellets have a MC of 10% and a lower heating value (LHV) of 16.5 MJ/kg. The total electrical power required for all processing machinery, screening, and auxiliary processes was estimated at 180 kWh/t pellets. This could be produced from an onsite generator, though in this case it was imported from the grid.

12.2.2 Fossil Reference System

Wood pellets are an ideal feedstock for co-firing with coal. They are made from hammermilled wood, which are reformed into small, cylindrical pellets under high pressure and temperature [17]. The pellets can easily be re-broken and fed into coal pulverisation lines and combusted via direct injection [18], presenting an inexpensive method of reducing carbon dioxide emissions from coal-fired power stations [19]. For example, Drax, in Selby, United Kingdom, co-fired coal with biomass between 2003 and 2012, but in 2012 and 2014, they fully converted their first and second boilers to dedicated biomass combustion. They currently consume around 4 million t of biomass pellets annually [20]. Drax has a reported electrical efficiency of almost 40% for coal [21], though it is not reported what the equivalent is for biomass. This case study assumes the same electrical efficiency; therefore, the two can be compared on an energy-content basis in which to calculate the GHG emission savings.

12.2.2.1 Counterfactuals

The 'counterfactual' refers to the alternative fate of the biomass in the case study, if not used for bioenergy [22]. This approach is more suited to consequential LCA (CLCA), which accounts for emissions that have not yet occurred [23], but considers the possible external consequences of utilising biomass for energy. This identifies additional impacts where alternative products may need to be sourced or highlights where potential emissions can be saved. For example, Grade A waste wood can be used to produce chipboard. One must assume that the demand for chipboard is the same if wood is diverted to bioenergy or not. This would mean that the waste wood pellet LCA would incur the consequences of sourcing chipboard material from other woody residues, virgin timber, or any suitable alternative nonwood products [22]. This is examined as one of the counterfactual scenarios.

Other alternatives include chipping waste wood for use in *animal bedding*, which could be substituted with cereal residues, or *horticultural mulching*, which could be substituted with bark from saw log processing. In either case, these are residues from other supply chains, and for simplicity, are not examined here.

Landfill is a good example of a counterfactual system that may have a negative or positive effect on the net CLCA results. This is a controversial counterfactual for waste wood, as there are existing markets that depend on this resource. It was selected in many early LCAs because it was assumed that the landfill tax introduced in 1996 would divert wood waste away from landfill [24], so it could be made available for bioenergy [25]. Since then, however, recycling rates have generally increased in the United Kingdom. Yet, although particleboard industries do consume a great majority of United Kingdom waste wood, United Kingdom waste statistics suggest that still a significant proportion of Grade A waste is not recovered and may end up in landfill [26].

If waste wood is then used for bioenergy production, the process of landfilling wood and its associated emissions are 'avoided' in the pelleting system. The process of landfilling involves the mechanic compression of waste where it is covered with a clay cap to restrict oxygen availability and methane emissions [27]. Data on the fuel requirements by landfill machinery is derived from [12]. There is some uncertainty over the avoided methane emissions from landfill attributed to wood, as these are highly sensitive to the rate at which it breaks down. Typically, wood is composed of cellulose (50%), hemicellulose (25%), and lignin (25%). Lignin acts both as a microbial and physical barrier to decay [28] and does not generally degrade in anaerobic conditions [29]. Therefore, in landfill lignin may hinder the decomposition of up to 50% of other carbohydrates integrated within the woody structure [30]. In this case, landfill could effectively act as a 'sink' of carbon. A number of studies have examined this. One hypothesised that 35% of the carbon in woody material is degraded in landfill, while the remaining 65% is sequestered [30]. Experimental studies in Australia, however, found evidence that the loss of carbon was actually very low (around 3%–4%) after 19 and 29 years, and around 17%–18% after 43 years after landfilling [31]. Another study suggests that a maximum of 3% carbon lost in landfill, though a timescale is not clearly stated [32]. A study in Korea found a similar degradation rate (17%–18%) for wood in landfill; however, again the timescale is not clear [33].

To examine this counterfactual system, a simple carbon-accounting model was created to estimate the net carbon balance from landfill (Fig. 12.1). The model considers fugitive emissions of methane, displaced grid electricity from energy generation, and carbon sequestration in landfill sites. Table 12.1 lists some model assumptions.

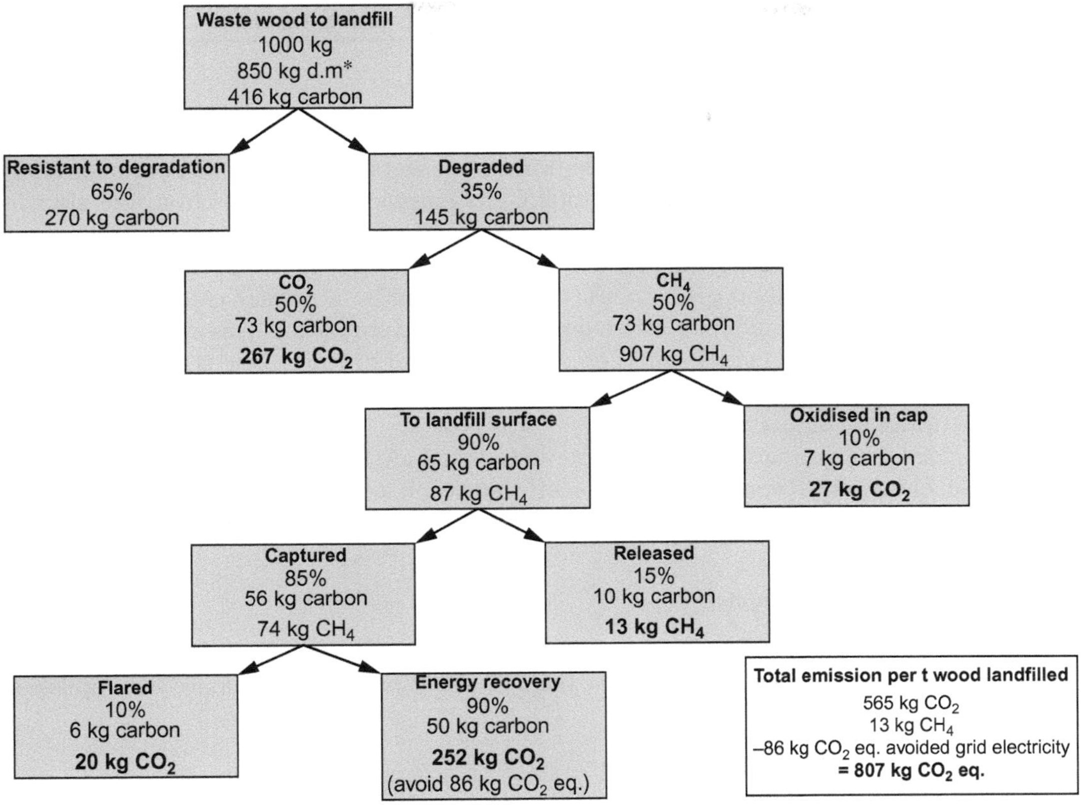

FIG. 12.1 Example of outputs from landfill model used in this study, based on a 35% degradation rate [30,31].

TABLE 12.1 Default Parameters in the Landfill Counterfactual Model Used in this Study

Aspect		Value	Unit	Notes
Biomass characteristics	Carbon content	50	% (dry basis)	[30]
	Moisture content	10–25	% (wet basis)	Estimated
Fugitive emissions	Degradation of carbon	0–100	% carbon	
	Landfill gas capture efficiency	85	%	[27]
	Proportion oxidised on exit	10	%	[34,35]
Energy recovery	Landfill gas methane content	60	%	[36]
	Proportion of sites with energy recovery	90	%	[37]
	Electrical efficiency	42	%	[27]
	Energy content	21–25	MJ/m^3	[38]
		18–21	GJ/t	Assume landfill gas has same density as air ($1.2\,kg/m^3$)

12.3 RESULTS: NET GREENHOUSE GAS EMISSIONS

12.3.1 Base-Case GHG Emission Results

Table 12.2 shows the breakdown of GHG emissions for the production of 1 t of waste wood pellets, to the factory gate. Pellet production from waste wood incurs a GHG burden of 111 kg CO_2 eq./t. The largest proportion of the total GHG emissions originates from the electricity requirements for pelleting and auxiliary power (72%). The GHG intensity of this input could be reduced if a natural gas generator was located onsite, as this would have a lower GHG emission intensity than grid electricity, although incur a greater capital investment. Otherwise, the diesel required for waste wood delivery and shredding has similar burdens on the total GHG emissions. A small emission of methane and nitrous oxide is caused from the combustion of biomass used for drying [39], and hence, this parameter is usually excluded from LCA studies on biomass combustion.

Overall, the GHG emission savings from waste wood are 93%, assuming that they can be converted to electricity at similar conversion efficiencies. If a coal power station was co-firing at rate of 10% (by energy content), it would be reducing its overall GHG emissions by 9.3%.

12.3.2 Sensitivity Analysis

The total GHG emissions in Table 12.2 assume the base-case parameters described in Section 12.2. Fig. 12.2 shows impact that the following have on the final GHG emission savings:

- Collection distance
- Processing energy (diesel)
- Pelleting and auxiliary power
- Parasitic drying demand
- Proportion contamination by plastic/metals

TABLE 12.2 Breakdown of GHG Emissions From Producing 1 t of Waste Wood Pellets, Including GHG Emissions Savings Compared to Coal, on an Energy-Content Basis and Including Combustion of Coal

Stage	Input	Carbon Dioxide Emissions (kg CO_2)	Methane Emissions (kg CH_4)	Nitrous Oxide Emissions (kg N_2O)	Total Greenhouse Gas Emissions (kg CO_2 eq.)
Delivery	Diesel fuel	14.7	0.0	0.1	14.9
Shredding	Diesel fuel	13.1	0.0	0.1	13.2
Drying	Waste wood	0	0.01	0.01	3.0
Pelleting	Electricity	79.6	0.1	0.5	80.2
Total		**107.5**	**0.1**	**0.7**	**111.3**
Total per GJ		*6.6*	*0.00*	*0.04*	*6.6*
% Saving (vs coal)		**93**	**93**	**97**	**93**

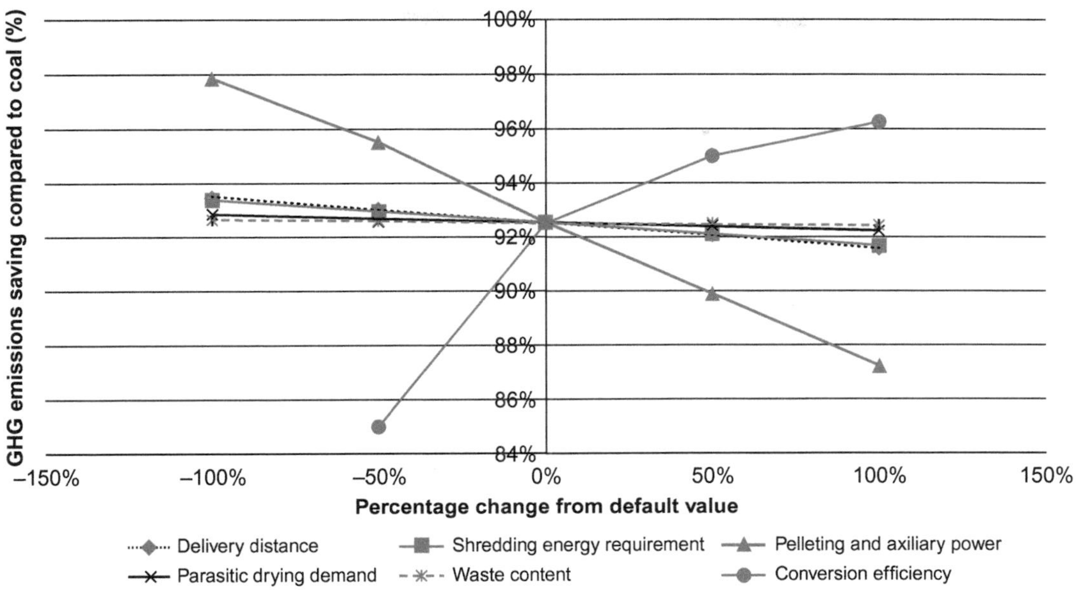

FIG. 12.2 Sensitivity analysis of the effect of several parameters on the GHG emission savings of waste wood pellets compared to coal.

Fig. 12.2 shows that the most sensitive parameter tested is the electricity requirement for pelleting and auxiliary power, which was also the highest contributor to the total GHG emissions. This was mainly because of the higher emission factor multiplier of grid electricity.

Other parameters having a less sensitive impact on the GHG emission savings include shredding power requirements, the drying demands, and waste content of the waste wood stream. The collection distance also had a small effect on the results: in this study, doubling the collection distance to 120 km increased the total GHG emissions by 13%, or a 2.4% increase for every additional 10 km in distance.

12.3.3 Effect of Counterfactuals: Chipboard and Coal

The counterfactual system considers what would have occurred if the wood waste was not collected and used for pellets. This is shown in Table 12.3, where coal would have been used for 'energy generation' (for example) and chipboard would be produced from the waste wood. In the pellet system, the energy is now supplied by pellets and the chipboard is supplied from virgin round wood.

The GHG emissions for the forest operations involved in growing and harvesting round wood are shown in (see Table 12.3) [40]. In this case, using wood waste for pellets will require an additional emission of 21.1 kg CO_2 eq./t pellets, which could be added as a 'penalty' to the pellet emission factor. In this case, the emissions are increased to 134.06 kg CO_2 eq. and the GHG savings (compared to coal) are decreased to 90%. This assumes that 100% of the waste wood has been diverted away from chipboard manufacture, whereas in real market conditions the amount displaced would depend on competition for resources. It is possible for a

TABLE 12.3 GHG Emissions From the Counterfactual System Where Waste Wood is Used for Chipboard Manufacture and Energy is Provided From Coal

Parameter	Counterfactual System	Pellet System
Energy carrier	Coal	Wood pellets
Amount (t)	0.8	1
Amount (GJ)	16.5	16.5
GHG emissions from production and combustion (kg CO_2 eq.)	1507.6	134.06
Material	Chipboard from waste wood	Chipboard from primary forest
Amount (t)	1	1
Production/collection of feedstock	16.50	21.1
Total for system	1524.1	155.2
% Saving		90

The following assumes a density of 593–801 kg/m^3 for chipboard, a ratio of 1:1 for pellets to chipboard equivalents, and an energy content of 21.3 GJ/t for subbituminous coal.

small amount of lower quality wood (Grades B-C) to be recycled into chipboard [41], so long as the levels of contaminant concentrations of heavy metals are within the specified voluntary limits set in the particle board industry [8]. Therefore, if this could be used instead of virgin timber, then the penalty would be negligible, assuming that an increase in demand for wood waste leads to an increase in waste recovery rates.

12.3.4 Effect of Counterfactuals: Landfill

Fig. 12.3 shows the impact of the landfill counterfactual system, assuming different proportions in which wood breaks down to carbon dioxide and methane in landfill. In this case, the timescale of breakdown is not specified; however, this is assumed to occur during a reporting period of 100 years. The rate of decomposition will depend on climate, landfill composition, and particle size of the landfilled material. At an extreme case where there is zero breakdown of waste wood in landfill, all of the carbon in the landfilled wood would be stored indefinitely. Therefore, using the waste wood for pellet production means that this sequestration no longer occurs, and the wood pellet supply chain receives a 'penalty' for that forgone sequestration. In this case, the CLCA GHG emission savings are reduced to 84% (vs coal). Therefore, the avoided GHG emissions from burning coal have a greater effect on the GHG emission savings than forgone carbon sequestration in landfill.

Fig. 12.3 shows how the net CLCA GHG emissions savings steadily increase as the total breakdown proportion increases in the counterfactual system, and this is due to increasing emissions of methane from landfill, which is the major source of GHG emissions from landfilling wood. When attributing this to pellet system as an avoided emission, if 25%, 50%, 75%, and 100% of the wood breaks down in landfill, the GHG emission savings change from 121%, 157%, 194%, and 231%, respectively. As the breakdown proportion increases, there are

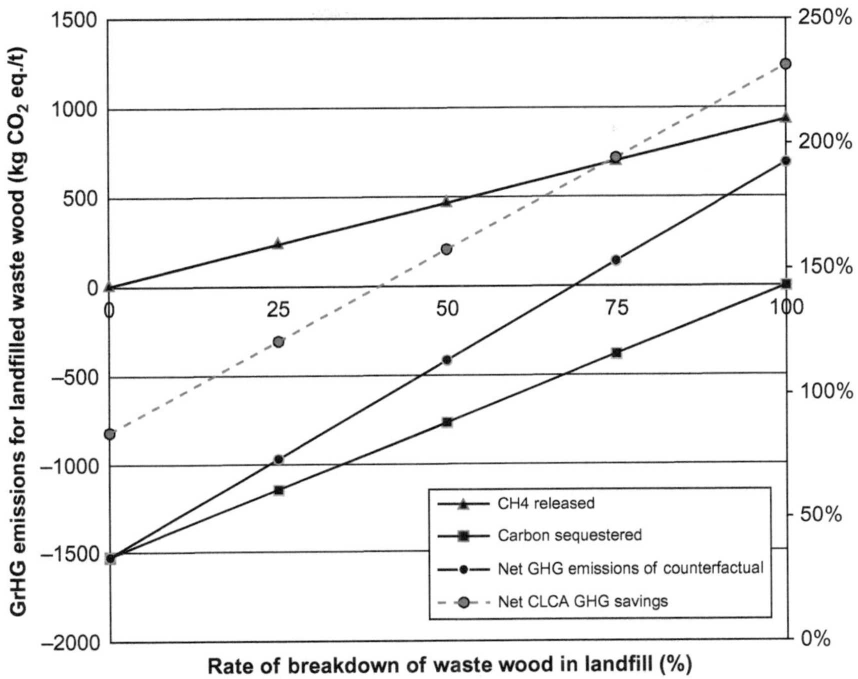

FIG. 12.3 The landfill counterfactual: shows how the breakdown rate of wood in landfill affects the methane emissions, sequestered carbon, and net GHG emission balance of utilising waste wood pellets to displace coal.

increasing credits for avoiding grid electricity through capturing the released methane, but this has a very small impact on the final results. The majority of the GHG emission impacts are attributed to the methane emissions from wood breakdown.

12.4 CONCLUSION

From this study, we have seen that the production of wood pellets from waste wood has a high (over 90%) GHG mitigation potential when used to displace coal. The most sensitive parameters affecting the emission savings are the electricity requirements for pelleting and auxiliary processes and the conversion efficiency of biomass compared to the fossil fuel reference. If the pellets are co-fired with coal, one must ensure that the power plant efficiency is not compromised.

The counterfactual systems considered two alternative fates for the waste wood used for wood pellet production: chipboard manufacture and landfill. Having to source virgin timber from forestlands has a small impact on the GHG emission savings from waste wood pellets. Diverting waste wood away from landfill will lead to GHG emission savings when the wood is used to displace coal. Although the foregone potential from carbon sequestration in landfill means that the savings could be reduced by, at most, to 84%, this depends on the proportion of wood that decays in landfill.

KEY POINTS

- Waste is an interesting resource and there are many potential conflicts with alternative markets that utilise them. Diverting wood away from landfill is beneficial to mitigating GHG emissions when used to displace fossil fuels.
- Producing wood pellets from waste wood releases around 111 kg CO_2 eq./MJ. Assuming this is combusted at the same conversion efficiency of coal, they have the potential to achieve GHG emission savings of over 90%. If the use of waste wood stimulates the use of more virgin timber for products such as chipboard, this incurs a small penalty on the waste wood pellets (−3% penalty). If wood was diverted from landfill, then in some circumstances where there is evidence to believe the wood does not effectively decay, carbon could be sequestered in landfill. In this case, the worst case scenario gives an 84% saving for waste wood pellets.
- Understanding how waste wood markets respond to changes in the values of waste wood is crucial for understanding the potential competition that biomass sector could have with the wood products and waste industries and help identify realistic counterfactual products. Understanding the breakdown of wood in landfill sites would help refine the GHG emission calculations for this aspect.

References

[1] WRAP. Realising the value of recovered wood: Market situation report. WRAP; 2011.

[2] Howes P, Beardmore L, Bates J, Heaume H, Godley A, Veryan E, et al. An assessment of the environmental impact of management options for waste wood. Didcot: AEA Technology, Oxford Economics and Forest Research; 2011.

[3] DEFRA. Waste wood as a biomass fuel. London: Department for Environment, Food and Rural Affairs; 2008.

[4] Tolvik. Briefing report: The UK waste wood market. Dursley: Tolvik Consulting Ltd; 2011.

[5] Beardmore L, Bates J, Edberg O, Godley A. An assessment of the environmental impact of management options for waste wood. In: Howes P, editor. Didco: AEA Technology; 2010.

[6] BSI. PAS 111:2012. Specification for the requirements and test methods for processing waste wood. London: British Standards Institute; 2012.

[7] EC. Directive 2000/76/EC of the European Parliament and of the Council of 4 December 2000 on the incineration of waste. Brussels: European Commission; 2000.

[8] CLA. Identification and assessment of types and levels of chemical contamination in wood wastes. In: WRAP, editor. Leamington Spa: CL Associates; 2006.

[9] Nikitas C, Boulos S, Shokar P, Leach B. Reference document on the status of wood waste arisings and management in the UK. Banbury: M.E.L Research Ltd. and WRAP; 2005.

[10] Aylott M, Higson A, Evans G, Hamer A, Mortimer N. What is the most appropriate LCA method for measuring greenhouse gas emissions from bioenergy? Biofuels Bioprod Biorefin 2011;5(2):122–4.

[11] Nuffield Council on B. Biofuels: Ethical Issues. London, UK; 2011.

[12] AEA Technology, Associates NE. Biomass Environmental Assessment Tool. 2.1 ed. Oxford and Sheffield; 2010.

[13] ECN. Phyllis2, database for biomass and waste, Energy Research Centre of the Netherlands; 2012.

[14] Salzmann R, Nussbaumer T. Fuel staging for NOx reduction in biomass combustion: experiments and modeling. Energy Fuels 2001;15(3):575–82.

[15] EPC. European Pellet Council: Handbook for the Certification of Wood Pellets for Heating Purposes. European Pellet Council; 2013.

[16] Alakangas E, Paju P. Wood pellets in Finland - technology, economy and market. OPET Report 5. Jyvaskyla: VTT Processes; 2002.

[17] Karkania V, Fanara E, Zabaniotou A. Review of sustainable biomass pellets production – a study for agricultural residues pellets' market in Greece. Renew Sustain Energy Rev 2012;16(3):1426–36.
[18] Li Y, Liu H. High-pressure densification of wood residues to form an upgraded fuel. Biomass Bioenergy 2000;19(3):177–86.
[19] Stelte W, Sanadi AR, Shang L, Holm JK, Ahrenfeldt J, Henriksen UB. Recent developments in biomass pelletisation: a review. Bioresources 2012;7(3):4451–90.
[20] Drax. Biomass Supply. Selby: Drax Group plc; 2015.
[21] Henderson C. Upgrading and efficiency improvement in coal-fired power plants. London: IEA Clean Coal Centre; 2013.
[22] Stephenson AL, MacKay DJC. Life cycle impacts of biomass electricity in 2020. London: Department of Climate Change; 2014.
[23] Brander M, Wylie C. The use of substitution in attributional life cycle assessment. GHG Measure Manag 2011;1(3-4):161–6.
[24] EC. Council Directive 1999/31/EC of 26 April 1999 on the landfill of waste. Brussels: The Council of the European Union; 1999.
[25] DfT DECC, DEFRA. UK biomass strategy. London: DEFRA; 2012.
[26] DEFRA. UK statistics on waste. London: Department of the Environment, Food and Rural Affairs; 2015.
[27] Jardine C, Boardman B, Osman A, Vowles J, Palmer J. Methane UK. Oxford, UK: Enviromental Change Institute, University of Oxford; 2001.
[28] Barlaz MA. Forest products decomposition in municipal solid waste landfills. Waste Manag 2006;26(4):321–33.
[29] Klimiuk E, Pokój T, Budzyński W, Dubis B. Theoretical and observed biogas production from plant biomass of different fibre contents. Bioresour Technol 2010;101(24):9527–35.
[30] Mann M, Spath P. A life cycle assessment of biomass cofiring in a coal-fired power plant. Clean Prod Processes 2001;3(2):81–91.
[31] Ximenes FA, Gardner WD, Cowie AL. The decomposition of wood products in landfills in Sydney Australia. Waste Manag 2008;28(11):2344–54.
[32] Micales JA, Skog KE. The decomposition of forest products in landfills. Int Biodeter Biodegr 1997;39(2–3):145–58.
[33] Park J-K, Lee W-J, Kim Y-J, Yoo M-H, Lee N-H. Assessment of greenhouse gas emission factors for wood wastes in a solid waste landfill. J Korea Soc Waste Manag 2014;31(8):811–9.
[34] Kuang X, Shankar TJ, Bi XT, Sokhansanj S, Lim CJ, Melin S. Characterization and kinetics study of off-gas emissions from stored wood pellets. Ann Occup Hyg 2008;52(8):675–83.
[35] Pipatti R, Alves JWS, Gao Q, Cabrera C, Mareckova K, Oonk H, et al. Biological treatment of waste. IPCC Guidelines for National Inventories. United Nations: Intergovernmental Panel on Climate Change; 2006.
[36] Brindley T. Landfill gas industry code of practice: the management of landfill gas. Codicote: Tokeda; 2012.
[37] Fisher K, Collins M, Aumonier S, Gregory B. Carbon balances and energy impacts of the management of UK wastes. Oxford: ERM; 2006.
[38] DECC. Estimated average calorific values of fuels 2014. In: Stastics N, editor. London: Department of Climate Change; 2015.
[39] Hagberg L, Sarnholm E, Gode J, Ekvall T, Rydeberg T. LCA calculations on Swedish wood pellets—according to the renewable energy directive. Stockholm: Swedish Environmental Research Institute; 2009.
[40] Puettmann M, Oneil E, Wilson J. Cradle to gate life cycle assessment of U.S. Particleboard Production. Corvallis: American Wood Council; 2013.
[41] TRADA. Enviros. Options and risk assessment for treated wood waste. In: WRAP, editors. Banbury: WRAP; 2005.

Further Reading

[1] Tillman DA. Biomass cofiring: the technology, the experience, the combustion consequences. Biomass Bioenergy 2000;19(6):365–84.

CHAPTER

13

Second-Generation Ethanol from Lignocellulose

*Sophie Parsons**, *Marcelle C. McManus**, *Caroline M. Taylor*†

*University of Bath, Bath, United Kingdom †EarthShift Global, Kittery, ME, United States

13.1 INTRODUCTION

Advanced processes, such as lignocellulosic ethanol production, have the potential for reduced life cycle GHG emissions when compared with conventional grain-based ethanol, and indeed fossil-based fuels. Sustainability assessment tools and energy analysis have been used to help identify impacts along supply and processing chains and reduce impact. This has helped target necessary and/or beneficial improvements and optimizations, providing valuable sustainability strategies. Key levers have emerged particularly in the use of crop and process residues, the selection of technologies, and the optimization of processes.

While there are both biological and thermochemical routes to lignocellulosic ethanol, the biological route has seen far more study and significantly more commercial effort. Likewise, biomass-derived hydrocarbon fuels ('drop-ins'), though their higher energy densities and full compatibility with existing infrastructure ensure their continued appeal and study, have been outpaced so far by lignocellulosic ethanol. As a result, its GHG performance is particularly scrutinised.

Focus on GHG emissions reduction is driven in part by various policy frameworks. By stimulating resource and processing efficiency improvements and encouraging circularisation of resource use that supresses demand, reducing GHG emissions can improve economic competitiveness as well.

The relative environmental impact of lignocellulosic ethanol compared with conventional fuels and/or grain ethanol is commonly assessed through life cycle assessment (LCA). LCA's ability to quantify resource and process improvements has contributed to its popularity as an important tool in the evaluation of bioenergy systems; the number of such studies has dramatically increased in the last decade [1]. These evaluate a variety of feedstocks and a more limited number of production technologies.

GHG emissions associated with lignocellulosic ethanol production can range between −1.1 kg CO_2 eq/km travelled to 0.28 kg CO_2 eq/km for E10, −1.15 Kg CO_2 eq/km to 0.79 kg

Greenhouse Gas Balances of Bioenergy Systems
https://doi.org/10.1016/B978-0-08-101036-5.00013-6

CO_2 eq/km for E85, and −1.25 Kg CO_2 eq/km to 0.84 kg CO_2 eq/km for E100, in comparison to 0.26 Kg CO_2 eq/km from conventional gasoline [2].

The majority of bioethanol production to date has been concentrated in the United States and Brazil, with Brazil producing over 30% of global ethanol, mostly consumed domestically to replace 40%–45% gasoline [3]. Brazil's sugarcane-derived ethanol is highly efficient, with crop yields between 80 and 85 t/ha and over 90% industrial sugar recovery; it is economically and energetically high performing, generally with a correspondingly low environmental impact compared to conventional gasoline and other bioethanol technologies [3]. It illustrates some clear examples of GHG reduction-driven development at scale and provides a glimpse of the potential for lignocellulosic fuels in integrated biorefineries to improve ethanol yield and reduce GHG emissions.

While this chapter focuses on GHG emissions, the wider bounds of sustainability assessment encompass many other factors; they may be well-aligned or require decisions about trade-offs. GHG emissions reduction and biodiversity conservation, for example, are influenced by common variables: species of crop grown, land brought into production, land management practises adopted, and ecological communities affected, *interalia*. Mixed prairie grasses are generally lower-yielding than energy grasses, but with a higher biodiversity index [4]; whether to use a higher-yielding crop to affect the smallest possible area or to use a lower-yielding system providing higher local biodiversity and habitat, but over a larger area, to achieve the same amount of fuel will always require a context-specific analysis [5].

This chapter summarises some of the LCA literature on the production of ethanol from lignocellulose, highlighting key areas where assessing the GHG impacts has or may lead to production and process efficiency improvements, lowering GHG emissions and supporting policy goals. We also address some challenges associated with assessment and touch on future outlooks.

13.2 KEY ASPECTS OF LIGNOCELLULOSIC ETHANOL PRODUCTION

Biofuel production from agricultural products is comprised of two main stages: feedstock production, including selection, agricultural production, and management, and harvest and delivery; and conversion to fuels, pre-processing, hydrolysis, fermentation, and recovery. Fuel blending and use in vehicle follow, to take the study through from well-to-wheel (WTW). Together, these are the system components considered in the life cycle of lignocellulosic ethanol. The magnitude of benefit to the overall system depends on the blend amount, for example ranging from nearly 100% reduction for E100 (100% ethanol) to <10% for E10 (10% ethanol, 90% gasoline) for wheat straw [6].

Most LCAs have addressed the two production stages, which have distinct challenges and needs for LCA and GHG emissions assessment that fall into several groups, relating to LCA application and practise, or to the uptake of LCA as an assessment methodology for bioenergy [1].

Decisions relating to either feedstock and processing technology can substantially affect total GHG calculations, making comparative analysis challenging and the use of LCA models to inform scaled process design potentially misleading. There remain unresolved issues relating to the LCA of lignocellulosic ethanol, among them are the challenges associated with feedstock-related emissions, system boundaries, attributional versus consequential assessment, and allocation methods [7]. Developments in both attributional and consequential LCA are needed to better inform LCA used in a design or regulatory context.

13.2.1 Issues in Lignocellulosic Ethanol LCA Modelling and Application

Consistency across studies is essential for LCA-based comparisons, and the large variations among published studies make such comparisons, particularly to guide policy, fraught. A meta-analysis of 100 different second-generation bioethanol pathways over the past four years [6] identified system boundary and functional unit as key to comparison, with co-products being a frequent differentiator, in keeping with other analyses [8].

The proliferation of models for calculating GHG emissions from bioenergy production and range of databases available makes integration of different modelling systems challenging. Here the overall complexity and level of detail in the analysis can mask areas where boundaries conflict even for relatively narrow studies. The wider implications for goal and scope setting are often not taken into account as data when a study is taken forward to inform wider decision-making, particularly in relation to policy [8].

To support policy, modelling advances in agricultural productivity are needed [8]. Projections of productivity changes over time are speculative, but important for long-term planning. For example, carbon payback times decrease 30%–50% when crop yields reach the 90th percentile in global yield, representing crop productivity increases and/or substantial management changes across global averages [9], in effect identifying the benefits of addressing the global yield gap. This also has important implications for the modelling of market-mediated effects such as indirect land use change.

Impacts attributed to biomass cultivation are highly sensitive to the way in which soil emissions (particularly Nitrogen as N_2O) are modelled. This in turn can significantly affect GHG emissions for the whole process as the global warming potential of N_2O is large. Other variables affecting soil carbon include the type of crop used in the assessment and the assumptions made around land management practises. Prior land use is the primary differentiator [10].

GHG emissions estimates' high sensitivity to net emissions of N_2O comes from fertiliser application and the decomposition of organic matter within soils, significant for both dedicated energy crops and crop residues. The many variables which can affect the fate of nitrogen within the environment and the conversion to N_2O make modelling emissions challenging. Coupled with N_2O's far higher global warming potential than CO_2, the Nitrogen contributes a large uncertainty associated with global warming impact [11].

Another GHG issue with bioenergy in general is that the carbon within bioenergy carriers is biogenic, meaning that the combustion of biomass releases the same amount of CO_2 as was captured by the growing plant. In contrast to the emission of CO_2 from fossil fuels which has been stored for millions of years, the time period within which the biogenic carbon is released and captured is short, and historically sequestered carbon is not released to the atmosphere. This point is key in modelling bioenergy as biogenic carbon emissions do not themselves increase overall CO_2 levels. Studies differ in how, and how clearly, this is handled.

The significance of co-products in differentiation reflects in part the importance of allocation. As discussed in Chapter 3, allocation in multi-product processes is a challenging issue in LCA. The ISO 14040/44 standard recommends avoiding allocation through sub-division or system expansion. If this is not possible, allocation can be done through physical relationships like energy, mass, or other variables such as economics. The use of different allocation procedures affects results [12]. Integrated biorefineries, in which multiple products are produced simultaneously, rely upon that allocation to apportion both costs and impacts. When an integrated biorefinery-producing ethanol, acetic acid, lactic acid, electricity, and heat from wheat straw, forest residue, poplar, and miscanthus is credited for avoided impacts, seven of 11 impact categories return environmental savings [13]. However, different allocation methods result in different rankings: in this case, using substitution favours ethanol from poplar, but allocation to products based on their market value (economic allocation; 81% of impact to ethanol, 17% to propanol, 2% butanol) favours ethanol from forest residue [13]. However, ultimately land-use change had a far greater effect on the overall results than allocation method.

13.3 FEEDSTOCK

Commonly assessed lignocellulosic ethanol feedstocks are agricultural residues (straw, stover, or cane leaves and tops) and forestry residues, forestry by-products, and dedicated energy crops (Miscanthus, switchgrass) that can be grown on marginal land, and bagasse, the lignocellulosic residue from sugarcane ethanol production [14].

When compared with both first-generation bioethanol and gasoline, producing advanced ethanol from a dedicated energy crop, biomass residue, or biomass waste can produce a fuel with lower GHG emissions. Residues, particularly, have large potential to mitigate emissions: in one large meta-analysis, residues could provide almost up to double the GHG reduction of dedicated crops on a distance-travelled basis (82%–91% for wheat straw or stover, versus 53%–93% for switchgrass or 52%–62% for wood, compared to gasoline)[6].

Significant impacts in the agricultural system component relate to land use, carbon sequestration, fertiliser use, the effect of removing biomass residues from the soil, and transportation of biomass to the ethanol production facility [15]. The relative importance of these differs depending on what type of feedstock is being used: energy crop, biomass residue, or biomass waste. The ideal energy crop has efficient solar energy conversion, requires minimal agrochemical inputs, has a high water use efficiency and low water requirements, and has a low moisture content at harvest [14]. Energy crops can have a positive impact on ecosystem services through providing carbon sequestration, biodiversity enhancement, and improvements in soil and water quality [11]. The negative environmental impacts associated with energy crops are generally based on management decisions, through land and water use, and impacts relating to fertiliser or pesticide use. Their magnitude depends on the type of crop and land being used, and the assumptions made around land management [11].

13.3.1 Feedstock Production and Agricultural Management Practise

Details of management practises are crucial. The scenario selected, and parameter and model uncertainty, plays a large part in constructing an LCA model for bioethanol. Both projected values and actual data cover a wide range. In stover, for example, actual harvest rate, farming energy for collection, and nitrogen fertiliser use can vary by an order of magnitude or more [7], influencing productivity, yield, and direct emissions. This in turn has a significant effect on feedstock production's contribution to total emissions.

Impacts of management decisions can change a system from net positive to net negative for GHG emissions: an estimate of management swing potential for Miscanthus including prior land use, harvest timing and technique, and fertilisation changes from 470 to −50 gCO_2eq/MJ, a swing of −520 gCO2eq/MJ [16]. Even changes in harvest timing are significant, for Miscanthus, e.g., highest yields are found in autumn immediately around senescence, but harvesting in early spring reduces GHG emissions the most [16]. Opportunities for improvements can be found across numerous sections of the chain; however, low energy intensity activities per hectare and high biomass yield crops reduced overall global warming potential relative to conventional gasoline in alfalfa stems, poplar, flax shives, hemp hurds, and by up to 145% (E85) for Ethiopian mustard [12].

Fertilisation is particularly challenging, both because of the mechanistic uncertainties and magnitude mentioned earlier and because real-world practise and outcomes vary so widely. For example, Miscanthus' response to fertilisation depends on stage of growth, showing no yield benefit during establishment, but increased emissions versus some response including longer stand yields after establishment [17], but response is strongly site-specific [18]. This lack of generalizability of application, uptake, and emissions across sites [18] and management regime [19] is consistent.

Likewise, harvest and collection equipment and practise can change the analysis dramatically. Automated and optimised cane harvest systems in Brazil had 40% lower GHG emissions when compared to conventional [20]. Some of this is driven by explicit GHG reduction policy that changes actual management practise: improvements in cane harvesting efficiency arising from the transition from manual to mechanised harvesting as a result of São Paulo state's ban on burning for manual harvesting on slopes up to 12% by 2021[1] reduced GHG emissions by ~46% at current efficiencies relative to traditional, manual harvest systems [21].

13.3.2 Feedstock Selection

Lignocellulosic feedstocks can be broadly categorised as energy crops, biomass residue, or biomass waste. These feedstocks are generally treated differently in LCA depending on whether they are a dedicated crop or a waste/residue. Crop residues may not have the emissions from agricultural production allocated to them, only including impacts from replacement fertiliser or energy use in harvesting the residue. However, a number of studies have chosen to allocate environmental impact based on economic value of the crop and crop residue [22–26], or mass and/or energy [2,26,27]. Feedstock allocation method has been shown to have an impact on total ethanol GHG emissions [7,23,25,26,28]. This shows that a consistent, transparent method is key; but most important is that we use these findings to help identify the areas where improvements can be made; requiring a true systems approach to feedstock

[1]São Paulo (2002) Lei Estadual no. 11.241/2002: Dispõe sobre a eliminação gradativa da queima da palha da cana-de-açúcar e dá providências correlatas. Available at: http://www.al.sp.gov.br/repositorio/legislacao/lei/2002/lei-11241-19.09.2002.html

and waste management. For example, moving to a perennial cropping system from more intensive row cropping decreases physical disturbance of soils through tillage and hence can increase the ability for soils to store carbon. Understanding these impacts and where they lie has been instrumental in encouraging change in the feedstocks used for bioenergy. For example, encouraging research and development into alternative feedstocks, and the reallocation of existing agricultural products leading to emphasis on residues use and high yield dedicated crops on marginal land.

13.3.3 Crop Residues

Corn stover is the most studied lignocellulosic feedstock for advanced ethanol production [7,14,26,27,29–32], although work around use of wheat straw, corn stover, and sugarcane harvesting residues (leaves and tops, 'trash') continues to grow as recognition of the footprint implications of fuel yield per land area grown has expanded.

The manner of stover production is significant, but needs to be considered alongside the wider system. The balance between stover removal and added nutrients to replace that in the harvested stover is one example. For two stover harvest scenarios, one with 40% stover harvested with nutrient replacement, the other 25% stover harvested but with no need for nutrient replacement, GHG emissions were lower with lower harvesting fractions, but including the additional transportation needed for the larger catchment area at lower harvest fraction raised emissions above the penalty for adding nutrients [31]. Similarly, the use of sugarcane straw and tops has complex effects on emissions, through both collection yield [33] and highly localised sensitivity to biomass remaining around the stalks [34]. Thus, while more efficient harvesting and collection systems can decrease emissions, it is challenging to generalise the yield impacts of utilising the resource.

Wheat straw has also been covered significantly using LCA, with complex results [2,22,23]. Consequential effects of straw removal are important in the bioethanol life cycle, contributing significantly to global warming, eutrophication, and ecotoxicity potentials [22]. In comparison with the energy crops miscanthus and poplar, Jeswani et al. [23] found wheat straw to be the least preferable option for all impact categories excluding land use (despite differences in impact values caused by allocation method). Cherubini and Ulgiati [35] compared corn stover and wheat straw biorefinery systems which showed a higher global warming (GWP 100), acidification, and eutrophication potential for corn stover, but lower corn stover impacts compared with wheat straw for other impact categories such as eco and human toxicity (these impacts are influenced by fertiliser application during cultivation).

Forestry residues for bioethanol are investigated through LCA, with some indications they could give significant reductions in GHG emission. Environmental impacts of pine residues for fuel have been assessed to be lower than the other forestry products pine, eucalyptus, unmanaged hard wood, and switchgrass Daystar, Reeb [36]. Additionally, differences in other impact categories may change the relative appeal among a set of feedstocks. While forest residues had lower acidification and eutrophication potential impacts when compared against wheat straw, poplar, and miscanthus, for example, they had higher photochemical smog impacts [13].

Sugarcane produces two lignocellulosic residues: straw and leaves, and bagasse, the partially pretreated residue from sugarcane ethanol production. About ~15 Mg/ha (dry mass) of straw is produced from cane, generally left in the field, [37] which can also be used for CHP generation or ethanol, but the impacts from collection are complex [38].

13.3.4 Energy Crops

Energy crops are typically woody or herbaceous plants, or grasses. Of these, high yield perennial crops such as willow, switchgrass, and miscanthus have been commonly assessed using LCA [13,36,39]. Wagner and Lewandowski [40] investigate the relevance of different impact categories to willow and miscanthus cultivation, highlighting several key factors that affect perennial crop performance: yield; the emissions factors and calculation models used; assumed CO_2 sequestration; ability to demonstrate through quantitative assessment the positive biodiversity, and soil quality impacts brought about by using perennial crops; indirect land use change caused by non-marginal land use; and the modelling of emissions from utilisation. Their study showed that perennial crops can contribute positively to reducing GHG emissions, but also identified cases, such as some ecotoxicity and human toxicity impact categories for willow and miscanthus where adverse effects may be found simultaneously with benefits in other categories. This highlights the need for a full spectrum of impact categories to be considered when evaluating perennial crops [40].

Rotation times for energy crops significantly alter the emissions estimates. Daystar et al. [36] found total fossil fuel use for pines, eucalyptus, and unmanaged hardwoods was lower than that of switchgrass. This is due to longer rotation lengths assumed in the model of 4 years for eucalyptus, 12 for pine and 50 for unmanaged hardwood. In comparison with switchgrass which is harvested annually and hence requires greater soil management than forestry products, Janssen et al. [24] found that using spruce wood chip was comparable to wheat straw, except for eutrophication potential which was lower for wood chips given that fertiliser is used in the cultivation of the straw.

13.3.5 Other

By-products or wastes from other processes can also be used to produce ethanol. The lack of competition with food crops makes them particularly appealing (although can compete with other uses such as soil conditioner or fertilisers which is an alternative use for the by-product or waste).

Sugarcane bagasse, particularly in Brazil, is a commonly discussed by-product in LCA [41–43]. Bagasse is already used to supply process heat and power, as well as for export to the grid, and is a major component in the overall sustainability of sugarcane ethanol. Watanabe, Chagas [44] also assessed first-generation ethanol from sugarcane against utilising bagasse alongside sugarcane. They found that optimising first-generation technology reduced GHG emissions to $14.6\,gCO_2$ eq/MJ. This can be reduced further to give a negative emissions value of $-10.5\,gCO_2$ eq/MJ when the system boundary is extended to include all surplus electricity generated from the bagasse. A combined first- and second-generation technology (producing less bioelectricity and utilising the bagasse for ethanol production) yields GHG emissions of $4.6\,g\ CO_2$-eq/MJ [44].

Ebner, Babbitt [45] utilises food waste to produce ethanol, compost, and an animal feed product. They use food waste from the industrial and retail food sector which is source-separated—in contrast with some MSW studies where total GHG emissions can vary depending on waste classification. Bozorgirad, Zhang [46] found that MSW incineration with electricity generation or MSW-to-ethanol was preferable to incineration of MSW with heat recovery. They also found MSW composition, along with facility location and size, to be factor affecting the environmental impact of the system evaluated.

13.4 CONVERSION TECHNOLOGY

Most lignocellulosic ethanol analyses focus on new facilities. However, incorporating lignocellulosic technologies with first-generation bioethanol production is attractive, particularly in the Brazilian context, and provides an intermediate picture of potential improvements.

13.4.1 Pre-Treatment

For all types of biofuel production, pre-treatment of biomass serves as a way to improve sugar yields for downstream processing and reduce the formation of by-products such as furfurals which inhibit downstream hydrolysis and fermentation [47]. Wiloso, Heijungs [15] show the majority of studies have chosen to assume an acid pre-treatment for different feedstock options. Borrion et al. 2012 [2] found this prehydrolysis stage to be the greatest contributor to environmental impact—particularly relating to the use of acid, steam, and electricity to treat the biomass. The production and pre-processing stages are also found by Murphy and Kendall [26] to correspond to 60% of total GHG emissions (using a corn stover feedstock and mechanical grinding pre-treatment).

Prasad et al. [47] reviewed the impacts for different pre-treatment options on corn stover to produce 1 kg fermentable sugar. Their findings showed that dilute acid (2% H_2SO_4; impact will vary with acid type) had the highest CO_2 emissions at 385 kg CO_2-equivalent, followed by 14.3 kg CO_2-equivalent (steam explosion), 9.23 kg CO_2-equivalent (organosolv), and 0.94 kg CO_2-equivalent (liquid hot water). Overall for eutrophication, acidification, and water depletion, dilute acid pre-treatment performed consistently badly against the other treatments, whereas organosolv performed consistently well (with the exception of aquatic eutrophication where it was the worst).

Kumar and Murthy [48] evaluated the sensitivity of assuming different pre-treatment types in bioethanol production. They assumed steam explosion would be used to substitute fossil energy requirement for the process, reducing overall energy demand through the use of process steam to produce electricity. This reduced total process energy demand (assuming steam explosion is used) to below that of gasoline and of corn ethanol. A more recent study by Pourbafrani, McKechnie [49] showed dilute acid to give the highest ethanol yield compared with ammonia fibre expansion and autohydrolysis. Zhu and Pan [50] estimate the mechanical pre-treatment of wood biomass to be 0.54–2.88 MJ/kg wood, with disk milling treatment achieving energy consumption values of 0.07–0.18 MJ/kg. Overall they found the lignin-preserving pre-treatment SPORL to be promising for woody biomass, particularly for softwood species. The treatment involves treating the wood chips with a solution of sodium bisulfite before mechanical milling. It is designed so that the lignin can be more easily recovered. Other physical pre-treatments include microwave treatment. Energy consumption for this can be between 5.5 and 169.2 kJ/g glucose, depending on microwave power output and exposure time [51].

The selection of pre-treatment option is dependent on feedstock, but these results show that the LCA analysis can show practitioners where to focus optimization efforts, and within technical boundaries, which pre-treatment options to select. Improving the combined efficiency of pre-treatment and conversion by about 25% against the NREL model [52] decreases GHG by about 25% and results in an energy balance of about 50% better [53]. This is complicated by the fact that these improvements in sugar release and conversion are generally accompanied by an uptick in fossil fuel use in the process as total ethanol produced increases, since decreasing final residue for co-firing decreases electricity yields [54].

13.4.2 Hydrolysis and Ethanol Conversion

Wiloso, Heijungs [14] found of the hydrolysis options investigated through LCA only 6 assumed acid hydrolysis, whereas 15 studies incorporated enzyme hydrolysis. Sebastiao et al. [55] found enzyme hydrolysis and neutralisation of $CaCO_3$ (neutralisation impact is specific to the pulp and paper sludge feedstock used which contained ~30% $CaCO_3$); they found they were able to decrease these impacts through co-fermentation of glucose and xylose.

Dunn, Mueller [56] present a useful paper on energy consumption and GHG emissions relating to enzyme and yeast manufacture for ethanol production. For cellulosic ethanol from switchgrass, they calculate contribution to GHG emissions to be 4.6 g CO_2eq./MJ ethanol for enzymes and 0.95 g CO_2eq./MJ ethanol for yeast. They found that, for cellulosic ethanol, the ingredients in cellulase production can contribute significantly to its energy intensity, meaning enzyme production should be included in the scope of any LCA analysis. Similar conclusions are made for yeast, where often yeast production is not included within the scope of assessment. This conclusion is also confirmed in earlier work by MacLean and Spatari [53] who point to exclusion of enzymes and hydrolysis chemicals on the grounds that they are present in insignificant amounts (as is commonly done when assessing first-generation ethanol production) as not holding true for second-generation technology. They determine the GHG emissions for switchgrass to be 3.3 (AFEX)–3.6 (Dilute acid) g CO_2eq./MJ ethanol.

Process configuration options addressed through LCA include: SHF (separate hydrolysis and fermentation), SSF (simultaneous saccharification and fermentation), SSCF (simultaneous saccharification and co-fermentation), and CBP (consolidated bioprocess). Typically, most LCA studies have assumed a co-fermentation approach or SSCF [15].

A common source of LCA data for the enzymatic hydrolysis and fermentation of lignocellulose is from the National Renewable Energy Laboratory (NREL). Their biochemical conversion presents a process design and economic analysis for corn stover under dilute acid pre-treatment and enzymatic hydrolysis [52]. Daystar, Venditti [39] adapt NREL's thermochemical gasification model [56] to model the impact of using different feedstock types on the ethanol conversion process. Overall, they found a low ash content and moisture content as correlating to a higher yield of ethanol, with alcohol yields from woody biomass significantly higher than non-woody biomass. It is important to note that biochemical paths may not always be preferable. Daystar, Reeb [36] compare their gasification model with biochemical conversion, finding environmental impact to be lower for thermochemical conversion.

LCA estimates of GHG emissions may also lead to new or unusual approaches to help lower GHG emissions and improve overall process efficiency. Janssen, Tillman [25], for

TABLE 13.1 Summary of GHG Emissions for Lignocellulosic Ethanol LCAs

Feedstock	GHG Emissions (gCO_2eq/MJ)	References
Petrol	86.4	[22]
Energy crop		
Wood	2–4	[6,62,63]
Switchgrass	−70 to 6	[6,64]
Miscanthus	11–15	[13,65]
Crop residue		
Agricultural waste	13–72	[6]
Forestry residue	11–23	[13,38]

example, use high gravity hydrolysis to process wheat straw and add polyethylene glycol (PEG) to increase the alcohol yield. Comparing several processing strategies, they find the addition of PEG lowers impacts from increased dry matter content, in turn decreasing environment impact as it is dominated by enzyme production, very sensitive to dry matter content (Table 13.1).

13.4.3 Integration

Increasingly, LCAs are evaluating integrated approaches producing ethanol alongside other products, primarily integrated biorefineries because the benefits are notable. For example, integrated production of bioethanol, bioenergy, and biochemicals from switchgrass decreased GHG emissions by 79% [57]. Ethanol from integrated biorefineries is found to be more environmental sustainable for most impact categories apart from when using wheat straw [23]. However, producing several co-products or utilising by-products for energy purposes creates issues around allocating impact among products, as discussed above.

Brazilian sugarcane ethanol is an established biorefinery, producing at least ethanol and electricity, and sometimes also sugar [58]. Scenarios of first- and second-generation ethanol in integrated biorefineries, producing ethanol, sugar, molasses, and electricity in Brazil, suggest that maximal ethanol yields may be economically favoured, while generating more products gives the lowest environmental impact [59]. That is, integration of first- and second-generation technology had a positive impact on GHG emissions, but was less economically desirable. Notably, biorefinery integration may reduce global warming impact, but show less reduction in other more localised impacts such as ecotoxicity, human toxicity, and freshwater eutrophication compared with conventional gasoline [59]. This is likely due to the nature of agricultural production compared with petroleum oil production, and the additional yield of fuel from integrating second-generation technology decreases overall environmental impact compared with conventional ethanol production.

Sugarcane ethanol systems deploy both integration and maturation of 1G and 1G2G systems, with large improvements in efficiency and decreases of impacts [60], and have long co-produced electricity from bagasse [61]. Rocha, Capez [3] evaluate two studies finding the Net Energy Ratio (NER) to be 12.37 on average—meaning the $KJ/MJ_{ethanol}$ out is far higher than energy. Da Silva et al. [41] evaluated a number of scenarios based on the use of sugarcane residues for electricity (for minimum tillage and no tillage cultivation) and the use of sugarcane residues for ethanol with reduced electricity output (for minimum tillage and no tillage cultivation), finding reduced environmental impact of sugarcane ethanol (compared with current good practise technique) across all impact categories assessed, with best performance from the no tillage scenarios. However, aside from reducing GHG emissions against gasoline, all scenarios were worse than gasoline for other impact categories, including human toxicity, ecotoxicity, acidification, and eutrophication, which was related to nematicide and pesticide use [41].

13.5 CONCLUSIONS AND PERSPECTIVES

Determining the environmental performance of bioethanol systems is complex. Key aspects such as GHG emissions, land use and biogenic carbon, eutrophication, and acidification impact are highly dependent on assumptions made about the system under investigation. The majority of LCA studies conducted to date have shown that lignocellulosic ethanol can have a positive environmental impact when compared with conventional gasoline. However, a lack of available data from commercial second-generation ethanol plant and the uncertainties in technology performance have made LCAs of lignocellulosic ethanol conversion process particularly difficult and challenging.

For dedicated energy crops, agricultural production is often the main contributor to environmental impact, with scale of impact sensitive to location and land use, crop productivity and harvest rate, fertiliser use, and land management. The extent to which agricultural production accounts for environmental impact often depends on the number of impact categories chosen and whether the study considers indirect land use. For crop residues and wastes, impact may be more closely linked with ethanol production, but this is dependent on how primary crop production impacts are allocated to the residue. Many LCAs point towards biological process pre-treatment and hydrolysis stages as being key contributors to environmental impact for crop residues.

The integrated biorefinery approach to ethanol production is increasingly discussed in LCA. This approach produces a number of products alongside ethanol and often utilises first- and second-generation technology. The production of bioelectricity and it's substitution into the process has been shown to reduce environmental impact, but the production of different products again raises challenges in allocation of impact between them.

Despite the complexities and uncertainties associated with lignocellulosic ethanol LCA, the method remains a valuable tool in assessing the environmental impacts relating to second-generation biofuel technology. Future LCA investigation must seek to improve modelling of agricultural systems and land use and develop more dynamic and consequential modelling for feedstocks and co-products. Model sensitivity to aspects such as allocation should be discussed as part of analysis.

KEY POINTS

- GHG and LCA have identified areas for overall improvement in the production of bioethanol
- Leveraging LCA and GHG reduction-based improvement opportunities requires transparent, robust studies
- System boundaries and allocation method are key to understanding result consistency
- The use of lignocellulosic waste material enables reduction in feedstock impacts
- Adoption of precision agricultural management techniques and improved residue recovery technologies provides key opportunities for improvement in the feedstock production phase
- Technology-feedstock matching and increased sugar to fuel yields provide key opportunities for improvement in the conversion phase
- Integrated biorefineries allow capturing additional efficiencies and outputs and provide key opportunities across the production chain.

Acknowledgements

The authors would like to thank their funders for this work: the BBSRC UK/US partnership grant (BB/JO20184/1), the EPSRC Supergen Bioenergy Hub (EP/JO17302/1), and the Integrated energy efficient microwave and unique fermentation processes for pilot scale production of high value chemicals from lignocellulosic waste (EP/N013522/1).

References

[1] McManus M, Taylor C. The changing nature of Life Cycle Assessment. Biomass Bioenergy 2015;82:13–26.
[2] Borrion AL, McManus MC, Hammond GP. Environmental life cycle assessment of bioethanol production from wheat straw. Biomass Bioenergy 2012;47:9–19.
[3] Rocha MH, et al. Life cycle assessment (LCA) for biofuels in Brazilian conditions: a meta-analysis. Renew Sust Energ Rev 2014;37:435–59.
[4] Dauber J, et al. Yield-biodiversity trade-off in patchy fields of Miscanthus×giganteus. GCB Bioenergy 2015;7(3):455–67.
[5] Meyer MA, Chand T, Priess JA. Comparing bioenergy production sites in the southeastern US regarding ecosystem service supply and demand. PLoS One 2015;10(3). e0116336.
[6] Morales M, et al. Life cycle assessment of lignocellulosic bioethanol: environmental impacts and energy balence. Renew Sust Energ Rev 2015;42:1349–61.
[7] Gerbrandt K, et al. Life cycle assessment of lignocellulosic ethanol: a review of key factors and methods affecting calculated GHG emissions and energy use. Curr Opin Biotechnol 2016;38:63–70.
[8] McManus MC, et al. Challenge clusters facing LCA in environmental decision-making—what we can learn from biofuels. Int J Life Cycle Assess 2015;20(10):1399–414.
[9] Gibbs HK, et al. Carbon payback times for crop-based biofuel expansion in the tropics: the effects of changing yield and technology. Environ Res Lett 2008;3(3). 034001.
[10] Duval BD, et al. Predicting greenhouse gas emissions and soil carbon from changing pasture to an energy crop. PLoS One 2013;8(8). e72019.
[11] Cherubini F, et al. Energy and greenhouse gas-based LCA of biofuel and bioenergy systems: key issues, ranges and recommendations. Resour Conserv Recycl 2009;53:434–47.
[12] Gonzalez-Garcia S, et al. Comparative life cycle assessment of ethanol production from fast-growing wood crops (black locust, eucalyptus and poplar). Biomass Bioenergy 2012;39:378–88.
[13] Falano T, Jeswani HK, Azapagic A. Assessing the environmental sustainability of ethanol from integrated biorefineries. Biotechnol J 2014;9:753–65.
[14] Somerville C, Youngs H, Taylor C, Davis SC, Long SP. Feedstocks for lignocellulosic biofuels. Science 2010;329(5993):790–2.

[15] Wiloso EI, Heijungs R, de Snoo GR. LCA of second generation bioethanol: a review and some issues to be resolved for good LCA practice. Renew Sust Energ Rev 2012;16:5295–308.
[16] Davis SC, et al. Management swing potential for bioenergy crops. GCB Bioenergy 2013;5(6):623–38.
[17] Davis MP, et al. Effect of nitrogen addition on Miscanthus × giganteus yield, nitrogen losses, and soil organic matter across five sites. GCB Bioenergy 2015;7(6):1222–31.
[18] Arundale RA, et al. Nitrogen fertilization does significantly increase yields of stands of miscanthus × giganteus and panicum virgatum in multiyear trials in illinois. Bioenergy Res 2013;7(1):408–16.
[19] Maughan M, et al. Miscanthus × giganteus productivity: the effects of management in different environments. GCB Bioenergy 2012;4(3):253–65.
[20] Chagas MF, et al. Environmental and economic impacts of different sugarcane production systems in the ethanol biorefinery. Biofuels Bioprod Biorefin 2016;10(1):89–106.
[21] Galdos M, et al. Trends in global warming and human health impacts related to Brazilian sugarcane ethanol production considering black carbon emissions. Appl Energy 2013;104:576–82.
[22] Wang L, Littlewood J, Murphy RJ. Environmental sustainability of bioethanol production from wheat straw in the UK. Renew Sust Energ Rev 2013;28:715–25.
[23] Jeswani HK, Falano T, Azapagic A. Life cycle environmental sustainability of lignocellulosic ethanol produced in integrated thermo-chemical biorefineries. Biofuels Bioprod Biorefin 2015;9:661–76.
[24] Janssen M, Xiros C, Tillman A-M. Life cycle impacts of ethanol production from spruce wood chips under high-gravity conditions. Biotechnol Biofuels 2016;9(53). https://doi.org/10.1186/s13068-016-0468-3.
[25] Janssen M, et al. Influence of high gravity process conditions on the environmental impact of ethanol production from wheat straw. Bioresour Technol 2014;173:148–58.
[26] Murphy CW, Kendall A. Life cycle inventory development for corn and stover production systems under different allocation methods. Biomass Bioenergy 2013;58:67–75.
[27] Canter CE, et al. Policy implications of allocation methods in the life cycle analysis of integrated corn and corn stover ethanol production. Bioenergy Res 2016;9:77–87.
[28] McKechnie J, et al. Exploring impacts of process technology development and regional factors on life cycle greenhouse gas emissions of corn stover ethanol. Renew Energy 2015;76:726–34.
[29] Daylan B, Ciliz N. Life cycle assessment and environmental life cycle costing analysis of lignocellulosic bioethanol as an alternative transport fuel. Renew Energy 2016;89:578–87.
[30] Dunn JB, et al. Land-use change and greenhouse gas emissions from corn and cellulosic ethanol. Biotechnol Biofuels 2013;6(51). https://doi.org/10.1186/1754-6834-6-51.
[31] Zhang Y, Kendall A. Life cycle performance of cellulosic ethanol and corn ethanol from a retrofitted dry mill corn ethanol plant. Bioenergy Res 2017;10(1):183–98.
[32] Zhao L, Ou X, Chang S. Life cycle greenhouse gas emissions and energy use of bioethanol produced from corn stover in China: current perspectives and future prospectives. Energy 2016;115:303–13.
[33] Cardoso TF, et al. A vertical integration simplified model for straw recovery as feedstock in sugarcane biorefineries. Biomass Bioenergy 2015;81:216–23.
[34] Fortes C, et al. Contribution of nitrogen from sugarcane harvest residues and urea for crop nutrition. Sci Agrár 2013;70(5):313–20.
[35] Cherubini F, Ulgiati S. Crop residues as raw materials for biorefinery systems - a LCA case study. Appl Energy 2010;87:47–57.
[36] Daystar J, et al. Environmental life cycle impacts of cellulosic ethanol in the Southern U.S produced from loblolly pine, eucalyptus, unmanaged hardwoods, forest residues, and switchgrass using a thermochemical conversion pathway. Fuel Process Technol 2015;138:164–74.
[37] Dias MOS, et al. Integrated versus stand-alone second generation ethanol production from sugarcane bagasse and trash. Bioresour Technol 2012;103(1):152–61.
[38] Carvalho JLN, et al. Agronomic and environmental implications of sugarcane straw removal: a major review. GCB Bioenergy 2016;.
[39] Daystar J, et al. Impact of feedstock composition on alcohol yields and greenhouse gas emissions from the NREL thermochemical conversion process. Bioresources 2013;8(4):5261–78.
[40] Wagner M, Lewandowski I. Relevance of environmental impact categories for perennial biomass production. GCB Bioenergy 2017;9(1):215–28.
[41] da Silva C, et al. Long-term prospects for the environmental profile of advanced sugar cane ethanol. Environ Sci Technol 2014;48:12394–402.

[42] Palma-Rojas S, Caldeira-Pires A, Nogueira JM. Environmental and economic hybrid life cycle assessment of bagasse-derived ethanol produced in Brazil. Int J Life Cycle Assess 2015;22(3):317–27.
[43] Souza SP, Gopal AR, Seabra J. Life cycle assessment of biofuels from an integrated Brazilian algae-sugarcane biorefinery. Energy 2015;81:373–81.
[44] Watanabe M, et al. Hybrid input-output life cycle assessment of first and second generation ethanol production technologies in Brazil. J Ind Ecol 2016;20(4):764–74.
[45] Ebner J, et al. Life cycle greenhouse gas (GHG) impacts of a novel process for converting food waste to ethanol and co-products. Appl Energy 2014;130:86–93.
[46] Bozorgirad MA, et al. Environmental impact and cost assessment of incineration and ethanol production as municipal solid waste management strategies. Int J Life Cycle Assess 2013;18(8):1502–12.
[47] Prasad A, et al. Life cycle assessment of lignocellulosic biomass pretreatment methods in biofuels production. Int J Life Cycle Assess 2016;21:44–50.
[48] Kumar D, Murthy GS. Life cycle assessment of energy and GHG emissions during ethanol production from grass straws using various pretreatment processes. Int J Life Cycle Assess 2012;17(4):388–401.
[49] Pourbafrani M, et al. Impact of pre-treatment technologies and co-products on greenhouse gas emissions and energy use of lignocellulosic ethanol production. J Clean Prod 2014;78:104–11.
[50] Zhu JY, Pan XJ. Woody biomass pretreatment for cellulosic ethanol production: Technology and energy consumption evaluation. Bioresour Technol 2010;101:4992–5002.
[51] Lu X, et al. Microwave pretreatment of rape straw for bioethanol production: focus on energy efficiency. Bioresour Technol 2011;102:7937–40.
[52] Humbird D, et al. Process design and economics for biochemical conversion of lignocellulosic biomass to ethanol. In: NREL, editors. CO, USA: US Department of Energy; 2011.
[53] MacLean HL, Spatari S. The contribution of enzymes and process chemicals to the life cycle of ethanol. Environ Res Lett 2009;4:1–10.
[54] Mu D, et al. Comparative life cycle assessment of lignocellulosic ethanol production: biochemical versus thermochemical conversion. Environ Manag 2010;46(4):565–78.
[55] Sebastiao D, et al. Life cycle assessment of advanced bioethanol production from pulp and paper sludge. Bioresour Technol 2016;208:100–9.
[56] Dunn JB, et al. Energy consumption and greenhouse gas emissions from enzyme and yeast manufacture for corn and cellulosic ethanol production. Biotechnol Lett 2012;34:2259–63.
[57] Cherubini F, Jungmeier G. LCA of a biorefinery concept producing b ioethanol, bioenergy, and chemicals from switchgrass. Int J Life Cycle Assess 2009;15(1):53–66.
[58] Conab. Perfil do Setor do Açúcar e do Álcool no Brasil—Volume 5—safra 2011/2012. In: Monitoramento agrícola – Cana-de-açúcar. Brasilia: Companhia Nacional de Abastecimento; 2013 [p. 88].
[59] Gnansounou E, Vaskan P, Pachón ER. Comparative techno-economic assessment and LCA of selected integrated sugarcane-based biorefineries. Bioresour Technol 2015;196:364–75.
[60] Cavalett O, et al. Environmental impacts of technology learning curve for cellulosic ethanol in Brazil. Ind Crops Prod; 2017. https://doi.org/10.1186/s13068-017-0722-3.
[61] Khatiwada D, et al. Optimizing ethanol and bioelectricity production in sugarcane biorefineries in Brazil. Renew Energy 2016;85:371–86.
[62] Dutta A, Philips SD. Thermochemical ethanol via direct gasification and mixed alcohol synthesis of lignocellulosic biomass. CO, USA: NREL; 2009.
[63] Woods J, Bauen A. Technology status review and carb\on abatement potential of renewable transport fuels in the UK. London, UK: Department of Trade and Industry, Imperial College London; 2003.
[64] Huang J. Life cycle analysis of hybrid poplar trees for cellulosic ethanol. Mechanical Engineering. MA, USA: Massachusetts Institute of Technology; 2007.
[65] Groode TA, Heywood JB. Ethanol: a look ahead. MA, USA: Massachusetts Institute of Technology: Laboratory for Energy and the Environment; 2007.

Further Reading

[1] Hudiburg TW, et al. Bioenergy crop greenhouse gas mitigation potential under a range of management practices. GCB Bioenergy 2015;7(2):366–74.
[2] Scown CD, et al. Life cycle greenhouse gas implications of US national scenarios for cellulosic ethanol production. Environ Res Lett 2012;7(7):1–10.

CHAPTER

14

Electricity From North American Forest Residues

Mirjam Röder

The University of Manchester, Manchester, United Kingdom

14.1 INTRODUCTION

This chapter discusses the greenhouse gas (GHG) impact from electricity generated in the United Kingdom from wood pellets imported from North America (South East United States). Electricity from biomass—bioelectricity—is considered as a feasible option to reduce GHGs in the United Kingdom's energy sector to support the country's climate change targets reducing its overall emission by 80% by 2050 compared to 1990 levels [1–3].

With the political support of large-scale bioelectricity [4,5], the United Kingdom has become the largest importer of forest residue-based wood pellets globally. About 6.5 million tons of wood pellets are imported annually which is about one third of the global trade [6]. Research suggests that GHG reductions of 60%–90% are possible compared to a coal reference [7–14].

To identify sources of uncertainties, this work evaluates key parameters for GHG emissions of forest-based wood pellet supply chains applying lifecycle assessment (LCA) with sensitivity analysis. The focus of interest is imported wood pellets from the South-East United States used in large-scale electricity generation in the United Kingdom. The aim is to evaluate emissions and GHG impacts of imported wood pellets for an improved understanding of biomass supply chains processes to maximise the emission reduction potential.

Two baseline options are investigated:

(A) Large-scale bioelectricity from forest residues
(B) Large-scale bioelectricity from sawmill residues

The term forest residue is used for different types of woody biomass and can cover different types of materials such as sawmill residues (e.g. off-cuts, sawdust, chips, bark) and parts of trees left in the forest after timber harvest (e.g. tree tops, branches, bark, thinnings) [7,8,15]; and sometimes also untreated waste wood from the wood processing industry (furniture, construction, joinery). Sawmill residues, in particular sawdust, are currently the main

Greenhouse Gas Balances of Bioenergy Systems
https://doi.org/10.1016/B978-0-08-101036-5.00014-8

raw material for producing wood pellets [16]. With an increasing demand for pellets, forest residues like tree branches, tree tops, bark, and early thinnings are increasingly used [7,15]. This can already lead to emission variations between these two supply chains as different processing steps are involved as described in the following section.

14.2 WOOD PELLET TO ELECTRICITY SUPPLY CHAIN

In the United States, about 7.4 Mt of wood pellets were produced in 2015 [6]. The South-East United States is one of the major regions for wood pellet production from sawmill residues globally [7,8,11,17,18]. Wood pellets are mainly produced from sawmill and to a small extent forest residues. The most common tree species in this region are mixed loblolly pine (Pinus taeda) and shortleaf pine (Pinus echinata) stand, which make up about 25% of the forest area [19,20]. The forests are privately managed focussing on high-quality timber production and are regulated by sustainability policies [7,21,22]. These are planted and managed following medium-intensive cultivation including land preparation, planting of new seedlings, fertilising, herbicide application, and thinning [19,20,23] and are then harvested by clear cut after 45 years [20,24]. The trees are harvested in a whole-tree logging system [25,26], with a minimum of 35% of the residues suggested to be left in the forest to maintain nutrient levels [8,15,21,22,27–29]. The tress are then cut to length and left for seasoning at the landing site. After that the logs are taken to the sawmill where they are processed.

As described above, two different residue options are considered:

(A) In the case of the forest residue option, thinnings and residues are collected at the landing site and left for natural drying for about 3 months. During that time, the moisture content reduces from 50% to about 30% [25]. After that the residues are chipped at the landing site and transported to the pellet mill for processing [25].

(B) In the case of the sawmill residue option, the residues are collected and stored at the sawmill where the logs are processed. Then the sawdust and cut-offs get transported to a local pellet mill. A moisture content of 30% is considered [25].

At the pellet mill, the chipped forest residues and the sawmill residues are produced to pellets. During this process, the biomass is ground, dried, and pressed into pellets. The pellets will then have a moisture content of 10% and a lower heating value of 16.5 MJ kg^{-1} [30]. The pellets are transported by rail to a port and shipped in a bulk vessel to the United Kingdom. At the United Kingdom port, the pellets get loaded to rail, which takes the pellets to the power plant where they are then combusted in a dedicated biomass boiler to generate energy [31]. The considered combustion unit is a 670 MW boiler with a load factor of 80% and efficiency of 40% [32]. That means that, to generate 1 kWh of electricity, 0.54 kg wood pellets are required [31]. The biomass (as dry matter) required at forest stand level for sawmill residues was significantly higher (3.1 kg kWh^{-1}) compared to the forest residue system (0.610 kg kWh^{-1}) as a tree has to be grown as a whole considering all the parts, which are related to sawdust, such as the saw log. The proportions of the different sawmill products (lumber, timber, chips, sawdust, shavings, and bark) were considered to identify the share of each product and then price allocation was applied to evaluate the GHG emissions. The system boundaries and processes included in the LCA of these two supply chains are provided in Fig. 4.1.

14.3 LIFE CYCLE ASSESSMENT AND SENSITIVITY ANALYSIS

The goal and scope of this research was to identify and evaluate supply chain processes that are subject to significant emission uncertainties that could limit real GHG reductions. An attributional LCA was conducted to evaluate the emissions of the two wood pellet options including the processes of forest establishment, management, harvesting feedstock processing, transport, and energy conversion. The supply chains were selected according to existing pathways of large-scale electricity production in the United Kingdom from biomass. The functional unit (FU) of the LCA was 1 kWh of generated electricity. The full inventory is provided by Röder [31]. The impact category highlighted in this chapter is global warming potential (GWP). The results account for carbon dioxide (CO_2), nitrous oxide (N_2O), and methane (CH_4) expressed as CO_2 equivalent (eq) with a 100-year time horizon [33] and a final unit of measurement of g CO_2eq kWh^{-1}. The fossil fuel reference considered is coal-based electricity generated in a similar application unit as the wood pellets.

LCA methodology suggests to avoid allocation and extend the system boundaries to include impacts of co-products. Forest production was therefore included in the emission boundaries and price allocation was applied.

Sensitivity analysis was applied for the parameters with potentially a significant impact on the overall GHG emissions. The following parameters were tested:

(a) Fuel for producing heat to dry feedstock during pelleting: During the pelleting process, the biomass feedstock has to be dried to reduce the feedstock's moisture content from about 30% to 10%. Many pellet mills use biomass to generate this heat (e.g. bark or sawdust) and, with this, the pelleting process contributes about 30% to the total GHG emissions of the selected supply chain. Nevertheless, diesel or electricity is also common to generate this heat. While drying with biomass-based heat is considered as the baseline option, variations with diesel and electricity-based heat were explored additionally.

(b) Dry matter losses: At various supply change stages, dry matter is lost as the feedstock is handled, processed, or stored [10,34–36]. Research by others showed that these losses can range from 7% [17] up to 20% [9], with a high degree of uncertainty in the literature [34–37]. For the baseline options, total losses of 9.5% and 10.5% for forest residues and sawmill residues, respectively, were assumed. For the sensitivity analysis, dry matter losses of 20.5% and 21.5% for forest residues and sawmill residues, respectively, were assumed following values provided in the literature [10,34–40]. Values for losses varied in particular for supply chain processes like chipping, during storage, transport, and handling [31].

(c) Storage of feedstock: during storage of wood chips and sawdust, CH_4 is released from micro-bacterial processes [41,42]. Considering GHG emissions from wood chips and sawdust piles found by others [43,44], sensitivity analysis has been conducted considering an estimated CH_4 emission of $24\,g\,d^{-1}\,m^{-3}$ from wood chips and sawdust when stored outside to allow natural ventilation [45]. For the baseline, just-in-time delivery with no storage has been considered and variations of 1–3 months storage at the different potential stages of the supply chain within the system boundaries have been included in the sensitivity analysis.

14.4 LCA RESULTS AS GREENHOUSE GAS EMISSIONS FROM BIOELECTRICITY

14.4.1 Baseline Emission and Emission Savings

Fig. 14.1 presents the emission profile of the baseline options for forest and sawmill residues and the variations when drying heat is generated from fossil fuels. The overall GHG emissions producing 1 kWh of electricity are 132 g CO_2eq kWh^{-1} for forest and 140 g CO_2eq kWh^{-1} for sawmill residues. The main emission contributors are transport with 39% for forest and 36% for sawmill residues. The main share of transport emissions are released from the transatlantic shipping of the pellets from the United States to the United Kingdom, while all land transport along the supply chain takes up about one fifth of the overall transport emissions. Transport emissions are followed by emissions from pelleting with 31% for forest and 29% for sawmill residues. The emission from the conversion process caused 18% of the total emissions of forest residues and 17% for sawmill residues, while the wood production (forest establishment, management, and harvesting) created 12% and about 18% of forest and sawmill residues, respectively. Additional to this, direct emissions from soil from fertiliser application are considered, which are relatively low (less than 1%) for forest residues, but as high as 5% for sawmill residues as at forest side level a larger amount of wood per unit of bioenergy feedstock is required.

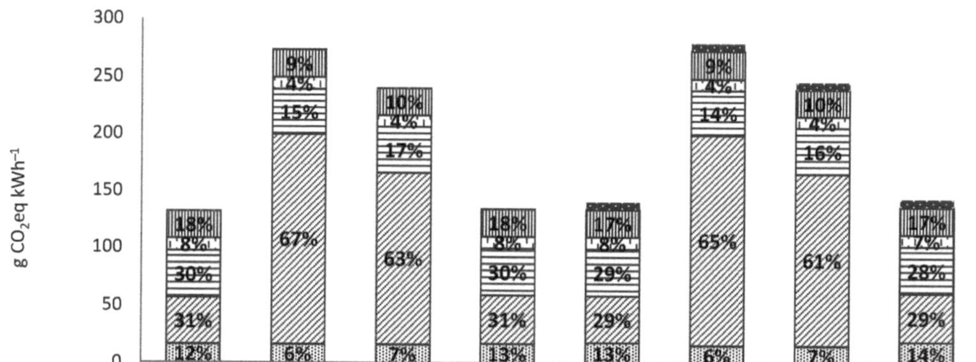

	Forest residues				Sawmill residues			
	Drying with biomass	Drying with diesel	Drying with electricity	High losses	Drying with biomass	Drying with diesel	Drying with electricity	High losses
Direct soil emissions	0.1	0.1	0.1	0.2	6.5	6.4	6.4	7.1
Energy conversion	23.9	23.9	23.9	23.9	23.9	23.9	23.9	23.9
Transport (land)	10.9	10.2	10.2	10.4	10.7	10.0	10.0	10.4
Transport (trans atlantic)	40.1	40.1	40.1	40.1	40.1	40.1	40.1	40.1
Pellet processing	40.7	182.8	149.7	42.4	40.7	182.8	149.7	42.4
Wood production	16.2	16.2	16.2	17.2	18.3	16.1	16.1	19.8

FIG. 14.1 Supply chain emissions of forest and sawmill residues. Baseline options and sensitivities for alternative drying fuels and including losses in terms of volume.

The emissions from generating electricity from coal vary between about 750 and 1000 g CO_2eq kWh^{-1} [46–48]. In 1990, the United Kingdom average electricity mix had an emission intensity of 718 and fell to 412 g CO_2eq kWh^{-1} in 2016. As coal reference system for this calculation, 752 g CO_2eq kWh^{-1} [49] is considered as this was the average carbon intensity of the comparable power station generating electricity. This means that the two baseline cases achieve emission savings of 82% for forest and 81% for sawmill residues.

14.4.2 Sensitivity Analysis

The emissions from the different drying fuel options are included in Fig. 14.1. When drying with diesel or electricity, the supply chain emissions double in case of diesel (to 271 g CO_2eq kWh^{-1} for forest residues and to 279 g CO_2eq kWh^{-1} for sawmill residues) and increase by about 88% (to 238 g CO_2eq kWh^{-1} for forest residues and to 246 g CO_2eq kWh^{-1} for sawmill residues) in case of electricity. Additionally, the proportion of emission from pelleting increases to over 60% of the total emission profile. This also means that the savings compared to the coal reference case decrease for forest residues to 64% when drying with diesel and 68% when using electricity and for sawmill residues to 63% and 67%, respectively.

A higher amount of losses (increase from about 10% to about 20% losses at supply chain processes like chipping, storage, handling, and transport) increased the total emissions compared to the two baseline scenarios from 132 to 135 g CO_2eq kWh^{-1} for forest residues and from 140 to 144 g CO_2eq kWh^{-1} for sawmill residues (see Fig. 14.1). This means an increase of total supply chain emissions of 2.3%–3% for forest and sawmill residues, respectively.

In addition to variations in drying fuels and increased losses, storage emissions in the form of CH_4 have been considered for 1–3 month of storage at different supply chain stages. Fig. 14.2 presents the findings in monthly steps, showing that the supply chain emissions increase compared to the baseline options from 132 to 317 g CO_2eq kWh^{-1} within the first month and to a total of 670 g CO_2eq kWh^{-1} when storing for 3 months in the case of forest residues. Storage-derived CH_4 emissions were considered even higher in the sawmill residue option, as it was assumed that sawmill residues have an additional storage stage prior to arriving at the pellet mill and are more prone to micro-bacterial activities than piled and unchipped forest residues [35,36,38,40,42,50]. With this, the supply chain emissions from sawmill residues increased from 140 to 495 g CO_2eq kWh^{-1} after 1 month and reach 859 g CO_2eq kWh^{-1} after 3 months, latest exceeding the GHG emissions from the coal reference case. The detailed analysis is provided by Röder et al. [31] Nevertheless, work by others [41] has indicated that micro-bacterial process is not linear during storage and linked to a number of various emission and dry matter losses through the flux of CO_2, CO, CH_4, and also N_2O under given storage and environmental conditions. Nevertheless, so far no real emission data is available on the storage sawdust and forest wood chips at different lengths and given weather and seasonal conditions.

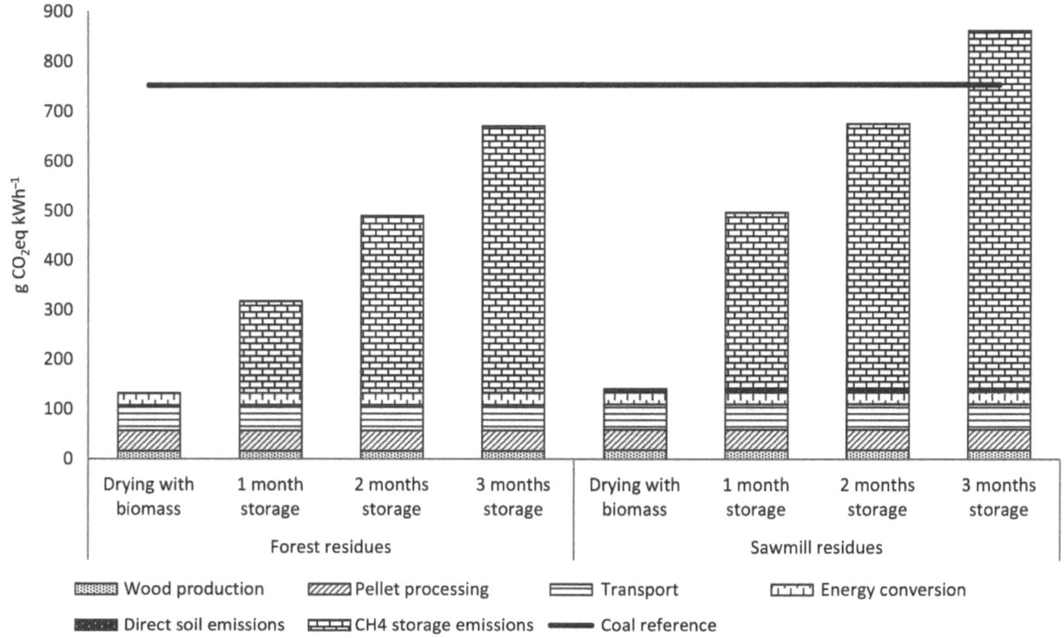

FIG. 14.2 Supply chain emissions including CH_4 emissions during feedstock storage in monthly steps and compared to the fossil fuel reference.

14.5 CONCLUSIONS

The results show that bioelectricity from forest and sawmill residues can achieve emission savings of over 80% compared to coal-based electricity. Nevertheless, the management of feedstock and practices of certain supply chain processes can diminish these savings significantly, and in some cases, even lead to GHG emissions higher than the fossil fuel reference. Using renewable energy for processing, and as the example has shown, biomass for supply chain processes such as feedstock drying should be encouraged as it can increase the emission reduction potential. This is, in particular, a valid option if the biomass for drying low-quality feedstocks like bark and forest residues is currently not used and left unutilised at landing sites [28,51] Micro-bacterial activities in stored wood chips and sawdust could be minimised if biomass is stored in a form less prone to decomposition and fungi, e.g. unprocessed wood and forest residues, as pellets, torrified, in dry and weather-sheltered environment [31,41]. Nevertheless, reliable data on such micro-bacterial activities is limited and, to date, sawdust and wood chip storage of forest residues under given conditions and over the considered length of time is lacking. Dry matter losses during processing did not significantly increase the supply chain emissions, but need to be considered from an economic and sustainability point of view. If dry matter is lost in the forest during storage and chipping, this might not be considered as a loss but could potentially have sustainability benefits [28,29].

This research presents a snapshot of two specific supply chains and applying LCA allowed evaluation of these by producing a set of robust numbers. Nevertheless, the sensitivity analysis indicates that there can be various sources of uncertainties, which may differ when considering a different setting and characteristics along the supply chain. This should be taken into account when considering the emission reduction potential of biomass-based electricity.

References

[1] DECC. UK bioenergy strategy. London: Department of Energy & Climate Change; 2012. p. 86.

[2] DECC. UK renewable energy roadmap update 2012. London: DECC; 2012.

[3] CCC. The fifth carbon budget—the next step towards a low-carbon economy. London: CCC; 2015.

[4] DECC. Interaction between the RHI and renewables obligation. London: DECC; 2013.

[5] DECC. The renewables obligation (RO). London: DECC; 2014.

[6] FAO. Forest statistics; 2016. Available from: http://www.fao.org/forestry/statistics/80938@180724/en/ [cited 10.02.2017].

[7] Aebiom. Forest sustainability and carbon balance of EU importation of North American forest biomass for bioenergy production. September 2013. p. 77.

[8] Colnes A, Doshi K, Emick H, Evans A, Perschel R, Robards T, et al. Biomass supply and carbon accounting for southeastern forests. Montpelier, VT: Biomass Energy Resource Center; 2012. February, p. 132.

[9] Forsberg G. Biomass energy transport: analysis of bioenergy transport chains using life cycle inventory method. Biomass Bioenergy 2000;19:17–30.

[10] Hamelinck CN, Suurs RAA, Faaij APC. International bioenergy transport costs and energy balance. Biomass Bioenergy 2005;29:114–34.

[11] Puneet D, Madhu K, Robert B, Adrian G. Potential greenhouse gas benefits of transatlantic wood pellet trade. Environ Res Lett 2014;9. 024007.

[12] Repo A, Känkänen R, Tuovinen J-P, Antikainen R, Tuomi M, Vanhala P, et al. Forest bioenergy climate impact can be improved by allocating forest residue removal. GCB Bioenergy 2012;4:202–12.

[13] Smith P, Bustamante M, Ahammad H, Clark H, Dong H, Elsiddig EA, et al. Agriculture, forestry and other land use (AFOLU). In: Edenhofer O, Pichs-Madruga R, Sokona Y, Farahani E, Kadner S, Seyboth K, Adler A, Baum I, Brunner S, Eickemeier P, Kriemann B, Savolainen J, Schlömer S, von Stechow C, Zwickel T, Minx JC, editors. Climate change 2014: mitigation of climate change. Contribution of working group III to the fifth assessment report of the intergovernmental panel on climate change. Cambridge, New York: Cambridge University Press; 2014.

[14] Laganière J, Paré D, Thiffault E, Bernier PY. Range and uncertainties in estimating delays in greenhouse gas mitigation potential of forest bioenergy sourced from Canadian forests. GCB Bioenergy 2015;.

[15] Kuiper L, Oldenburger J. The harvest of forest residues in Europe. Postbus 253, 6700 AG Wageningen. Biomassa-upstream stuurgroep. February 2006; p. 6. D15a.

[16] Cocchi M, Nikolaisen L, Goh CS, Junginger M, Henimo J, Bradley D, et al. Global wood pellet industry market and trade study. Trade IBTSB; 2011; p. 190.

[17] Jonker JGG, Junginger M, Faaij A. Carbon payback period and carbon offset parity point of wood pellet production in the South-eastern United States. GCB Bioenergy 2013;6:371–89.

[18] ITC. Trade map—4401 fuel wood; 2013. Available from: http://www.trademap.org/Index.aspx [updated 10.03.2015; cited 01.11.2013].

[19] Oswalt SN, Smith WB, Miles PD, Pugh SA. Forest resources of the United States, 2012. Washington, DC. U.S. Department of Agriculture. Forest Service. October 2014; p. 228. WO-91.

[20] Smith WB, Miles PD, Perry CH, Pugh SA. Forest Resources of the United States, 2007. Washington, DC. U.S. Department of Agriculture. Forest Service. 2009; p. 336. WO-78.

[21] McKechnie J, Colombo S, Chen J, Mabee W, MacLean HL. Forest bioenergy or forest carbon? Assessing trade-offs in greenhouse gas mitigation with wood-based fuels. Environ Sci Technol 2010;45:789–95.

[22] Walker T, Cardellichio P, Colnes A, Gunn J, Kittler B, Perschel B, et al. Biomass sustainability and carbon policy. Boston, Massachusetts. Manomet Center for Conservation Sciences. June 2010; p. 182. NCI-2010-03.

[23] FAO. Global forest resources assessment 2010. Rome: FAO Forestry Department; 2010. p. 75. FRA2010/223.

[24] Matthews R, Mortimer N, Mackie E, Hatto C, Evans A, Mwabonje O, et al. Carbon impacts of using biomass in bioenergy and other sectors: forests. Forest Research. North Energy Associates Limited. 23/04/2012; p. 178. URN 12D/085.

[25] Leinonen A. Harvesting technology of forest residues for fuel in the USA and Finland. Helsinki. VTT Tiedotteita. Espoo 2004; p. 132 + app. 10 p. Research Notes 2229.
[26] Mälkki H, Virtanen Y. Selected emissions and efficiencies of energy systems based on logging and sawmill residues. Biomass Bioenergy 2003;24:321–7.
[27] Gan J, Smith CT. Co-benefits of utilizing logging residues for bioenergy production: the case for East Texas, USA. Biomass Bioenergy 2007;31:623–30.
[28] Paré D, Thiffault E. Nutrient budgets in forests under increased biomass harvesting scenarios. Curr Forestry Rep 2016;2:81–91.
[29] Thiffault E, Béchard A, Paré D, Allen D. Recovery rate of harvest residues for bioenergy in boreal and temperate forests: a review. Wiley Interdiscip Rev Energy Environ 2015;4:429–51.
[30] Verhoest C, Ryckmans Y. Industrial wood pellets report. Brussels: European Commission; 2012. p. 30.
[31] Röder M, Whittaker C, Thornley P. How certain are greenhouse gas reductions from bioenergy? Life cycle assessment and uncertainty analysis of wood pellet-to-electricity supply chains from forest residues. Biomass Bioenergy 2015;79:50–63.
[32] Drax. Annual report and accounts 2013. Selby: Drax Group plc; 2014. p. 156.
[33] IPCC. 2006 IPCC guidelines for national greenhouse gas inventories. In: Eggleston HS, Buendia L, Miwa K, Ngara T, Tanabe K, editors. Prepared by the national greenhouse gas inventories programme. Japan: IGES; 2006.
[34] Filbakk T, Høibø OA, Dibdiakova J, Nurmi J. Modelling moisture content and dry matter loss during storage of logging residues for energy. Scand J For Res 2011;26:267–77.
[35] Nurmi J. The storage of logging residue for fuel. Biomass Bioenergy 1999;17:41–7.
[36] Thörnqvist T. Drying and storage of forest residues for energy production. Biomass 1985;7:125–34.
[37] Lehtikangas P. Storage effects on pelletised sawdust, logging residues and bark. Biomass Bioenergy 2000;19:287–93.
[38] Jirjis R. Storage and drying of wood fuel. Biomass Bioenergy 1995;9:181–90.
[39] Sikkema R, Junginger M, Pichler W, Hayes S, Faaij APC. The international logistics of wood pellets for heating and power production in Europe: costs, energy-input and greenhouse gas balances of pellet consumption in Italy, Sweden and the Netherlands. Biofuels Bioprod Biorefin 2010;4:132–53.
[40] Eriksson LN, Gustavsson L. Biofuels from stumps and small roundwood—costs and CO2 benefits. Biomass Bioenergy 2008;32:897–902.
[41] Whittaker C, Yates NE, Powers SJ, Misselbrook T, Shield I. Dry matter losses and greenhouse gas emissions from outside storage of short rotation coppice willow chip. Bioenergy Res 2015;1–15.
[42] Alakoski E, Jämsén M, Agar D, Tampio E, Wihersaari M. From wood pellets to wood chips, risks of degradation and emissions from the storage of woody biomass—a short review. Renew Sust Energ Rev 2016;54:376–83.
[43] Ferrero F, Malow M, Noll M. Temperature and gas evolution during large scale outside storage of wood chips. Eur J Wood Wood Prod 2011;69:587–95.
[44] Pier PA, Kelly JM. Measured and estimated methane and carbon dioxide emissions from sawdust waste in the Tennessee Valley under alternative management strategies. Bioresour Technol 1997;61:213–20.
[45] Wihersaari M. Evaluation of greenhouse gas emission risks from storage of wood residue. Biomass Bioenergy 2005;28:444–53.
[46] Parliament. Carbon footprint of electricity generation. 2011.
[47] DECC. UK biomass and biogas carbon calculator. Available from: http://www.ofgem.gov.uk/Sustainability/Environment/RenewablObl/FuelledStations/bbcc/Pages/bbcc.aspx; 2012.
[48] AEA, Associates NE. Biomass environmental assessment tool (BEAT2). 2011; Available from: http://www.biomassenergycentre.org.uk/portal/page?_pageid=74,153193&_dad=portal&_schema=PORTAL.
[49] Drax. Personal communication, Drax Group plc. 2014.
[50] Eriksson L, Gustavsson L. Comparative analysis of wood chips and bundles—costs, carbon dioxide emissions, dry-matter losses and allergic reactions. Biomass Bioenergy 2010;34:82–90.
[51] Thiffault E, Smith CT, Junginger M, Berndes G. Mobilisation of forest bioenergy in the boreal and temperate biomes. London: Academic Press; 2016.

CHAPTER

15

Agricultural Lessons

Ian Shield

Rothamsted Research, Harpenden, United Kingdom

15.1 SUMMARY

Ruminant livestocks are the biggest single emitters of greenhouse gases on farms. Reductions in animal numbers open the opportunity for land use change to energy crops, with a big impact on climate change through reduced emissions from ruminants and fossil carbon substitution.

All land use change must be carefully managed; the soil is a valuable sink for carbon, but that carbon is vulnerable to oxidation due to soil aeration. Zero and very minimal cultivation technologies are key to protect that sink during land use change.

Nitrogen fertiliser is responsible for N_2O emissions from soils and increasing vulnerability of soil organic carbon to oxidation. Ammonia-based fertilisers are energy- and greenhouse gas-intensive to manufacture. Yet nitrogen is the one crop input that is most readily justified in terms of crop response. Robust data is required for all crops with which to predict response to N fertiliser in terms of finances, energy, and greenhouse gases. In many cases, the balance is positive.

Despite being a lesser source of greenhouse gases, the direct combustion of fuels in agriculture is important to consider so that supply chains are as efficient as possible. As for nitrogen fertiliser, the optimum management of a crop may not be immediately obvious and robust data sets are crucial to making the best decisions. For example, densification of an energy crop may not be justified until transport distances become quite large, determining how large requires access to a suitable data set.

15.2 THE NATIONAL PICTURE

Bioenergy derived from agriculture must tackle the big issues faced by farmers more generally. In 2015, as per the latest statistics available, agriculture was responsible for 49.1 Mt CO_2e or ~34% of the UK greenhouse gas emissions. The greatest contributors were CH_4 from enteric fermentation by ruminants (particularly cattle) and their manures (46%) and

https://doi.org/10.1016/B978-0-08-101036-5.00015-X

N_2O emissions from soils associated with nitrogen fertiliser applications (29%). Reduced emissions from agriculture have only made a small contribution to overall United Kingdom emission reductions 1990–2015. This can be attributed to increased efficiency in the beef and dairy sector, leading to fewer animals and improved crop and fertiliser management. Adding bioenergy to farming systems has great potential to further those emission reductions, albeit requiring a more comprehensive accounting system that crosses sectors and time spans and credits agriculture for savings elsewhere.

15.3 LIVESTOCK

Declining animal numbers impact bioenergy through the supply of slurry and dairy and abattoir waste which is widely used in anaerobic digesters and in potentially making land available for dedicated energy crop production. Although much of that land is not well-suited to arable cropping, it may be very productive when forage grasses (e.g. for anaerobic digestion) or woody crops (e.g. for more general thermochemical conversion) are grown. Availability of land avoids issues of agricultural intensification associated with maintaining overall production on a farm or in a region and known as indirect land use change. Achieving any land use change without increasing greenhouse gas emissions is a challenge that needs to be addressed. At the extremes of land quality (steep slopes, waterlogged, rock outcrops), the only realistic method of harvesting the sun's energy is by grazing ruminants and so they maintain a role in United Kingdom agriculture and research into reducing their impact must remain a priority.

15.4 LAND USE AND LAND USE CHANGE

This has been one of the most controversial aspects of greenhouse gas inventories and the bioenergy debate. In some cases quite justifiably so, but in the case of the United Kingdom and other similar land use regions, it is of little consequence and in some cases a positive outcome. Overall in the United Kingdom, land use has been a nett sink for greenhouse gases in recent years. Photosynthesis remains the dominant mechanism by which carbon dioxide (the principle greenhouse gas) may be removed from the atmosphere. Marine and inland water organisms contribute somewhat to the process, but land-based plants are the principle actors in fixing atmospheric carbon. Storage of that carbon in soils is a potentially significant sink, but largely it is in flux and annual reporting does not fully account for the temporal scale of those fluxes.

Ley Arable farming remains a common land use in mixed farming regions. On an annual basis, the 5.8 Mt CO_2e emitted by the conversion of land to cropland (arable) is balanced by the −4.6 Mt CO_2e fixed by conversion of land to grass leys, as a proportion of the arable land is sown to grass and a proportion of the grass is destroyed to return to arable. These two land use changes may not be perfectly in balance as market forces affect the mix on a farm. Additionally, the oxidation of soil organic carbon fixed during the ley period is more rapid during arable cultivations than the fixation of new soil organic carbon by newly sown leys which take time to establish. This is reflected in the −4.9 Mt CO_2e fixed by grassland remaining as grassland, whereas crop land remaining as such is an emitter. Soils have a diminishing ability to retain soil organic carbon as it accumulates, but this operates on a decadal scale. The addition of nitrogen fertiliser to arable and grass crops increases the vulnerability of the soil organic carbon to breakdown and oxidation.

Ironically, sowing of those leys is driven by the needs of ruminant livestock, the greatest emitters of greenhouse gases in United Kingdom agriculture. Switching from ruminant livestock production to energy crop production on suitable grassland could achieve the double win situation of reduced emissions from the animals and of fossil carbon substitution in the energy sector. It should be possible to achieve such land use change without cultivations that aerate the soil and drive soil organic carbon oxidation. Unfortunately, early attempts at planting SRC without cultivation were inconclusive.

Energy crops such as miscanthus and SRC are similar to grass leys in that they facilitate the fixation of soil organic carbon, but that carbon remains vulnerable to subsequent oxidation. Theory suggested a productive life span of ~20 years for both crops, at which point re-planting or land use change would occur. However, market conditions have seen crops removed early and recent research around those crops being removed has demonstrated the vulnerability of the carbon fixed once the crop was destroyed. Drewer, Howard, and McNamara (Chapter 6, this Volume) describe the gas fluxes from energy crops in detail.

Research continues into forms of carbon in soils, identifying and increasing those recalcitrant forms resistant to oxidation. The spatial location of carbon can also be an important factor in retention within the soil system; position within aggregates in the soil matrix and depth in the profile are important factors. Management practises of such planting over winter cover crops on arable land introduce the opportunity to seek particularly deep rooting plants that place carbon below the cultivation depth where oxygen is less freely available. Deep rooting grasses have been developed for environmental engineering (soil stabilisation, flood alleviation, etc.) as well as forage that have the potential to add carbon to depth in soils.

The greatest contribution to negative emissions from land use comes from forestry where not only is soil carbon accumulated on a long timescale (60–80 year), but the standing crop of wood represents a significant carbon stock. These carbon stocks are also vulnerable to flux as forests reach maturity and are harvested or degenerate. Most re-planting methods involve some oxidation of soil organic carbon. Wood used for construction may extend the carbon cycle to a scale of hundreds of years, but ultimately the carbon will be oxidised back to atmospheric CO_2. The area of woodland in the United Kingdom has increased by 400,000 ha since 2004 and a significant area of woodland planted prior to that time matured to maximal productivity in recent years.

All agricultural and forestry land use involves a significant component of carbon flux; however, all operate on a timescale that is realistic to impose sustainable management upon. Fossil fuels are not in flux on a timescale that can be subjected to sustainable management.

15.5 SOILS AND FERTILISERS

Emissions due to natural soil processes are not large; when nitrogen fertilisers are added to soils, the greenhouse gas emissions rise considerably. Inorganic fertilisers, manures, and slurries and direct additions from grazing animals increase emissions to 14.1 Mt CO_2e as N_2O

from arable and grassland soils. Manures and slurries in storage prior to application to the land also emit a mixture of CH_4 and N_2O equivalent to 4.9 Mt CO_2. By comparison, human waste water treatment emits 4.1 Mt CO_2e and is relevant to agriculture, and bioenergy in particular, as a proportion is utilised as a fertiliser. Anaerobic digestate will behave similarly. Drewer, Howard, and McNamara (Chapter 6, this Volume) describe the gas fluxes from bioenergy crops in detail.

However, before nitrogen fertiliser is applied to the soil, it has caused the emission of large quantities of greenhouse gases in the industrial process of manufacture. The majority of nitrogen fertilisers are based upon ammonia produced via the Haber Bosch process in which CH_4 (mainly fossil gas) is used as the source of hydrogen, with the carbon ultimately being emitted as CO_2. The hydrogen is reacted with nitrogen from the air, but nitrogen being relatively unreactive, this requires high temperatures (450°C, energy-intensive) and pressures (maximum 18 MPa, large embodied energy in pressure vessels plus an energy requirement in operation) in the presence of a catalyst. So little ammonia is produced in the United Kingdom that it appears as a small emission on the United Kingdom inventory. Most N fertiliser is imported.

Of all inputs to crop production, the application of nitrogen fertiliser is the most responsive in terms of yield and return on investment. Most crop types have financial response curves produced for different nitrogen fertiliser and crop prices. More recently, alternative response curves have been produced based upon energy and greenhouse gas responses. The yield response to application of nitrogen fertiliser is often so strong that an energy crop may easily fix more energy into harvestable biomass than is expended in manufacturing and distributing the fertiliser. For wheat, the optimum nitrogen application for energy balance was below the optimum for financial balance, resulting in recommendations that did not optimise energy balance. In miscanthus (where response to N fertiliser is not always apparent), the financial and energy optima were co-incidental. Similarly, the CO_2 fixed in that biomass may be greater than the CO_2 equivalence of manufacturing and distributing fertiliser and emissions from the land. However, those data have not yet accounted for emissions from the land at the end of crop life when cultivations may release soil organic carbon and with it N_2O (see Section 15.4). Such data has only just become available in sufficient quantity and quality in recent years.

15.6 ENERGY CONSUMPTION

The consumption of fuels (primarily mineral oil) by internal combustion engines and some heat demand (gas and oil) in agriculture is a relatively minor source of greenhouse gases at the national scale (4.5 Mt CO_2e), yet is worthy of consideration when assessing supply chains, especially for bioenergy. Short rotation coppice (SRC) is a very low emission crop to grow until the harvest and post-harvest processes are considered. As Whittaker and Shield (Chapter 7, this volume) show, multiple harvesting options are available for energy crops such as SRC and there is much debate regarding the more efficient. Assuming that the product required by the end user is wood chip at a certain specification, the crop may be chipped during cutting in the field or harvested as larger billets or whole stems. Chips may be force-dried or left in heaps to ambient dry; both involve financial, energy, and greenhouse gas costs. A fuel source is required to force dry and significant losses occur during ambient drying. Larger harvest products, billets and whole stems, dry under ambient conditions with only ~5% of the losses from chips. However, they require a secondary handling and chipping when dry, which has been shown to require greater energy in the case of willow. A recent study, thought to be the first to analyse financial, energy, and greenhouse gas data from the same supply chain, found that the difference between the three options (chips, chips force dried, and whole stems) was small. Therefore, small changes in any one component part of the process could shift the balance. Further studies may reveal the probable variance that may be encountered in each component. One factor that could shift the balance greatly was the emission of methane from the heap of ambient drying chip. Methane was not detected at all heaps, and its production is thought to be related to consolidation of the heap during construction. Whole stem heaps produced no methane, and as stated, lost very little dry matter.

Densification is another finance, energy, and greenhouse gas-intensive process. It is the nature of bioenergy feedstocks that they are dispersed and often long distances from the point of final use. Transport is not only expensive, but also energy and greenhouse gas intensive. As with harvest management of SRC, densification and transport efficiency is a delicate balance. This was demonstrated by a study for a miscanthus pellet plant in France, where it was shown that short distance transport (up to 15 km from the plant) could be efficiently achieved by chipping the miscanthus as it was harvested, a low density product for transport but that was convenient for the plant to receive. Longer distance transport required densification using a baler. Chipping and then baling to achieve high-density bales did not produce a robust product that withstood handling.

Pelleting and briquetting can achieve greater product densities than balers, but do not always result in greater overall transport efficiencies when the densification process is also accounted for (see Whittaker and Shield, Chapter 7, this volume). In the United Kingdom, transport distances needed to be several hundred kilometres before pelletization was justified. A recent consideration of the sugar beet industry pointed out that access to good quality roads increased the efficiency of transport over small rural roads. Other forms of biomass for energy are often dispersed and in less accessible locations than sugar beet. Often, pellets are formed to produce a more uniform product and to mill the material to a fine particle size in specialised mills in cases where end-user facilities cannot produce such specification.

Adding greater certainty to these aspects of crop management is a crucial part of demonstrating the best practise for bioenergy and ensuring that the greenhouse gas savings and climate impact are maximised.

CHAPTER

16

Engineering Lessons—Using Engineering Design to Minimise GHG Emissions From Bioenergy Production

Paul Adams

University of Bath, Bath, United Kingdom

16.1 BACKGROUND

In previous chapters of this book, we have described different issues that need to be considered when assessing the greenhouse gas (GHG) emissions that arise from bioenergy systems. Several of these factors relate to the natural environment and ecological systems, waste management, resource efficiency, land management, and agricultural systems; however, there are also many aspects where engineering can play a key role in minimising GHG emissions. Engineering can be described as the art or science of making practical application ('know how') of the knowledge ('know what') of pure sciences, as physics, biology, or chemistry, to invent, design, develop, construct, improve, and maintain physical products and systems [1]. This means that engineering is critical in almost all aspects of bioenergy and its sustainable development and implementation.

Historically, engineering is applied in all stages of the bioenergy life cycle, for example:

- Cultivation and harvesting—agricultural machinery and agronomic inputs have been engineered to maximise yields, increase the speed of production, and reduce the physical human effort involved in agriculture and forestry;
- Transportation—sophisticated transport methods have been developed through the engineering of tractors, trucks, trains, boats, and mass freight transport;
- Processing—engineering has developed techniques that allow biomass to be processed to maximise the energy potential, reduce logistic costs, and improve the efficiency of subsequent conversion processes;
- Conversion—there are many different conversion routes that can be taken to produce useful energy from biomass as described in Chapter 8. Engineering is applied to all of

https://doi.org/10.1016/B978-0-08-101036-5.00016-1

these whether it is biological, physical, or chemical, so that bioenergy systems can be adapted to match the biomass input and required energy output, while maintaining high levels of availability and reducing maintenance and replacement costs;
- End-use—whether it is direct heating, electricity generation, or biofuel use in vehicles, engineering has been applied to maximise the efficiency of biomass or biofuel to useful energy.

It could be argued that traditionally engineering design has been used to develop the different stages of bioenergy production in order to maximise energy production and economic return. Therefore, the focus to date has to a large extent been on increasing feedstock resource and improving the biomass conversion efficiency. There are also regulations to adhere to such as emissions to air; therefore, mitigating and managing emissions from combustion to meet regulatory limits is a focus. Engineering has been applied to control emissions from combustion often with a focus on air quality emission sources such as particulates, carbon monoxide, and NOx. There is perhaps a risk therefore that non-CO_2 GHG emissions from combustion are often overlooked in LCA as the common assumption is that biomass combustion is carbon- neutral. Methane (CH_4), black carbon, and nitrous oxide (N_2O) are formed in combustion processes under certain conditions, all of which can have greenhouse gas impacts [2,3]. The amount of CH_4 emitted depends on the efficiency (completeness) of the combustion process. The amount of N_2O emitted depends on the nitrogen content of the fuel and the combustion temperature.

Strategies implemented to increase bioenergy production are pursued by biomass suppliers and developers to maximise energy output for economic reasons. This has meant that, without limits on GHG emissions from supply chains farmers may, for example, cultivate crops to maximise the energy yield (and therefore economic return) per hectare. In these cases, it may be more economical to apply high amounts of Nitrogen-based fertiliser in order to increase crop yields. With this comes increased GHG emissions but from an economic view point it is worthwhile so as to increase the benefit from the available Government incentives. Similarly, while wastes and residues often have lower GHG emissions, they are also commonly harder to obtain, difficult to secure finance, and can be more expensive to operate. This again means that the economics may push operators towards feedstocks with higher GHG emissions. For the engineer, this presents a challenge that is to continuously look at ways to design, innovate, and improve the different aspects of bioenergy systems so that GHG emissions and the costs of production can be unceasingly reduced. The key is to ensure that the available supply of biomass is utilised efficiently using appropriate conversion processes to achieve the maximum possible energy output without compromising GHG emissions.

The focus of this chapter is therefore to assess some of the engineering challenges at each stage of production in different bioenergy value chains. A number of different examples are described with existing solutions that have been developed in order to reduce GHG emissions. Due to the diversity of biomass supply and conversion routes, this chapter attempts to consider different biomass sources and conversion technologies, but does not aim to address all of these. A selection of engineering lessons that have been learnt from current and historical bioenergy operators and research is presented. Finally, some of the future challenges and constraints are evaluated so as to inform engineering research and development in the context of existing and future bioenergy systems. The chapter is structured so as to follow the main stages of production with common bioenergy production systems.

16.2 BIOMASS SUPPLY—CULTIVATION, HARVESTING, COLLECTION, AND TRANSPORTATION

16.2.1 Emissions From Fertiliser Production and Use

For annual crops, the largest GHG emissions arise from the production of fertiliser and soil emissions from fertiliser application (see Fig. 16.1). It can be observed that Nitrogen applied has the largest influence on total emissions per hectare of land cultivated. These are therefore the 'hot-spots' and have been a primary focus for engineering development. The key lesson is that the type of fertiliser used and how this is applied is important in terms of GHG emissions. For example, emission factors collated by Fertilizers Europe for different types of fertiliser are presented in Table 16.1 and demonstrate how fertilisers can be selected to ensure minimum upstream GHG emissions from production [4]. The timing of application is also important as the more Nitrogen that is taken up by the crop, the lower the potential for N_2O emissions. The management approach is crucial as traditionally farmers will apply N at a rate to ensure the optimal economic return on N-application in relation to yield. This thinking is now starting to develop to consider optimal GHG balance as well as the economics. This change to fertiliser application strategy has primarily been driven by biomass sustainability criteria as emissions from feedstock supply are critical in overall GHG balance.

In terms of agricultural engineering, the focus is to develop low-emission fertilisers and fuels; develop fuel-efficient machinery and equipment; advance methods that can lower fuel use in field operations; apply fertilisers in a way that minimises application emissions;

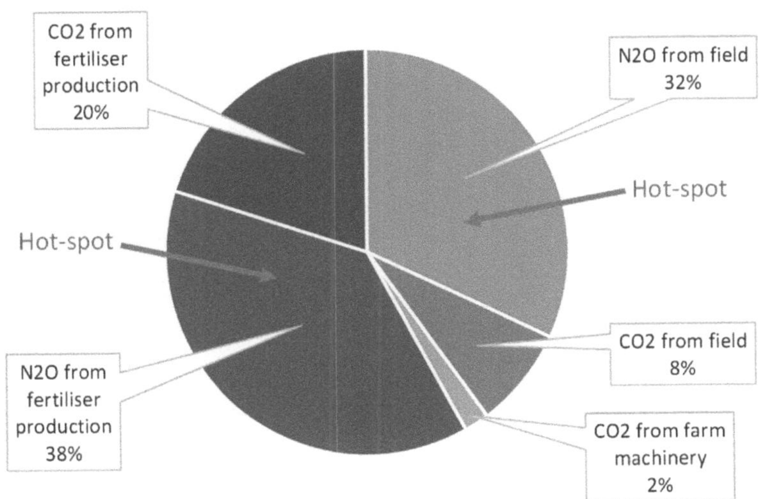

FIG. 16.1 Typical carbon footprint of commonly used annual biomass crops. *Adapted from Fertilisers Europe. Energy efficiency and greenhouse gas emissions in European nitrogen fertiliser production and use. Brussels: Fertilisers Europe; 2015.*

TABLE 16.1 Fertilizers Europe Carbon Footprint Reference Values for European Mineral Fertiliser Production and Use [4]

			Fertiliser Production	Fertiliser Use (Soil Effects)					Fertiliser Production and Use	
			Plant Gate	CO_2 From Urea Hydrolysis	Direct N_2O From Use	Indirect N_2O Via NH_3	Indirect N_2O Via NO_3^-	CO_2 From Liming and CAN	Total	Total
Fertiliser Product	Abbreviation	Nutrient Content	kg CO_2-equiv./kg Product						kg CO_2-equiv./kg Product	kg CO_2-equiv./kg Nutrient
Ammonium nitrate	AN	33.5% N	1.18	0.00	1.26	0.01	0.35	0.27	3.06	9.14
Calcium ammonium nitrate	CAN	27% N	1.00	0.00	0.89	0.01	0.28	0.20	2.40	8.88
Ammonium nitrosulphate	ANS	26% N, 14% S	0.83	0.00	1.10	0.02	0.27	0.40	2.62	10.09
Calcium nitrate	CN	15.5% N	0.68	0.00	0.65	0.00	0.16	0.00	1.50	9.67
Ammonium sulphate	AS	21% N, 24% S	0.58	0.00	0.98	0.02	0.22	0.50	2.30	10.95
Ammonium phosphates	DAP	18% N, 46% P_2O_5	0.73	0.00	0.76	0.01	0.19	0.34	2.03	11.27
Urea	Urea	46% N	0.91	0.73	2.37	0.28	0.48	0.36	5.15	11.19
Urea ammonium nitrate	UAN	30% N	0.82	0.25	1.40	0.10	0.32	0.24	3.13	10.43
NPK 15-15-15	NPK	15% N, 15% P_2O_5, 15% K_2O	0.76	0.00	0.56	0.01	0.16	0.12	1.61	10.71
Triple superphosphate	TSP	48% P_2O_5	0.26	0.00	0.00	0.00	0.00	0.01	0.27	0.56
Muriate of potash	MOP	60% K_2O	0.25	0.00	0.00	0.00	0.00	0.00	0.25	0.43

and develop approaches to mitigate emissions from soil. GHG emissions from N-fertiliser production are mainly from two sources, as explained by Fertilizers Europe [4]:

These are CO_2 from the use of fossil energy sources (mainly natural gas) as feedstock and fuel in ammonia synthesis and N_2O emitted from nitric acid production. Approximately, 70% of the natural gas feedstock (methane) used to make ammonia provides 60% of the hydrogen (H_2) required for the reaction with nitrogen (N_2) from the air for ammonia synthesis. The other 40% of the H_2 is derived from water in modern steam-reforming ammonia plants. These reactions are almost at their maximum theoretical efficiency. The other 30% of the methane is used as fuel to heat the processes. Fig. 16.2 gives the levels of GHG emissions from ammonium nitrate production at different levels of technology and demonstrates how engineering has significantly reduced emissions over time. Improvement in energy efficiency is now the main target of developments in N-fertiliser production technology. More recently, N_2O abatement technology is also being used to reduce N_2O emissions from nitric acid production. Catalytic systems are being developed which break down N_2O under high temperature into harmless nitrogen (N_2) and oxygen (O_2). This process enables a reduction in N_2O emissions of up to 70%–85%.

Another more radical option would be to produce ammonia for fertiliser from renewable biomass sources. This could reduce greenhouse gas emissions associated with ammonia by 65% [5]. Assessing economically viable carbon reductions for the production of ammonia from biomass gasification is close to being economically competitive per unit of output but requires a radical, high-risk change to the existing infrastructure and there are significant challenges around matching the scale of feasible biomass supply with the conventional scale of the Haber–Bosch process for ammonia production.

N_2O from fertilised soils is a crucial area for GHG mitigation. Nitrogen, whether it is from organic or inorganic sources, is subject to various natural microbial conversion processes in the soil, some of which may produce N_2O [4]. The main inorganic forms of nitrogen in the soil are ammonium (NH_4^+) and nitrate (NO_3^-) [6]. Ammonium originates either directly from

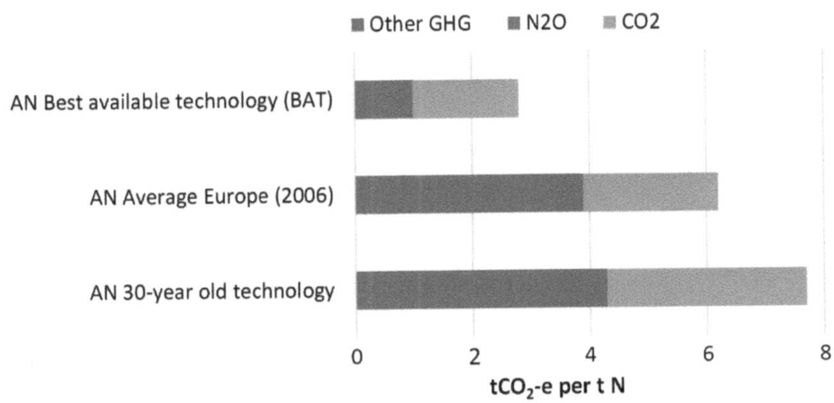

FIG. 16.2 Greenhouse gas emissions of ammonium nitrate production at different levels of production technology. *Adapted from Fertilisers Europe. Energy efficiency and greenhouse gas emissions in European nitrogen fertiliser production and use. Brussels: Fertilisers Europe; 2015. Data derived by Fertilizers Europe from Jenssen and kongshaug, 2003 for '30 years old tech' and Fertilisers Europe data for 'Average Europe 2006' and 'BAT today'.*

mineral fertilisers containing NH_4^+ (e.g. ammonium nitrate or ammonium sulphate), from the conversion of organic nitrogen (e.g. manure or crop residues), or from urea fertiliser (see Table 16.1). Nitrate is either directly applied as nitrate mineral fertiliser (e.g. ammonium nitrate or calcium nitrate) or results from the microbial oxidation of ammonium [4]. Nitrate is dissolved in the water in the soil and is unable to be stored in the soil over the long term. Nitrate is taken up at high rates during the period of crop growth, though at times of low or zero crop demand, nitrate can be lost either to the air via denitrification or to water by leaching [7]. Ammonium is not mobile and most of it has to be converted into nitrate before crops can take it up. Losses of ammonium from the soil occur via volatilisation of ammonia (NH_3). Nitrification is the oxidation of ammonium to nitrate. This natural process supplies energy to the nitrifying bacteria. During the oxidation of ammonium to nitrite, N_2O is produced as a by-product.

As growers are unable to influence the climatic conditions, efforts should be concentrated towards maintaining a good soil structure that enables good drainage and avoids waterlogging. The choice of the right N-fertiliser product under the given conditions (e.g. nitrate-based products applied on nonwaterlogged soils) can help minimise N_2O emissions from the soil [4].

Another development in farming is the use of precision agriculture, an example application being where fields are analysed for soil composition. This allows agronomists to assess nutrient requirements and farmers to apply pesticides and fertilisers at different rates in different parts of the field, thereby optimising application rates and minimising potential emissions. The goal of precision agriculture is to develop decision making for whole farm management, so returns can be optimised against inputs while preserving resources [8]. The practical implementation of precision agriculture has been facilitated by the introduction of GPS and satellite mapping. Farmers can now locate precise positions in a field which allows for the creation of maps of the spatial variability of as many variables as can be measured (e.g. crop yield, terrain features/topography, organic matter content, moisture levels, nitrogen levels, pH, etc.) [9]. Data can also be collected by crop yield monitors mounted on GPS-equipped combine harvesters, which collects a range of information from real-time vehicle mountable sensors. This data is then used by variable rate technology (VRT) including seeders, sprayers, etc. to optimally distribute resources [8]. In terms of GHG emission reduction, this engineering development offers significant opportunities.

There are also options available to directly control GHG emissions from fertilisers including nitrogen inhibitors [9]. With moderate temperature and soil water content, nitrification occurs on most soils within a few days or weeks after application of ammonium sources. The rate of nitrification can be controlled by preserving N as ammonium which can be achieved through applying a compound known as nitrogen inhibitors. This delays nitrate production and has been shown to reduce direct N_2O emissions from soil between 40% and 70%, with a mean, nonsignificant reduction of 56% [6].

16.2.2 Diesel Use in Feedstock Supply

Diesel used in field operations and harvesting can make a reasonable contribution to the emissions from feedstock supply. In principle, biodiesel could be used to reduce the GHG emissions; however, it is often difficult to access a dedicated supply of appropriate quality/provenance.

Engineering in recent years has improved the efficiency of engines and cultivation; harvesting and collection methods have also improved, as discussed in Chapters 7 and 17.

In agricultural systems, the key engineering lessons are to implement crop management techniques that reduce the need for heavy machinery; examples include irrigation systems for distributing fertiliser, no-till or min-till cultivation, or harvesting techniques that maximise the yield obtained per unit of process energy input. Genetic engineering innovations have facilitated no-till or low-till, herbicide-resistant crops and could be applied to other crops if there were demand.

In forestry systems, a constraint on truck movements is the bulk density and moisture content of the wood product. It is fairly standard practise to allow wood to dry naturally in the forest after harvesting; this reduces moisture content and therefore a greater quantity can be transported for the same amount of diesel energy. Depending on the end product, chipping often occurs close to harvest as this can reduce the bulk density and make the feedstock more suitable for pellet production. Wood pellets are the most common form of woody biomass when it comes to importation due to the ease of handling, energy density, durability, and restrictions on importing unprocessed biomass such as wood chips. Transporting pellets by boat is now common practise in global supply chains and this reduces emissions per t-km in comparison to road transport.

16.2.3 Collection and Transportation of Biomass Feedstocks

All biomass has to be collected, regardless of whether this is cultivated and harvested as a crop or forestry resource, or if this is waste arising. However, emissions from collecting biomass are, in general, not significant and should not be a core focus for engineering development. Losses in collection or inefficient collection can increase net GHG emissions as lower tonages ultimately increase the emissions per ton and consequently per unit of energy produced. Therefore, ensuring efficient collection systems that minimise losses is important. Another key lesson from existing biomass supply chains is that in general the closer the feedstock is located to the conversion facility, the lower the net GHG emissions will be. The mode of transport also influences emissions as well as associated infrastructure required, for example, creating roads in forests or laying railway tracks.

Drying and moisture content should also be considered. Greenhouse gas emissions associated with the supply chain will be increased by transporting water long distances, so efficient drying early in the logistics process makes environmental sense. This is particularly so when we take into account the potential losses described in Chapter 7. Chapter 14 also showed that, even when actively drying forest residues, a good greenhouse gas balance can be achieved, but this is improved by drying with wood rather than fossil fuel and densification by pelleting actually reduces the subsequent transport costs and emissions per unit of feedstock. So, drying solutions using biomass where possible should be considered early in engineering logistics/integration.

16.2.4 Wastes and Residues

Biomass wastes and residues are materials of biological origin from agriculture, forestry, industry, and households and can be derived from a wide range of sources. In contrast to dedicated energy crops, they are not produced specifically as a feedstock for bioenergy and usually derive from an economic activity. Engineers have therefore developed various systems for energy recovery from wastes and residues. In most cases, this is a good use of resource in terms of GHG avoidance; however, caution is needed as there are some examples where utilising wastes and residues doesn't always reduce net GHG emissions. Cherubini et al. describes two such examples [10]:

- Removing forestry or agricultural residues from land can reduce carbon storage in carbon pools like soil, dead wood, or litter and can deplete soil nutrients.
- Creating markets for biomass residues can make production of the main product economically more attractive and therefore increase pressure on land use. For example, in forestry selling residues can make harvesting timber more profitable as it provides an extra income source from the harvesting operations.

The engineering lesson for wastes and residues is therefore to completely consider the feedstock being used and assess the alternative and competing uses in terms of GHG emissions. LCA practitioners therefore have a role in assessing the GHG emissions of different economic and waste/residue management scenarios. Considering the counterfactual and impact on existing systems is therefore important when developing engineering solutions for wastes and residues. There are a wide range of waste management LCAs in the literature that should be considered when undertaking an LCA of waste-derived bioenergy.

It is also important to realise that when using wastes or residues, there is little control over the physical and chemical characteristics of the incoming feedstocks. This means that systems need to be designed for a wide range of tolerance to, for example, variable moisture and ash contents and in some cases to cope with high variability in sulphur, chlorine, silica, or heavy metal contents. A general corollary of this is that there is a need to adapt the engineering design for a wide operating envelope of feedstock characteristics rather than to optimise performance for a specific feedstock specification. Essentially, engineering resilient plant performance for waste feedstock is more important than efficiency optimisation.

16.3 PROCESSING BIOMASS

Various techniques have been developed to process biomass either so it can be converted more easily into an intermediate fuel or so it can be converted directly into useful energy. Examples of the former are generally known as pre-processing and include techniques such as drying, pelleting, torrefaction, hydrolysis, and steam explosion [11,12]; examples of the latter are described in Chapter 8.

In terms of GHG emissions, there is often a trade-off between additional emissions from the processing stage and reduced emissions in the subsequent conversion to fuel and energy. This shows the importance of designing engineering solutions that don't increase the net GHG balance of the bioenergy system, and hence, LCA is required to assess the different processing options impact on the supply chain GHG emissions. An example is the requirement for wood chip to be dried prior to pellet production (see Chapter 14). There are different utility fuel options that can be used for drying including electricity, natural gas, and biomass [11,13]. In this example, electrical heat has much higher emissions where the electricity is produced using fossil fuels such as coal. Natural gas is therefore a more preferable option in terms of lower GHG emissions, but may not always be available in remote locations. Gas is also often preferred due to its ease of use and ability to control. Finally, biomass can be used to provide process heat which will deliver the lowest net emissions, but the facility will need to consider the increased logistics required and the additional land requirements. Engineers therefore need to be able to balance ease of use with economics and minimising GHG emissions from the system. The following description provides a brief introduction to different processing methods, the considerations, and trade-offs in terms of GHG emissions.

Biomass densification technologies can potentially add value to agricultural, forestry, horticultural, wood processing (i.e. sawmills), local authority (as an alternative to composting), and the waste industry [14]. Diversification and efficiency gains can be achieved through improved logistics, reduced infrastructure, and equipment requirements compared to untreated biomass systems. Densification is achieved through engineering techniques including converting the biomass to a pellet by pelletisation (see Chapters 12 and 14), producing 'bio-coal' by torrefaction, or to a liquid (bio-oil or biocrude) by pyrolysis or hydrothermal upgrading (see Chapter 8). The choice of processing technology depends on the context, e.g. feedstock characteristics, distance, and conversion technology and is therefore commonly assessed via process and techno-economic analysis.

Biomass pellets are commonly made from sawdust and wood residues, but can also be made from a number of comminuted biomass types, including wood, waste agricultural products such as straw, grasses, and other wastes. In use, biomass pellets provide a consistent feedstock which is attractive to end-users. They are dry and clean, have a specified ash content, and flow freely allowing them to be easily mechanically conveyed [14]. While there is an energy and emission cost in producing pellets when compared to using unprocessed wood, there can be substantial GHG savings when the transport logistics and combustion efficiencies are considered. Similarly for torrefaction, there is a trade-off that requires assessment on a case by case basis. From a GHG perspective, it is not always necessary to produce pellets or torrefied fuels, particularly if the transport distances are short or the end-use is relatively simple [11]. GHG benefits tend to be realised when there is larger scale of production, longer distances, and where more advanced conversion technologies are used.

16.4 CONVERSION TECHNOLOGIES

A key lesson learnt from both successful and failed biomass projects and the various GHG assessments performed through bioenergy LCAs is that the choice of conversion technology is crucial in developing sustainable bioenergy systems. Matching the most suitable conversion route and scale to the available feedstock and required end-use is a key challenge for engineers. There is little point in implementing a technology if the feedstock resource available is limited or there is a low end-user demand for the bioenergy produced. Several papers have reviewed the critical success factors for bioenergy projects and the reasons for failure [15–18]. Some of the main conclusions drawn include ensuring the technology is viable both technically and economically [19,20], that the technology is appropriate for the available biomass resource [21], and the scale and location are optimised for both feedstock supply and delivering the bioenergy output [22].

It should be obvious to most observers that bioenergy systems only deliver greenhouse gas reductions if they operate reliably during their lifetime and so engineering systems that have high levels of availability (e.g. thorough appropriate front-end handling, screening and exclusion; by incorporating parallel stream configurations and adequate redundancy; using robust devices and materials with generous wear and maintenance margins) are key to actually delivering the GHG reductions.

Bioenergy LCAs conducted have also provided useful insight into the conversion technology choice. The design of the whole system is therefore crucial when trying to minimise GHG emissions, while still developing a technically and economically feasible project. As Chapter 8 has shown, the conversion pathways for biomass are numerous and so the engineering challenge is to appreciate that bioenergy is not a 'one size fits all' technology and that the local conditions are the key determinant when developing and implementing a bioenergy project. The following sub-sections describe in further detail some example considerations for biomass engineering to consider in relation to technology choice.

16.4.1 Scale of Production

Scale is a key consideration in the design and operation of a biomass production facility. From the economic perspective, it is common to try and maximise the scale of output in order to maximise efficiency and minimise capital costs per unit of output and achieve economies of scale. Nonetheless, the scale needs to match the available feedstock as well as the end-user demand. Being able to operate consistently at a certain scale relies on maintaining feedstock supply and having the ability to utilise the energy output. For several bioenergy projects, feedstock has seasonal variability, for example due to the nature of harvest times or animals grazing outdoors at different times of year. In addition, end-user demand can be seasonal, for example heat demand is much higher in winter months. These seasonal variations mean that bioenergy facilities may have to operate well below the total installed capacity or (more commonly) incorporate significant storage/buffer supplies or feedstock diversification strategies. Hence, investigation of scale of production and associated efficiency variations are important to consider at the design phase.

Larger scale production benefits from increased efficiency of production and often the ability of larger scale facilities to mitigate and manage GHG emissions. However, small-scale production often has the advantage of emissions from feedstock supply being lower due to lower transport demand and less handling.

16.4.2 Location

The location of biomass facilities is of key importance in terms of GHG mitigation. From the engineering perspective, facilities should be located in a practical location for receiving and handling the feedstock supply. In general, this means being located as close as possible to the biomass resource in order to minimise emissions from transport and logistics. Where this is not possible, engineers have come up with methods to reduce the impact from the supply chain; a good example of this is densification and the use of pellets where the end-user is often located very long distances away from the unprocessed biomass. In this example, the pellet plant needs to be located as near as possible to the forest and transportation routes. The key consideration for engineers is to ensure that the system is designed to minimise GHG emissions and so locating the conversion process in the optimal location is important. This may be at a port or other transport node with appropriate infrastructure. Locating biomass conversion facilities at the point of use is important where on-site bioenergy use is required, such as boiler heat. If the form of bioenergy can be easily fed into the gas or electric grid then access to the distribution network is important.

16.4.3 Feedstock Quality

Feedstock variability has been one of the biggest challenges to the thermo-chemical conversion process routes due to the wide versatility of biomass feedstocks, e.g. moisture content, contaminants, morphology, etc. This creates several engineering challenges with the implementation of more sophisticated conversion technologies. A trade-off therefore arises when assessing if more complex conversion routes are utilised for inconsistent or unusual biomass feedstocks. Availability (discussed above) is perhaps the key driver of the GHG saving potential of a given biomass feedstock conversion process as achieving high operational hours will attain GHG reductions, hence poor quality feedstock is likely to impact on this.

16.4.4 Construction and Maintenance

GHG emissions associated with constructing bioenergy facilities have been assessed in a number of studies. The general conclusion is that while there are embodied GHG emissions associated with construction (e.g. steel, concrete, aluminium), these emissions are usually low in comparison to the operation of the facility. For the LCA practitioner, this typically means that the embodied impacts of capital equipment and infrastructure can be excluded from the LCA without further justification. There are however exceptions to this, for example production systems with a short economic life or low operating hours, or production systems requiring establishment of significant supporting physical infrastructure, such as dedicated roads, rail, pipelines, and intermodal change facilities. In these cases, the embodied emissions may contribute 5%–10% or more of total life cycle emissions, in which case capital equipment and infrastructure should be included at a scoping level in the LCA. A high level assessment can then be performed to estimate approximate embodied emissions in comparison to operational emissions.

Maintenance of plant, machinery, and equipment is regularly overlooked in LCA studies. This is often because LCAs are performed prior to a facility commencing operation or because the maintenance requirements are unknown until the production facility is fully operational. In most cases, emissions from maintenance activities and replacing equipment are unlikely to have a material impact on the GHG balance. Nonetheless, not all equipment has a long useful economic life and may require replacing several times in the operational lifetime of the whole facility. An example of this would be CHP engines which require regular maintenance and have limited run-time before needing replacement around every 8–10 years. Another example is wood chipping or hammermills used in pellet production that wear down which has an associated carbon cost of replacement. Plant associated with physical processing and comminution may experience severe wear and frequent replacement, e.g. blades in chippers and mechanical components of hammer and ball mills where these are used to reduce the size of hard/abrasive/high-ash material. Anaerobic digesters may also experience large methane fluxes during maintenance; for example, if the digester roof is removed thereby releasing biogas to atmosphere. In summary, construction and maintenance should be considered and assessed within the scope of a LCA.

16.4.5 Start-Up Fuel and Process Energy

Bioenergy systems typically require some fossil-derived energy for the production, transport, and conversion to bioenergy and are therefore crucial to consider in an LCA. The more fossil fuel that a bioenergy production chain requires, the less appropriate it becomes. Consequently, those supply chains requiring minimal fossil fuel input become more desirable. Several studies have assessed the cumulative energy demand (CED) or fossil energy demand (FED) which effectively calculates the energy balance or ratio of fossil energy required per unit of energy output. This type of data is crucial for engineering design so that ways of minimising associated GHG emissions can be mitigated and alternative energy sources considered.

For many conversion processes, start-up fuel is required to obtain the necessary temperatures and process conditions for the biomass conversion to commence. A good example of this is gasification where very high temperatures are required so that the thermo-chemical process is instigated. In terms of GHG emissions, the choice of start-up fuel is important as this often relies on fossil fuels to obtain the high temperatures quickly, hence the energy and GHG balance requires attention. Engineers also need to consider how regularly a system may start-up and shut down. A gasification system that only operates intermittently is likely to have much higher demand for start-up fuel in comparison to a process that runs for long periods of time. When new technologies are developed, start-up fuel and run-time are therefore important considerations so that GHG emissions can be minimised.

Process energy (or utility fuel) is required in a wide range of biomass conversion technologies. In many cases, there are different options for the supply of fuel or electricity, each with different considerations in terms of availability, practicality, user-friendliness, type of resource, and of course GHG balance. Some examples of process energy requirements in different biomass conversion technologies include:

- Pellet production—mechanical energy required for chipping, hammermill, and pressing, thermal energy required for drying;
- Gasification—thermal energy required to obtain the necessary process heat in the gasifier;
- Biomethane—electrical energy required for the biogas upgrading process including gas compression and grid injection;

Regulatory or policy incentives can play a significant role in determining start-up fuel requirements. For example, permits for waste plants may require pre-heating to minimum temperatures before feedstock is introduced which may require long periods of time on start-up fuel. However, renewable energy incentives sometimes specify limits on the amount of fossil fuel that can be used, providing an obvious incentive to reduce utilisation. In general, good management will result in start-up and supplementary fuel supplies being <1% of primary fuel consumption for most plants. However, where plants experience operational or maintenance problems and have frequent (e.g. greater than 10) start-ups per year this can increase and should then be accounted for in the LCA calculations. As above, the key objective is to maximise reliability and availability by appropriate engineering design.

16.4.6 Storage

Emissions from storage are a key consideration in engineering design. Chapter 9 showed that open digestate storage can lead to very high releases of methane. Chapter 14 showed that in some cases methane could be released from wood chip piles due to micro-biological activity. Bioenergy systems therefore need to not only focus on the feedstock supply chain or the conversion process, but also critical components of the system such as storage.

16.4.7 Load Factors and Efficiency

The load factor is the total actual output of a facility divided by the output it would have achieved operating continuously at maximum continuous rating. Therefore, the plant output and load factor is reduced both by failure to achieve maximum output and failure to operate continuously. Maximising this compound measure may require trade-offs between efficiency and availability whereby using a sub-optimal fuel may result in lower output, but maintain operational hours and avoid shutdowns and associated start-up fuel.

In a bioenergy system, there are many different steps, each of which has its own efficiency. There is a tendency to focus on the efficiency of the key conversion step to energy as the most significant engineering parameter and sometimes to seek to maximise this. However, in cases where the majority of GHG emissions are associated with feedstock production, this can be inappropriate. It makes much more sense in such scenarios to maximise the use of the biomass resource and this is done by maximising system or supply chain efficiency, not conversion efficiency. Thornley demonstrates that this rationale underpins the advantages of lingocellulosic/second-generation biofuels systems so that it is overall system efficiency and not conversion efficiency that should be considered [23]. Of particular importance is the ability of the conversion system to accept feedstocks with a wide range of feedstock characteristics, as this minimises losses along the supply chain.

A good example of where the load factor directly affects the conversion efficiency is in CHP. A key issue here is that the load factor is often determined by the heat demand and CHP plant may operate so that the overall load factor is modulated to match the heat demand on the system.

16.4.8 Trade-Offs

When it comes to technology choice and design, there are several trade-offs which need to be considered, which depend on the biomass conversion route in question. For instance, with a thermo-chemical conversion process such as gasification, it may be appropriate to preprocess the biomass so its composition is more suitable for gasification. Preprocessing often involves process energy and associated GHG emissions, therefore the trade-off for engineers to consider is whether the additional processing step is worthwhile in terms of the overall net GHG balance. If the improved energy output, quality of fuel, and increased conversion efficiency reduce the overall GHG emissions, then preprocessing is clearly worthwhile. However, there are also practical and economic factors to consider as it may be that there is a trade-off between a durable, easy to handle, and energy dense wood pellet which can be easily transported and un-processed wood chip. It could therefore be necessary to forgo some GHG savings in order to have a more economical and practical biomass supply chain.

16.4.9 Losses

Losses have a direct impact on supply chain efficiencies and therefore greater losses result in higher GHG emissions. Therefore, the challenge is not to minimise losses at all parts of the supply chain, but to focus particularly on those steps where significant losses can be mitigated, e.g. it has been shown in Chapter 7 that there may be significant losses in storage and Chapter 14 has shown the impact that such losses can have on the overall GHG balance.

16.4.10 Counterfactuals

In engineering terms, the counterfactual for a bioenergy system is important. Efforts should focus on those conversion routes which can achieve the largest GHG savings, but also those which are efficient and more difficult to replace with other technologies. Several studies have shown that the greatest GHG savings are obtainable from replacing coal-fired power stations with biomass electricity; however, alternative renewable technologies can produce electricity such as wind and solar. In contrast, off-grid heating systems or natural gas is more difficult to replace without using bioenergy. Therefore, the counterfactual needs careful consideration alongside the alternative options.

It should also be recognised that there is often more than one counterfactual option and the degree of GHG savings generated can be acutely sensitive to what fuel or technology would have been appropriate if bioenergy had not been used. Work that analysed over 100 different heat bioenergy pathways in the UK concluded that variations in GHG performance did not correlate closely with feedstock or technology choice, but were most sensitive to the inclusion of specific processing steps and displacement of certain counterfactuals, suggesting that policies and bioenergy implementation should target resources with high GHG counterfactuals [24].

16.5 END-USE

As noted above, the most significant GHG savings are incurred when targeting higher carbon intensity uses and this should be a key consideration when targeting the best use of biomass. For example if there are other equally economic sources of low carbon electricity, there is limited point in implementing bio-electricity in situations where there are likely to be requirements for liquid hydrocarbons. These are much more difficult to satisfy with other low carbon energy sources and so the delivery of fungible hydrocarbons is a more valuable end-use than electricity in many contexts. A similar logic applies to the trade-offs between use of biomass for energy and use of biomass for bio-products such as chemicals. Where biomass can be used to displace a product that is carbon-intensive and has a high volume market demand, it makes sense to consider whether it makes more environmental and economic sense to use biomass to replace natural gas or to use biomass to replace petroleum-derived bioplastics. This is the rationale behind many biorefinery concepts, but it is key to focus on products that have a significant demand and high carbon intensity.

16.6 SUMMARY

This chapter has highlighted and discussed some of the key issues for bioenergy engineering design. When the GHG balance of a given biomass conversion technology is considered it is important that engineers consider the whole system including supply chain and end-use demand. Key considerations in the engineering design of bioenergy systems include:

- Fertiliser production and application method is important when optimising crop yields while also balancing GHG emissions.
- Precision agriculture offers a significant opportunity for farmers to use modern technology to combine economic return with GHG reduction.
- Pretreatment technologies such as pyrolysis, torrefaction, and pelletisation should be considered when consistent feedstocks are required by the conversion technology, or when supply chain logistics require long distances.
- Producing liquid biofuels usually requires more fossil fuel inputs than the generation of electricity and heat from biomass. Consequently, electrical, heat, and CHP conversion pathways are likely to achieve larger GHG savings than liquid biofuels [10].
- Bioenergy systems utilising wastes and residues demonstrate the largest GHG savings due to avoided emissions associated with crop production and the emissions from waste management. Nonetheless waste feedstocks often have variable composition hence conversion technologies require engineering to accommodate biomass variability.
- While a wide range of conversion technologies exist, it is important to consider key factors such as scale, location, feedstock quality, maintenance requirements, start-up fuel, process fuel, storage, load factors, trade-offs, and losses in the engineering design stage.
- Counterfactuals are important for engineers to assess as the available resource and alternative replacement options should be considered alongside GHG saving potential.
- Concerns over competition with food production and limited available land resources mean that achieving high biomass yields are extremely important in achieving high GHG emission savings. Nonetheless, using chemical fertilisers to increase production can reduce the net savings.
- Alternative energy sources should be considered in bioenergy supply chains. Good existing examples include bagasse used in ethanol production, biogas CHP used in biomethane production, and biomass used for wood drying rather than gas or electricity.
- When using agricultural wastes and residues it is paramount that the effects of removing the material are accounted for. Straw use in ethanol and forest residues in pelleting where the removal of nutrients from soil needs to be assessed [10].
- High biomass conversion efficiency to energy products is a fundamental requirement for minimising GHG emissions from bioenergy production. Engineers therefore need to consider scale, start-up, downtime, maintenance, and typical operating hours.

References

[1] Anon. The universe of engineering—a UK perspective. London: Royal Academy of Engineering; 2000.

[2] Amous S. Non-CO2 emissions from stationary combustion. Good Pract Guid Uncertain Manag Natl Greenh Gas Invent; 2001. p. 41–54. Available from: http://www.ipcc-nggip.iges.or.jp/public/gp/english/2_Energy.pdf

[3] IEA. CO_2 emissions from fuel combustion online data service. 2017th ed. Paris: International Energy Agency (IEA); 2017. Available from: https://www.iea.org/statistics/relateddatabases/co2emissionsfromfuelcombustion/.

[4] Fertilisers Europe. Energy efficiency and greenhouse gas emissions in european nitrogen fertiliser production and use. Brussels: Fertilisers Europe; 2015.

[5] Gilbert P, Alexander S, Thornley P, Brammer J. Assessing economically viable carbon reductions for the production of ammonia from biomass gasification. J Clean Prod 2014;64:581–9. Available from: http://10.0.3.248/j.jclepro.2013.09.011.

[6] Sylvester-Bradley R, Thorman RE, Kindred DR, Wynn SC, Smith KE, Rees RM, et al. Project Report No. 548 Minimising nitrous oxide intensities of arable crop products (MIN-NO). 2015; (548). Available from: https://cereals.ahdb.org.uk/publications/2015/october/29/minimising-nitrous-oxide-intensities-of-arable-crop-products-(min-no).aspx.

[7] Ferchaud F, Vitte G, Machet J-M, Beaudoin N, Catterou M, Mary B. The fate of cumulative applications of 15N-labelled fertiliser in perennial and annual bioenergy crops. Agric Ecosyst Environ 2016;223:76–86. Available from: http://www.sciencedirect.com.ezproxy1.bath.ac.uk/science/article/pii/S0167880916301074 [cited 27 April 2017].

[8] JRC. Precision agriculture: an opportunity for EU farmers—potential support with the CAP 2014-2020. Brussels: Joint Research Centre (JRC) of the European Commission (EC); 2014.

[9] EC. Mitigation potential of green house gases in the agricultural sector—Final Report (COMM(2000)88). Brussels: European Commission (EC), Agriculture Directorate-General; 2001.

[10] Cherubini F, Bird ND, Cowie A, Jungmeier G, Schlamadinger B, Woess-Gallasch S. Energy- and greenhouse gas-based LCA of biofuel and bioenergy systems: key issues, ranges and recommendations. Resour Conserv Recycl 2009;53(8):434–47.

[11] Adams PWR, Shirley JEJ, McManus MC. Comparative cradle-to-gate life cycle assessment of wood pellet production with torrefaction. Appl Energy 2015;138:367–80. Available from: http://www.sciencedirect.com/science/article/pii/S0306261914011507 [cited 7 May 2017].

[12] Tabil L, Adapa P, Kashaninejad M. Biomass feedstock pre-processing—part 1: pre-treatment. In: Dos Santos Bernardes MA, editor. Biofuel's engineering process technology. InTech; 2011. https://doi.org/10.5772/17086. Available from: https://www.intechopen.com/books/biofuel-s-engineering-process-te.

[13] McNamee P, Adams PWR, McManus MC, Dooley B, Darvell LI, Williams A, et al. An assessment of the torrefaction of North American pine and life cycle greenhouse gas emissions. Energy Convers Manag 2016;113:177–88. Available from: http://www.sciencedirect.com/science/article/pii/S0196890416000224 [cited 7 May 2017].

[14] NNFCC. Techno-economic assessment of biomass "densification" technologies. York, UK: NNFCC Project 08-015 National Non-Food Crops Centre (NNFCC); 2008.

[15] Adams PWR, Hammond GP, McManus MC, Mezzullo WG. Barriers to and drivers for UK bioenergy development. Renew Sustain Energy Rev 2011;15(2):1217–27. Available from: http://www.sciencedirect.com.ezproxy1.bath.ac.uk/science/article/pii/S1364032110003266; [cited 7 May 2017].

[16] Roos A, Graham RL, Hektor B, Rakos C. Critical factors to bioenergy implementation. Biomass Bioenergy 1999;17(2):113–26. Available from: http://www.sciencedirect.com.ezproxy1.bath.ac.uk/science/article/pii/S0961953499000288 [cited 7 May 2017].

[17] McCormick K, Kåberger T. Key barriers for bioenergy in Europe: economic conditions, know-how and institutional capacity, and supply chain co-ordination. Biomass Bioenergy 2007;31(7):443–52. Available from: http://www.sciencedirect.com.ezproxy1.bath.ac.uk/science/article/pii/S0961953407000311 [cited 7 May 2017].

[18] Junginger M, van Dam J, Zarrilli S, Ali Mohamed F, Marchal D, Faaij A. Opportunities and barriers for international bioenergy trade. Energy Policy 2011;39(4):2028–42. Available from: http://www.sciencedirect.com.ezproxy1.bath.ac.uk/science/article/pii/S0301421511000504 [cited 7 May 2017].

[19] Strzalka R, Schneider D, Eicker U. Current status of bioenergy technologies in Germany. Renew Sustain Energy Rev 2017;72:801–20. Available from: http://www.sciencedirect.com.ezproxy1.bath.ac.uk/science/article/pii/S1364032117301089 [cited 7 May 2017].

[20] Patel M, Zhang X, Kumar A. Techno-economic and life cycle assessment on lignocellulosic biomass thermochemical conversion technologies: a review. Renew Sustain Energy Rev 2016;53:1486–99.

[21] Cherubini F, Strømman AH. Life cycle assessment of bioenergy systems: state of the art and future challenges. Bioresour Technol 2011;102(2):437–51.

[22] Souza GM, Ballester MVR, de Brito Cruz CH, Chum H, Dale B, Dale VH, et al. The role of bioenergy in a climate-changing world. Env Dev 2017. Available from: http://www.sciencedirect.com.ezproxy1.bath.ac.uk/science/article/pii/S2211464516302755 [cited 7 May 2017].

[23] Thornley P. Biofuels review: report for government office for science. Manchester, UK: University of Manchester; 2012.

[24] Welfle A, Gilbert P, Thornley P, Stephenson A. Generating low-carbon heat from biomass: life cycle assessment of bioenergy scenarios. J Clean Prod 2017;149:448–60. Available from: http://www.sciencedirect.com.ezproxy1.bath.ac.uk/science/article/pii/S0959652617302391 [cited 27 April 2017].

CHAPTER

17

Environmental Lessons—Making Bioenergy System Decisions That Benefit the Environment

Patricia Thornley

University of Manchester, Manchester, United Kingdom

17.1 ENVIRONMENTAL CONTEXT AND GREENHOUSE GAS MITIGATION

This book has focused throughout on the greenhouse gas balances of bioenergy systems. Climate change is widely recognised as the biggest environmental threat that humankind has ever faced. The economic impacts of climate change have been quantified [1] and it should be remembered that underpinning this is a set of environmental impacts that incorporate huge changes to land (changes in soil quality and condition leading to desertification, salination, etc.); increases in sea temperature leading to loss of marine species; increases in global mean surface temperature leading to reductions in biodiversity and changes in spatial distribution of species; reduced crop yields; loss of agricultural land; etc.

Bioenergy is one of several renewable, low carbon technologies that can help to reduce GHG emission and thereby help prevent the more extreme future climate scenarios. Therefore, it is entirely appropriate that this book should focus on the greenhouse gas balances of bioenergy systems, since it is absolutely critical to ensure that bioenergy delivers net GHG benefits when implemented. However, it must be remembered that the comparison point for judging the impact of low carbon energy systems is not the current world, but this radically climate-transformed world. In other words, the choice is not between continuing as we are or having a bioenergy system with its attendant impacts; it is actually a choice between a future with a catastrophically environmentally damaged planet with reduced biodiversity, land (and landscape) quality, reduced ecosystem provisioning, or a future with a bioenergy system or another renewable energy system and their associated impacts. It is critically important to remember when making environmental judgements that sufficient greenhouse gases have already been released to make staying within a 2°C temperature rise unlikely [2]. This has

Greenhouse Gas Balances of Bioenergy Systems
https://doi.org/10.1016/B978-0-08-101036-5.00017-3

two corollaries: first of all the correct comparison is with a world severely affected by climate change; secondly, there is a need not only to mitigate emissions, but also to adapt to the changing climate.

17.2 ENVIRONMENTAL CHANGE AND CLIMATE ADAPTATION

It is now widely accepted that a degree of irreversible environmental change has already been incurred and that there is a need to simultaneously mitigate, but also adapt to climate change impacts. There are many ways in which bioenergy can support adaptation strategies and so it is critical that this is also taken into account when considering bioenergy impacts, strategy, and planning. Adaptation can broadly be thought of as measures that support continued functionality within a changed ecosystem. So, these might include things like planting of perennial crops to improve soil (increasing agricultural crop options in the long term) or biodiversity or planting of hydrophilic short rotation crops to support flood protection. Both of these are viable adaptation options which would support adapting our practises to cope with existence in an altered climate and improve human resilience. The first is an action that would support reversal of negative environmental impacts associated with climate change; the second is an action that would support ecosystem recovery when climate impacts are experienced. A third option would be to support socio-economic resilience in the face of climate change and its impacts. Bioenergy also has a role to play here in providing alternative functions, activities, and income stream for farmers and other landowners: rural diversification. Finally, transpiration and the presence of vegetation can provide shade, cooling, and air purification effects that may be beneficial in situations where extreme high temperatures are being experienced as is likely with climate change projections.

17.3 ENVIRONMENTAL IMPACTS OF BIOENERGY SYSTEMS

As stated above, this book has focused primarily on the greenhouse gas balances of bioenergy systems. There are of course many other impacts and care must be taken to balance greenhouse gas reductions against other impacts. For example, Chapter 10 dealt with the greenhouse gas balance of biodiesel from soy cultivation in Argentina. It is possible to cultivate soy in different ways and genetic modification has facilitated no till or low till varieties that are herbicide-resistant and can drastically reduce greenhouse gas emissions. However, if, in practise, farmers then spray more herbicide, the ecotoxicity and human toxicity of the system will increase. So, greenhouse gas reductions are valuable, but sometimes need to be considered in context against the impacts of achieving them.

Throughout this book, we have considered a life cycle approach to greenhouse gas balance and exactly the same life cycle environmental approach can be used to catalogue the full environmental impact of a bioenergy system to consider these trade-offs. The approach to doing this is exactly the same as described throughout this book: starting with goal and scope definition, developing a full mass and energy inventory of all flows into and out of the system, and then ascribing impact factors to each component of the inventory. When considering the greenhouse gas impact, the factors used are the global warming potential, which may be averaged over different time horizons. Other environmental impacts commonly evaluated are:

(1) Fossil resource depletion
(2) Water depletion
(3) Terrestrial acidification potential
(4) Freshwater eutrophication potential
(5) Marine eutrophication potential
(6) Ozone depletion potential
(7) Photochemical oxidant creation potential
(8) Human toxicity
(9) Ecotoxicity

These are calculated by assigning a factor to every substance in the inventory that constitutes its impact in each of the nine categories above. This is commonly called a 'mid-point' evaluation as it combines the different substances which have the same environmental impact, but stops short of weighting these environmental impacts to carry out a net assessment of overall impact. Each of these topics could merit entire books in their own right, but for our purposes the important thing is to be aware of the potential for other impacts, sensitive to the relative importance of different environmental impacts in different contexts and that improving greenhouse gas emissions may make some of these impacts worse. It should also be noted that while many of these impacts are global, some impacts, e.g. smog resulting from photochemical oxidation, may have more local or regional consequences. It should also be noted that LCA has been developed as a 'context-blind' tool, which allows fair and even comparison across different settings by ignoring the local environment in which a bioenergy plant is sited. This is appropriate for calculation of global warming potential, since greenhouse gas balances and climate change are global issues. However, that is not so of all ecological impacts. For example, the impact of a high level of ecotoxicity could be very much more concerning if experienced in areas of high biodiversity or special scientific interest.

17.4 LAND-USE

Many of the ecological impacts associated with bioenergy systems are related to the occupation of land while biomass is growing. This is important to remember when comparing bioenergy impacts with those of other renewable energy systems and considering the appropriate goal and scope of the system. If a bioenergy system is compared with a photovoltaic system, it may be desirable to compare their net life cycle greenhouse gas balance to support decision making about the most appropriate renewable energy choice. However, the bioenergy system is likely to also incur negative environmental impacts associated with its land occupation: most commonly acidification and eutrophication. It could be argued that the comparison between the two technologies is not entirely consistent, since if the land were not used for biomass production, it would have been used for something else, which also would have incurred eutrophication and acidification impacts. Therefore, the assessment should not be to what extent has biomass growth incurred these impacts, but to what extent has biomass growth changed these impacts from what they would have otherwise been. Just as counterfactuals can be important in determining the greenhouse gas balance of a bioenergy system, they can also be important for wider environmental assessment.

For example, if the land was agricultural land which had, instead, been used for intensive cereal production, there would likely have been a greater set of impacts associated with this, which would need to be considered. So, careful consideration of goal and scope consistent with LCA objectives is needed to determine the extent to which wider ecosystem impacts should be taken into account.

17.4.1 Land-Use Change

A particularly important aspect of land use is that of land-use change and this has seen huge focus as a potential area for distorting the greenhouse gas balances of bioenergy systems. First of all, if land-use change is being caused by a new bioenergy system, the life cycle assessment and greenhouse gas balance should generally take that into account in their calculations. Sometimes, however, it can be difficult to determine the cause of a land-use change and what would have happened otherwise, leading to significant uncertainty potential.

Secondly, consideration should be given to whether it is also appropriate within the scope of system to consider reversion of the land, i.e. returning it to the state it was in before the bioenergy crop was planted. This decision has to be taken on a case by case basis with knowledge of the context. It may seem like a reasonable principle to return land to its original state, but perhaps only if that is desirable for the land owner or user and it should not reverse any benefits that have been incurred. So, for example, if a perennial bioenergy crop is planted in order to accumulate soil carbon and improve soil quality, it would not make much sense to return the land to its original state at the end, but the soil carbon benefit should generally be included in the LCA calculation, just as we would include a soil carbon deficit.

There has been considerable focus in recent years on the potential of indirect land-use change (ILUC) to magnify the greenhouse gas releases associated with bioenergy systems. The principle is that if land is converted to a bioenergy crop, it is displacing another land use, which needs to be accounted for. If that displaced land use results in other areas of land being converted to the original land use of the system, then it is reasonable to take that into account in the overall calculation of ecosystem impact. However, the challenge here is in determining what land-use changes have been caused by the bioenergy system and which ones would have happened anyway. Some researchers have attempted to develop factors to ascribe to this, known as ILUC factors. These may prove useful in some assessments, but it should be noted that there is very considerable regional variation and temporal variation in these, since these depend on the availability of land, market pressures and trends, etc. Therefore, extreme caution should be exercised in selecting the most appropriate factor, and where possible, a direct evaluation of the land impacts related to the specific planting being considered should be carried out.

17.4.2 Environmental Impacts of Waste Utilisation

Several of the examples examined in this book use waste materials or residues as a feedstock. Often, this can deliver a variety of environmental benefits. From a greenhouse gas balance perspective, much depends on whether the material is a waste or a co-product (often determined by whether there is another use or market for it) as this determines how the environmental burdens are allocated from an LCA perspective. Depending on the circumstances, it may be appropriate to either include a proportion of the impacts incurred in production of the main product, not assign any production impacts to the waste/residue, or credit the waste or residue with disposal impacts that have been avoided.

Where genuine waste materials are used and other more severe impacts are avoided, there are genuine ecosystem benefits associated with the bioenergy system, which can be evaluated via life cycle assessment. These could include, for example, avoided airborne pollution associated with burning of crop agricultural residues or avoided water pollution impacts associated with uptake of sewage or wastewater.

Ecologically sound management of waste is a global challenge and integration of this function into bioenergy systems can deliver genuine synergies and ecosystem benefits. One example is the use of perennial crops to remediate contaminated land sites, where heavy metals and other toxins may be removed from soil/land (improving its functionality and utilisation prospects), concentrated in biomass, and ultimately in a manageable biomass residue, such as ash from combustion or gasification. There may even be opportunities for chemical recovery of more valuable elements.

17.4.3 Ecosystem Impacts

17.4.3.1 Soil Carbon

Chapters 5 and 6 combine to give an extremely comprehensive overview of the soil carbon changes that occur when biomass grows on land. The reader should find there sufficient information to allow an assessment of the main potential changes in soil carbon that may occur in their specific bioenergy system and whether these are likely to be significant for the overall carbon balance.

This is an area of research which is evolving rapidly with current work focusing on evaluating the carbon changes in the short and long term associated with bioenergy system deployment. Therefore, it is good practise to seek experimental research outputs that have soil carbon data and are closest in context to the particular bioenergy system being considered or to model the closest appropriate system.

This is particularly important for perennial crop systems where there may be an expectation of a long-term increase in soil carbon. It is also important for systems where residues are being removed. The extent of residue removal may impact on long-term soil carbon contents and assumptions relating to this in an LCA should be justified with reference to the current knowledge base for similar levels of removal with similar species in similar ecological zones, where available.

17.4.3.2 Soil Nutrients

When plants and trees grow, they absorb essential nutrients from soil. These can be macro-nutrients, such as the nitrogen, phosphorus, and potassium present in commercial fertilisers or trace nutrients including metals. If biomass or residues are being removed to produce energy, this may result in a depletion in soil nutrients and this needs to be taken into account in an ecosystem assessment. However, care needs to be taken as the extent to which this occurs depends on the removal rate, species, soil, etc. It should also be noted that there are some potential synergies with the waste management function discussed above, in that waste materials, including sewage and waste water, may be used to provide macro- and micronutrients, thereby avoiding releases to watercourses or alternative treatment requirements.

17.4.3.3 Eutrophication and Acidification

As discussed above, eutrophication and acidification are common environmental impacts associated with production of biomass that are related to the use of land. Careful consideration needs to be given to whether it is appropriate to include these and any relevant counterfactuals in an assessment including consideration of what other use the land may have been ascribed if it were not producing biomass.

17.4.3.4 Fit Within Agricultural Rotation Patterns

Land use and management is constantly evolving to respond to human demand and farmers and land managers will generally design their land utilisation to rotate different land uses with different impacts. So, for example, a leguminous nitrogen fixing crop may be grown in rotation with intensive crop production that depletes nitrogen from soil. Such crop rotations are well-established and form part of most farm's environmental management plans so that they can optimise the long-term productivity and manage the condition of the land.

It is important to think about the role of biomass production within the relevant farming system, e.g. is this likely to be a long term land-use change or a temporary one; is it likely to interrupt, disrupt, or preclude other rotation options; could it provide benefits in opening up a wider set of rotation options or providing a particular role in a rotation cycle.

These are extremely difficult to quantify and depend very much on the context of the bioenergy system development, but should be taken into account in an overall assessment where feasible. They may be particularly useful in giving added insight into why bioenergy is being implemented and the appropriate scope of system given the geographical, temporal, and other interfaces.

17.5 AIRBORNE EMISSIONS

17.5.1 Air Quality Impacts

A key impact of bioenergy systems that should be considered in comprehensive environmental or life cycle assessments is the potential impact on air quality of biomass utilisation. As above, this may need to be seen in the context of a displaced alternative use depending on the goal and scope of the study.

Often, bioenergy system involves some form of combustion either of the solid biomass or a derived liquid or gaseous intermediate vector. No combustion system is ever 100% efficient and there will inevitably be release of carbon monoxide to atmosphere as well as the expected carbon dioxide and water. There may also be particulate emissions from solid combustion and nitrous oxide emissions. Particulates are important to characterise since the health and environmental impacts vary depending on the size of the particle, commonly described as PM2.5 (particles $<2.5\,\mu m$ in diameter), PM10, etc. Generally, finer particles have more significant toxicity and health impacts as do particles with adsorbed metals.

17.5.2 Greenhouse Gas Impacts

Particles also have a scattering effect on light, which can result in a greenhouse gas warming or cooling effect depending on the characteristics of the particle. There has been concern expressed in relation to the release of 'black carbon' (particles from combustion which absorb radiation and can influence the GWP of a system). It should be noted that the GWP impact associated with particulate emissions tends to be of much shorter term than that associated with carbon dioxide and methane as the species have a shorter atmospheric half-life. Data in respect of the impact of particulate emissions and black carbon in particular on the net greenhouse gas impact of bioenergy systems is beginning to emerge, but at the time of writing there is not a conclusive dataset that LCA practitioners could consult and use. The best recommendation would be to seek primary data for a combustion system as close to the one being considered as possible (device type, size, and feedstock) and evaluate the potential impact with appropriate sensitivity analysis where data is missing or uncertain.

17.6 RECONCILING TRADE-OFFS

Section 17.3 outlined a range of possible ecosystem impacts associated with bioenergy systems and pointed out that improving one impact may result in another one deteriorating. This is true not only with ecological impacts, but also when we look wider to social and economic impacts of bioenergy systems. It is very difficult, therefore, to define minimum thresholds or levels that are 'acceptable' for bioenergy systems as this can result in under-performance being tolerated or negative impacts in the systems that are not being so closely monitored or measured.

A holistic approach to bioenergy is therefore necessary that recognises and values all of the impacts (ecological, economic, and social). That is then compounded by consideration of the different bioenergy stakeholders and their perspectives. One might consider a particular impact to be of key importance, while another may disagree. It may not be possible to simultaneously deliver high levels of greenhouse gas savings with low costs of energy access. Trade-offs are almost inevitable and tools such as multicriteria analysis can help in identifying and investigating these to support decision making.

17.7 KEY MANAGEMENT LESSONS

This chapter has reviewed the main environmental impacts of bioenergy systems beyond the greenhouse gases which have been the core focus of the book. When these are taken into account, some recommendations emerge that should give general guidance for management of bioenergy systems to maximise environmental benefits (including greenhouse gas reduction). These are summarised below, but, as ever, context is important and the reader is urged to consider the individual situation in which a particular bioenergy system is being developed.

- Agricultural residue systems (including straw, cattle manure, etc.)—It is important to consider the impact of residue removal on soil organic carbon and long-term soil nutrient contents when evaluating the overall ecological impact and greenhouse gas balance.

- Purpose grown crops (including oil seed crops, e.g. rape, soy, palm, jatropha), sugar crops (cane and beet), starchy crops (e.g. wheat)—Often the agrochemical input over life cycle will be high and needs to be taken into account, including any releases of ammonia and nitrous oxide from soil when fertiliser is applied.
- Forestry systems—Usually it will be the lower value forest products that are used for bioenergy systems and there is a need to consider the origin of these, the alternative use they may be put to by considering wider markets for that component and the wider forestry system.
- Energy crops—There is a need to consider the land-type and rationale for using these, e.g. is there secondary ecosystem function such as improving soil carbon content, managing waste water, providing flood protection, or improving biodiversity

There is potential for bioenergy systems to be developed in a way that supports wider ecosystem benefits and integrates with other activities in a region. Sensitive recognition and development of these is important for successful bioenergy projects. From an impact assessment perspective, it is important to understand this wider development context and objectives as it may affect calculation methodologies or interpretation of results obtained in greenhouse gas balance assessments.

References

[1] Stern N. Economic impacts of climate change. London: HMSO; 2001.
[2] Bows A. Reframing the climate change challenge in light of post-2000 emission trends. Phil Trans A 2008;306:3363–882.

CHAPTER

18

Policy Lessons: The Role of Policy Regimes in Maximising GHG Savings in Bioenergy Systems

Patricia Thornley, Paul Adams*[†]

*University of Manchester, Manchester, United Kingdom [†]University of Bath, Bath, United Kingdom

18.1 INTRODUCTION

Bioenergy systems are being encouraged at different levels of Government from regional and country level through to wider continental and global policy regimes, targets, and mechanisms to support the development of renewable energy systems [1,2]. A key driver for this is the desire to reduce global greenhouse gas (GHG) emissions to avert the most damaging consequences of climate change. Consequently, bioenergy systems that are incentivised under policy mechanisms such as at the European level should deliver actual quantifiable GHG reductions. However, evaluating the net GHG reductions achieved by bioenergy systems can be challenging with assumptions, modelling, results, and the calculated emissions often disputed because of variations in scope of system, data inputs, or methodological approaches [3–7]. Life cycle assessment is now commonly used as the methodological basis for determining the global warming potential (GWP) of bioenergy systems and increasingly life cycle GHG calculations are required for eligibility under different policy schemes. For example, the EU has in place mandatory GHG reporting for biofuels and voluntary GHG reporting for solid biomass and biogas, although several countries have implemented minimum GHG thresholds.

The principle of life cycle assessment (LCA) is to assess the full impact of a product or system. Many of the challenges associated with bioenergy systems revolve around the fact that it is not a product, but a complex supply chain (system) with extensive physical and socioeconomic interfaces which actually affect their surroundings. Each of these variations (in system scope, data inputs, or methods) effectively amounts to asking a slightly different 'LCA question' which can be answered in different ways depending on such factors as the data

Greenhouse Gas Balances of Bioenergy Systems
https://doi.org/10.1016/B978-0-08-101036-5.00018-5

quality, methodological choice, system boundary, assumptions, and interactions with other systems (consequential impacts). This chapter examines the ways in which bioenergy differs from other forms of renewable energy provision and the consequences of that for effective policy development. To date, there have been a number of incremental changes where policy objectives have been set, instruments developed and then changed, modified, and adapted as new scientific evidence emerges, market conditions change, policy understanding develops, public acceptance evolves, and industry progresses [3,8,9]. The result is a complex policy landscape for bioenergy that does not necessarily always deliver against the key objective of minimising global greenhouse gas emissions.

18.2 RECENT DEVELOPMENTS IN BIOENERGY POLICY

Bioenergy has been explicitly encouraged by the European Union since 1990 [10]. At first, it was recognised as a renewable energy technology that would contribute to greenhouse gas reductions and therefore seen as of equal value to wind, tidal, and other systems. In addition, bioenergy is viewed as being able to directly substitute fossil-fuel vectors which other renewables can't do as easily due to storage issues and not being a primary fuel. However, various studies have emerged which showed examples of biofuel systems that were being rewarded as renewable, but actually incurred greater GHG emissions than mineral oil equivalents, particularly agrochemical-intensive production of annual crops [11].

Further questions around GHG savings were raised by Searchinger [12], who argued that the consequences of indirect land-use change needed to be taken into account and that these could completely offset greenhouse gas savings from biofuel systems. This is the case where land that would have otherwise been used to produce food is used for biomass production.

Biomass that originated from forestry systems therefore emerged as a more dependable way of achieving greenhouse gas savings as it avoided both intensive agrochemical application and land-use change. However, concerns were then raised that encouraging utilisation of forest products could influence forest management techniques (e.g. earlier thinning) to the extent that the standing stock of carbon in existing forest systems is threatened [13].

The cumulative impact of the above issues has been frequent changes in policy, which have created uncertainty in the market and a development hiatus. It could be argued that the root cause of this has been policy moving ahead of scientific knowledge and understanding. When the Renewable Energy Directive [14] was first put in place to encourage bioenergy systems at European level, there was little policy awareness of the potential impacts of agrochemical application, land-use change, or forestry management on the overall greenhouse gas balances of bioenergy systems. Scientists were aware that bioenergy systems did incur some GHG releases, but these were calculated as the emission associated with processing, transport, etc. and generally considered to not have a significant impact on the net greenhouse gas savings of the system.

However, the intense scrutiny encouraged a proliferation of studies that showed that, under certain conditions, there may be some systems that do not achieve significant greenhouse gas savings [5,6,12,15]. Policy-makers were forced to amend existing legislation to swiftly ensure that the mechanisms they had put in place did not inadvertently encourage bioenergy systems that had unacceptably high levels of GHG emission. A good example of this is the EU introducing land criteria and minimum GHG savings for biofuels and, more recently, the

United Kingdom and other European countries setting a minimum 60% GHG saving against fossil-derived electricity and heat.

The most pragmatic approach (for reasons of speed, ease, simplicity, competitiveness, and market reassurance) has been to maintain the essence of the existing policy instruments, but adapt them to preclude the worst bioenergy systems. This has resulted in moratoria on development [8], targets being revised downwards [9], prohibition of material originating from certain land types, awarding of multiple credits for advanced technologies, and requirements for minimum levels of demonstrated life cycle GHG savings. The result is a complex policy landscape for bioenergy as instruments have been adjusted to follow the latest scientific understanding on a frequent basis. This creates market uncertainty and therefore affects investor confidence which demonstrates the direct link between policy and economic development.

The situation is further complicated by the fact that bioenergy is the renewable technology that possibly interfaces most with the social, economic, and wider environmental spheres because of its use of land and interface with food [16]. This means that there are other attributes of bioenergy systems (such as job creation, soil remediation, biodiversity and landscape changes, storage and dispatch ability) that are also valued by those encouraging implementation, so the policy objective cannot be simply reduced to a one-dimensional problem of maximising greenhouse gas savings from bioenergy. Employment, social, and rural development are key factors in developing effective policy for bioenergy and these advantages cannot be understated. This is highlighted by the numerous interactions of bioenergy with the millennium development goals, and the fact that bioenergy often falls within the remit of many different Government departments. However, for the purposes of this chapter, we will focus on the single objective of maximising global greenhouse gas reductions and consider how policy mechanisms that actually incentivise these emission reductions could be framed.

18.3 REVIEW OF KEY CHALLENGES IN BIOENERGY POLICY DEVELOPMENT

A review of existing literature on bioenergy policy development indicates six key areas where bioenergy systems differ from other renewables and the significance of each of these for policy development is discussed below.

18.3.1 Uncertainty

There are various forms of uncertainty associated with calculation of the GHG balance of any bioenergy system. Conventionally, economic theory would indicate that the uncertainty drivers associated with bioenergy mean that hybrid instruments are likely to be required, e.g. where price instruments are used, but with an implementation ceiling or quantity instruments with quality constraints [17]. This approach can encourage renewable energy capacity that achieves specified minimum greenhouse gas savings, but has no available mechanism that allows us to actually maximise greenhouse gas savings. Just as greenhouse gases are 'counted' in the IPCC framework in order to promote reductions [18], they must be 'counted' in any policy mechanism that strives to reduce GHGs effectively.

Uncertainty is one of the biggest challenges in undertaking a GHG assessment of a bioenergy system or particular biomass-to-energy conversion pathway. When developing effective policy, this presents policy-makers with difficult decisions that encourage the most beneficial bioenergy supply chains in terms of GHG emission reduction. For example, attributional LCA is the required approach for individual operators to calculate GHG emissions in their biomass-to-energy product. Farmers and bioenergy developers/operators can collect data on the specific attributes or inputs required to cultivate, harvest, collect, process, and produce bioenergy from biomass, i.e. diesel use, fertiliser use, crop yield, transport distance and method, process energy, and conversion efficiency. However, within this there are often numerous uncertainties that are not captured by selected accounting methods. Examples include carbon sequestration, N_2O emissions from soil, losses in the supply chain, fugitive emissions, and so on. If policy and research does not address these uncertainties, then there is a risk that optimal biomass conversion pathways are not being implemented due to inaccurate accounting or GHG methodologies being implemented by policy-makers.

While these data and scientific uncertainties are important for the calculated GHG emissions in individual supply chains from the policy perspective, it is the uncertainties in the consequential impacts or counterfactuals that present the larger challenge. As described in previous chapters, considerations such as fossil-fuel comparator (Chapter 4), soil carbon (Chapter 5), land-use change (Chapter 6), externalities (Chapter 3), ecosystem services (Chapter 17), and emissions from counterfactual products or systems are critical for policy-makers to address and incorporate into policy.

18.3.2 System Interfaces/Boundaries and System Scopes

A second key issue is that of interfaces with other systems. Many published LCAs of bioenergy systems which show divergent results actually use quite different scopes of systems. Common areas of divergence are whether land-use change is included in an LCA assessment and whether account is taken of the full supply chain emissions where the biomass used for energy purposes is only one of multiple products (or co-products). Another example is accounting for avoided or negative emissions from feedstock supply such as waste that avoids going to landfill or is managed in a different way via bioenergy than it would be through conventional waste management. Consequential impacts are one of the main issues in system boundary definition that are difficult to address with accuracy and certainty. These factors can have a very significant effect on the reported GHG savings for a bioenergy system [12] and therefore demonstrate the importance in LCA of having a clearly defined goal and scope. The role of policy-makers is therefore to develop policy approaches that coherently and transparently undertake consequential LCAs for policy development, i.e. rationally assessing how bioenergy systems interface with other ecosystems and economic sectors. At the same time, policy-makers are required to develop and implement attributional LCAs for bioenergy operators that have a clearly defined system scope to ensure the data collected and reported is a fair reflection of the GHG balance of individual biomass supply chains.

18.3.3 Accounting for Land

Land-use change emissions are reported and managed within the IPCC framework in a section separate from that where energy system emissions are reported [19]. Essentially, recourse is made to data relating to land-use cover to assess the level of emissions from the land use. This incorporates emissions such as N_2O from soil during crop cultivation, which can be significant for some bioenergy feedstocks such as annual crops with high N-fertiliser demand. If a bioenergy policy instrument is to minimise global GHG emissions, it must actually include and account for these emissions from land. At present, this is done in United Kingdom and European legislation as part of a life cycle calculation that includes the land emissions from biomass supply in the bioenergy supply chain and compares this with the fossil-fuel energy equivalent energy provision. This approach does not necessarily result in a reduction of emissions from land compared to previous land use, nor does it appropriately incentivise a minimisation of those emissions.

18.3.3.1 Land-Use Emissions

Land emissions are highly variable due to geographical variation and depend on ecological zone, agronomy, topology, timing/weather, etc., making them challenging to measure and control. The best that can be done is to encourage best practise which needs to be done within the agricultural sector as part of a wider strategy focusing on land-use and agricultural emissions. In terms of annual crops, there is much that is already being done and recent evaluations in the United Kingdom show that lower emission levels are being reported, for example, for N_2O emissions levels from cereal production [20]. At the farm-level, land management and crop rotation are crucial in minimising GHG emissions, while encouraging wider ecosystem service benefits. Longer and more varied crop rotations are one such way that land-use emissions can be minimised, therefore the role of the policy-maker is to work closely with farmers, land managers, agronomists, and scientists so that policy instruments can be used to encourage best practise in farm and land management.

From a policy perspective, the question is how to encourage reductions in soil emissions when land is being used for bioenergy purposes. In theory, there should be no difference between emission levels whether the crop is used for bioenergy or other purposes. In practise, there are some nuances that are worth noting, e.g. wheat produced for bioenergy does not require the high protein content favoured for bread wheat and so there may be variations in how agrochemicals are applied, but the reality is that this component of the GHG emissions is highly variable, difficult to measure, and often beyond the control of the bioenergy interests in the chain. The only way of assuring that reductions are actually being achieved in feedstock production is to encourage such reductions across the agricultural sector. If farmers have to report GHG emissions for all crops, then that would encourage lower agrochemical application rates and best practise in application. Specific financial incentives could even support reduction measures like application of N inhibitors. However, the most effective mechanisms for reducing emissions from land should come from the agriculture rather than the energy sector. Otherwise, there is a risk that there is not a net overall reduction in global soil emissions, but rather the land chosen for bioenergy reporting is selected on the basis of the land with lowest emission under a farmers' jurisdiction.

18.3.3.2 Land-Use Change Emission

In bioenergy LCA calculations, it is possible to include a component that allows for direct greenhouse gas emissions related to land-use change. Requiring these to be included in any GHG calculation for bioenergy would reassure that at least these land-use change emissions were being accounted for and thereby there was some incentive to minimise them. When these are included, they can completely overwhelm any GHG reductions achieved, particularly in tropical regions or if forested or scrub land is involved [15]. There is also the potential for indirect effects resulting from land-use change to magnify this effect [12].Therefore, it is extremely important to ensure that bioenergy implementation does not incentivise inappropriate land-use change. This has been incorporated into current legislation by excluding bioenergy systems that have used biomass from high value carbon lands [21] and requiring land-use change to be incorporated into assessments of GHG balances, with only those reaching minimum thresholds being acceptable [22]. This does not necessarily incentivise minimisation of greenhouse gas emissions from bioenergy deployment, but does take some account of emissions when converting land to bioenergy use.

18.3.4 Spatial Governance

Climate change is a global phenomenon and global greenhouse gas emissions are of global concern. This presents very significant governance challenges. The UNFCCC framework addresses this by placing responsibility on individual countries to report and regulate the emissions within their territorial boundaries [18]. This makes sense from a policy objective perspective, since those emissions are under their direct control and, if all countries globally secure a reduction in territorial emissions, then a net global reduction should result.

Of course, all countries trade goods and products with each other, so the coal which is burnt in a United Kingdom power station may not originate from the United Kingdom but the direct (scope 1) emissions are counted in the country in which they are released. Likewise, any sequestration from crop or biomass growth is counted in the country in which the growth takes place. Therefore, under the UNFCCC framework, there is an incentive for a country to grow biomass as it receives sequestration credit for this, but there is no real incentive to use biomass, as this results in a similar level of emissions to fossil alternatives. However, when we consider the growth/sequestration and utilisation/emissions together, then the fuel could logically be described as low carbon, even though this is not apparent under a territorial reporting system.

Biomass is therefore ascribed a low, net carbon emission factor by virtue of sequestration at the location of growth and release at the location of utilisation. The policy objective of minimising greenhouse gas emissions therefore cannot be achieved by considering the energy system alone, but only by considering the supply chain. If the supply chain is in the same country as the utilisation plant, then it is relatively straightforward to gather data and there can be a direct incentive to reduce supply chain emissions and maximise sequestration since all those emissions are captured on that country's GHG balance account. However, where biomass is imported, governance is more challenging. First of all, the collation of relevant emissions balance data is more difficult. Secondly, the aim of minimising supply chain emissions is not under the direct control of the purchaser of the fuel or the energy producer. This essentially means that there is no incentive under the UNFCCC framework to minimise net emissions.

Additionally, we are effectively adding together GHG emissions that are normally distinct sectors: agriculture/forestry and energy. In the energy sector, there is an incentive to minimise emission with a long history of various governments' worldwide setting targets for GHG emission reduction and low-carbon energy provision, but this is less true in agriculture, and so there is no control or mechanism in place to control the emission from biomass production.

On one hand, it could be argued that it would be inappropriate to control such emissions, since biomass production is the same as any other agricultural commodity and, if we do not control the GHG emissions associated with rice production (a much more carbon-intensive process), why should we control biomass production. But the counter argument is that it is perverse to credit a bioenergy source as being 'low carbon' in one location if it is actually incurring significant emissions at another location.

Therefore, some consideration of the supply chain emissions incurred at another location does seem appropriate. Nevertheless, these remain emissions associated with production in the producer country and, so from a management and incentivisation perspective, they are really best treated as such. In other words, it makes sense to invoke responsibility for minimising emissions with the party that is best placed to implement such minimisation. In the case of biomass supply chains, there are many reported examples of perverse consequences of the failure to do this and many of the concerns relate to chain of custody issues.

If stakeholders are to have confidence in the ability of bioenergy systems to achieve greenhouse gas reductions, it is essential that there is confidence in the governance procedures at the point of production, not just those at point of use. Implementing such remote oversight and management is challenging and has mainly focused to date on certification procedures. These may be able to give confidence related to the manner of production and level of greenhouse gas emissions associated with a particular batch of material, but do not necessarily result in minimisation of global greenhouse gas emissions.

18.3.5 The Importance of Counterfactuals and Avoided Emissions

Bioenergy systems are generally encouraged as part of a transition to low-carbon, renewable energy. Therefore, when calculating a greenhouse gas balance for a bioenergy system, we are generally interested in how the bioenergy system performs compared to the previous energy service delivery. In order to actually achieve a reduction in greenhouse gas emissions, it is essential that the new system has a lower level of emissions than the previous one. Inevitably, the greatest greenhouse gas reductions will be achieved where previously a relatively high GHG alternative was in use. This introduces several policy challenges:

First of all, policy incentives and legislation can only influence the new system, not the previous one. So there may be scope to encourage the new system to have the lowest possible GHG emissions, with appropriate use of policy levers, but the figure for reductions is still usually more sensitive to what is considered to be the previous or incumbent use. This point of comparison is the counterfactual for any assessment of reduction and so will have a significant impact on the final GHG balance results.

Secondly, it is often argued that the actual point of comparison (particularly with a system that involves continuous activity/operation) should not be the previous system, but should rather be the one that would have otherwise existed at the time of energy generation. This argument is commonly invoked, e.g. when dealing with waste material. So, it is argued that bioenergy should be credited with the avoided emissions associated with landfilling of material that would have occurred if the material had not been diverted for energy purposes. That may make sense from an ideological perspective, but needs careful consideration. The circumstances under which a particular material is a waste or a product can change with market demand; so it can be difficult to distinguish the two under some circumstances. If a material has previously been treated as a waste and actually caused greenhouse gas emissions, then using it instead for bioenergy will patently reduce emissions. However, if that material previously had a use, it may be argued that there is now a changing market and so it should be credited as a waste with appropriate credit for avoided emissions. An example of this scenario is forest residue material. The main product of a forest is the timber from the saw log and this delivers most of the forest revenue, so residues are often left as waste, not economically worth recovering. However, introduction of a bioenergy market that values that material can result in a market price for residues. A different example would be organic household waste that previously went to landfill, but in more recent times would be source-separated and likely go to compost. The counterfactual is therefore different depending on whether the assumption is that landfill is avoided, or compost is replaced with bioenergy production.

A third consideration with counterfactuals and avoided emissions is whether policy-makers encourage the most efficient use of the limited biomass resource. For example, by comparing biomass electricity with coal electricity then larger GHG emission savings may be achievable than if, say, biomethane or solid biomass boilers replace natural gas. This is due to the relatively high emissions of coal combustion. However, this does not fully consider what is the best use of biomass as electrical efficiencies are around 35%–40% efficient, so unless the heat is utilised, a lot of the primary energy in the biomass is lost. In contrast, an 80%–90% efficient biomass boiler may achieve lower GHG savings, but produce a higher amount of renewable energy from the same feedstock. Considering the efficiency is therefore important, but the alternative technologies should also be considered. Biomass electricity may in many cases reduce GHG emissions by the largest amount; however, electricity can also be displaced with several other renewable technologies such as wind and solar. Heating is harder to displace with other renewables, so there is a trade-off between GHG savings and the alternatives available.

18.3.6 Temporal Aspects

In traditional life cycle assessment (LCA), dynamic aspects of when carbon releases take place are largely ignored and averaged over the time period being considered for the whole system. This approach has several failings when applied to bioenergy systems, outlined below.

18.3.6.1 Carbon Stocks and Flows

First of all, the problem of climate change is caused by transfer of carbon in the terrestrial ecosphere to the atmosphere. Terrestrial biomass therefore has many ecosystem service functions, but key among these are the maintenance of a solid carbon stock, which precludes the carbon being present as an atmospheric greenhouse gas and also maintains the capacity of the terrestrial ecosphere to continue to absorb carbon dioxide via photosynthesis. If the terrestrial carbon stock reduces over time (e.g. via deforestation), this results in a net exchange of carbon from the terrestrial ecosphere to the atmosphere; the greenhouse gas impact of which needs to be taken into account.

18.3.6.2 Biogenic and Fossil Carbon

As stated above, it is important to assess long-term changes in the phase of global carbon stock. This applies whether the terrestrial carbon stock is in the form of a fossil or a biological reserve.

However, when dealing with the biological or biogenic carbon, there are two key additional considerations: first the biogenic material effectively cycles or recycles between atmosphere and ecosphere on a cyclic basis and so the length of this cycle is important, and secondly, maintenance of the carbon stock (or planted area) provides opportunity for further sequestration of carbon dioxide in the future. Arguments that bioenergy systems are carbon-neutral are based on the assumption that all carbon present within a bioenergy material has actually originated in the atmosphere and so growth of the biomass has effectively removed carbon from the atmosphere [23]. Therefore, any thermochemical or biological conversion of the biomass which transforms the carbon in biomass to carbon dioxide in atmosphere does not contribute to any net additional greenhouse gases in the atmosphere.

Of course, it is important to ensure that conversion of biomass C to atmospheric C does accurately describe the full set of reactions going on. Any additional greenhouse gases generated by steps in the supply chain must be fully accounted for and life cycle assessment is a useful tool for identifying and quantifying these. Also, any carbon in the biomass that converts to carbon dioxide is effectively recycling atmospheric carbon, but if carbon converts to other products, e.g. methane or organic black carbon, it will have additional radiative forcing impacts that need to be taken into account.

This approach contains two implicit assumptions:

1. That the timescale of the recycling of the carbon is short in comparison to the timescale being considered for long-term greenhouse gas accumulation. This is discussed in Sections 18.3.6.3 and 18.3.6.4.
2. That the transference of a unit of carbon from biomass to atmosphere does not impact on the wider ecosystem in a way that has other greenhouse gas consequences. An example of where such transference may result in significant impacts is when the biomass growth or carbon sequestration rate of the terrestrial biomass is affected by removal of a portion of the standing biomass. Forestry systems are a focus of particular interest in respect to this and are discussed further in Section 18.3.6.5.

18.3.6.3 *Bioenergy Systems in a Cumulative Emission Framing*

Work by Anderson, Bows and Mander [24] focused on the total cumulative greenhouse gas emissions released over time to demonstrate that climate policy focus needs to be on the total emissions released over all time rather than on annual rates of emissions and that 2050 or similar long-term targets are not sufficient to ensure climate objectives, since it is the total cumulative emissions released over time that matter rather than the snapshot emissions in a particular year.

Logical analysis of this framing shows that the timing of GHG release is important and relevant to climate policy objectives. Failure to make sufficient progress in reducing net greenhouse gas emissions sooner in time means that the final, long-term, allowable annual emission rates will have to be lower and the rate of decarbonisation to reach that end-point much greater in the future. The corollary of this is that near-term reductions in emissions are more valuable than long-term reductions and so the timing of achieving net reductions does matter.

With many low-carbon energy systems, there is an initial release of greenhouse gases (often related to industrial production or installation of materials or devices) which is followed by low-carbon energy provision in the longer term. However, bioenergy is different in several respects:

First of all, the energy associated with the production of conversion plant has been demonstrated to be very low (to the extent that it can be ignored in most calculations).
Secondly, there is an ongoing GHG burden associated with fuel provision throughout the lifetime of energy provision.
Thirdly, sequestration of carbon dioxide from the atmosphere precedes release for bioenergy systems. Therefore, different bioenergy systems can have the same net greenhouse gas burdens over lifetime but very different temporal profiles. It is therefore possible to choose bioenergy systems that will deliver higher levels of GHG reductions earlier in the project time duration. This can be done by capitalising on the sequestration properties of biomass. So, for example, if we fix the anticipated future temperature rise that corresponds to a particular CO_2 concentration in the atmosphere and this is contributed to by all emission species released between now and the final stabilisation point. Despite this, we often frame targets as a 'permissible' annual emission level of greenhouse gases and attempt to evaluate the feasibility of achieving energy systems that have decarbonised to that extent.

However, if we decarbonise faster, the end-point carbon intensity that our energy system must reach to have the same chance of staying within a fixed temperature rise will be higher. So near-term decarbonisation allows a less stringent long-term target. In addition, this means that the rate of decarbonisation that must be achieved by the energy system is also less extreme if we decarbonise earlier in time.

Bioenergy offers significant advantages in this respect since it is a renewable technology which places sequestration first and emissions later. Therefore, if compared to another renewable technology with the same carbon intensity (measured in the conventional manner of GWP over a 25-year period), the actual emissions profile is very different; with the bioenergy system having lower emissions in the 'more valuable' earlier years. This could potentially ease the long-term end-point target and decarbonisation rate.

18.3.6.4 Timescales and Urgency

The requirement to decarbonise the energy system has arisen because of a transfer of carbon from the terrestrial ecosphere to the atmosphere. This had been kept in balance with fluxes between both spheres for many hundreds of years. However, anthropogenic activity in the industrial revolution significantly accelerated the rate of transfer from land to air and this has continued to accelerate to the present day. Scientists have calculated the cumulative remaining emissions space and formulated targets that tend to focus around a very significant (~90%) reduction in the next 35 years [24].

The need to address greenhouse gas releases and formulation of targets was first seriously addressed around 1990 and at that point the 2050 targets were two generations away and seemed 'a long way off'. With such a radical reduction required, yet over a relatively long time scale the question of the timing of emissions did not seem pertinent. It would make no noticeable difference to the net GWP if the emissions profile was slightly higher in the earlier years or if decarbonisation was deferred in time.

However, the lack of actual decarbonisation in the intervening 25 years has meant that the remaining timescale to achieve the transition is now very much more compact. It is still achievable, but the significance of the actual timing of the emissions is accentuated by the fact that we now have a relatively short period remaining.

18.3.6.5 Contribution and Role of Forest Biomass

As an example, consider a plantation forestry system (which is one of the systems most often referenced when concerns are raised over the timing of bioenergy emission profiles). Carbon dioxide is sequestered fastest during the early period of growth and the annual carbon sequestration (and biomass increment) gradually reduces until a steady value of carbon stock is reached with a mature forest. Harvesting of the planted forest releases a very significant proportion of the stored carbon so that the time-averaged figure for carbon stored over the plantation lifetime is around.

The vast majority of forestry material used in bioenergy plants comes from purpose grown plantation forests. These are planted as an economic enterprise with the main source of income originating from the high value, saw log product. The market value of this wood is generally too great for it to be attractive to the energy market. However, lower value roundwood and residues are increasingly being converted to feedstocks for bioenergy plants and concerns have been raised about the sustainability of this. If the removal of forest material leads to a long-term reduction in the forest carbon stock, that should be taken into account in any LCA calculations. However, the challenge associated with assessing the impact of a removal is establishing if the energy utilisation has actually caused the removal and can therefore be linked to it in any meaningful way. Considered from another perspective, we need to consider what would have happened if the bioenergy demand did not exist: would the forest have been managed differently and would the net impact have been a higher or lower level of greenhouse gas emissions?

18.4 OBJECTIVES OF AND RECOMMENDATIONS FOR EFFECTIVE POLICY INSTRUMENTS

It is worth remembering that no renewable energy system is free of greenhouse gas emissions. All incur some degree of energy and carbon emissions. For example, recent work on life cycle assessment of solar energy systems has shown that the avoided GHG impacts 'were more than five fold' the impacts of the solar system. This suggests a net greenhouse gas reduction of at least 83% [25] and other work has produced figures of 0.0402 kg CO_2/kWh for solar PV [26]. This is not very different to the sort of figures that bioenergy systems routinely achieve [27] and so it is not the case that bioenergy systems are 'worse' than other renewables in terms of greenhouse gas reductions. It is simply that different bioenergy systems produce different levels of GHG savings and the variability across the full range of GHG savings that may be achieved is very much larger than is generally the case for solar, wind, or PV systems.

The consequence of this is that it is insufficient, from a policy perspective, to simply state that the objective is to maximise the renewable energy produced. Instead, the focus must be on greenhouse gas reductions alongside renewables. At present, there is a requirement to achieve minimum levels of savings. This ensures that the worst offending systems are not developed, but it also encourages a 'lowest common denominator' or 'tendency towards mediocrity'. In other words, there is no incentive with the current policy regimes to actually maximise the GHG savings achieved, only to meet the current thresholds. No matter which policy instrument is used, the maximisation of GHG savings can only be encouraged if the key objective function being measured is the GHG savings and those are being remunerated in the policy mechanism. Therefore, it makes eminent sense to have a mechanism that rewards bioenergy on the basis of the actual GHG savings achieved. Support mechanisms such as feed-in tariffs or other incentives could therefore be tiered so that a higher rate of return is available for bioenergy production with larger GHG savings. An alternative policy approach is through the use of tax, and in particular, carbon tax could be used to encourage those systems that save or avoid the most carbon. Waste and residues as feedstock would benefit the most from this as there is the potential to generate carbon credits which can be sold on national or international carbon markets.

Although other considerations such as social and economic development are not the focus of this book, it is clear that these factors should also be taken into account as bioenergy policy is developed. As bioenergy systems affect people and places, the policies that are introduced should account for other facets of sustainability such as other environmental impacts (air quality, soil quality, water quality, resource depletion, etc.), economic prosperity (employment, industrial development, profitability etc.), and social factors (good jobs, energy affordability, well-being, etc.). Hence, there is a balance between solely focusing on GHG reductions and wider sustainability considerations.

When developing policy for bioenergy that focuses on GHG savings, policy-makers require robust scientific evidence about the emissions that arise from different biomass conversion routes. These assessments increasingly require data and quantified emission estimates of the consequential impacts of bioenergy production. For example, recent research has shown a wide range of values for GHG emissions depending on the system boundary and counterfactual assumptions. Policy-makers therefore need to work closely with scientists and LCA practitioners to ensure the models, goal and scope, data, and assumptions are robust. It is important that it is acknowledged that exact results are not always practical and so performing different types of LCA should be used to inform policy decision-making by highlighting emission hot-spots, identifying externalities, reducing uncertainty, and assessing the impacts of bioenergy policy on other sectors.

18.5 CONCLUSION

The review of policy issues and challenges above does not facilitate recommendations for alternative policy mechanisms that would incentivise greenhouse gas savings: that would require detailed econometric modelling. However, it does give some clear direction for development of policy instruments that could be more effective at encouraging real greenhouse gas reductions using biomass:

(1) It is essential that a framework designed to encourage low-carbon intensity bioenergy actually counts the greenhouse gas emissions associated with the bioenergy system and incentivises lower carbon intensity systems via higher rewards.

(2) Such instruments should be targeted on the currently worst performers, e.g. heat incentives that prioritise users who do not have access to the natural gas grid, electricity incentives where viable alternatives such as solar PV are not attainable, or AD incentives that focus on less well-managed manure and slurry, e.g. possibly with a focus on smaller farms to manage farm waste.

(3) Bioenergy users have little or no influence over the counterfactuals that can make such a huge difference to LCA calculations and, in reality, the emissions that 'might have occurred' if a user did not adopt bioenergy have no relevance to either a country's existing emission levels or the ones they will report when the bioenergy system is implemented. So, while LCA practitioners may often focus on what would have happened otherwise as the theoretical alternative, this cannot be effectively measured and should not have any place in policy incentives but can be used to inform policy decision-making. Therefore, crediting systems (e.g. for avoided landfill emissions) is inappropriate in an incentive mechanism, but modelling the avoided emissions helps determine the likely success of a policy incentive. The mechanism should instead focus on a scope of system that details what we know is actually happening. So it is legitimate to include an offset for sequestration that has already occurred (as long as this is in a relatively recent time frame, e.g. post the 1990 Kyoto protocol), but it makes no sense to include offsets for emissions that may or may not happen in future. It is, in some ways, unfortunate that the binding policy commitments focus on the concept of 'reductions', but the United Kingdom's Climate Change Committee has set carbon budgets for future time periods to achieve climate policy ambitions and other countries have done similar. Our focus really should be on ensuring we remain within these budgets rather than thinking of what might have happened if we had not implemented renewable energy. So, from an existing policy commitment perspective, it could be argued that it makes sense to look at reductions compared to what we were doing in 1990, but more recent understanding drives us towards a focus on what we are actually outputting year by year. This means that our policy focus should be minimising the actual outputs and so the greenhouse gases that are counted for a bioenergy system should be the actual physical output of the system without any adjustment for what has been avoided. This has the key benefit of minimising actual emissions in the coming years.

(4) However, such an approach fails to take account of the unique 'sequestration first' attribute of bioenergy, which it may make sense to exploit given the recent very stringent targets arising from the recent Paris COP [2] agreement.

(5) This is when timescales and temporal aspects must be taken into account. It makes no sense to assume that carbon emissions from an annual crop planted last year are increasing long-term atmospheric CO_2 concentrations. They are not. Terrestrial plants assimilate more than 95% of their carbon from aerial CO_2 [23] and so this is simply recycling of carbon on a short-term basis and makes no difference to net long-term concentrations. From a policy perspective, exactly the same argument can be presented for all carbon sequestered after 1990. The sequestration of that carbon contributed to a real atmospheric reduction in CO_2 after the point at which carbon targets were set and so it is entirely legitimate to ignore those emissions when they are returned to atmosphere during such a short time scale.

(6) Indeed, if we were dealing with the climate change problem in a more generic sense, it would always make sense to neglect emissions that have previously been sequestered from atmosphere. The only reason exceptions to this arise is because we are so close to potential tipping points based on the already accumulated greenhouse gases that releasing further GHG's into air even if only on a temporary basis (the equivalent of recycling) could result in the ecosystem crossing non-linear tipping points which may have consequences that cannot be repaired by the recycling philosophy. For this reason, it is important to look at the actual timing of releases and to consider them against the allowable carbon budgets and we should treat bioenergy releases the same as any other release in this respect.

(7) There is one caveat to this and that is we should also credit the bioenergy system for the sequestered carbon, but it only makes sense to credit what has physically been sequestered, not what might or might not be avoided.

All of this requires a subtly different scope of system from that conventionally used in existing legislative instruments. IPCC commitments record biomass sequestration against the land use and land-use change category, with releases on biomass combustion recorded against the energy use category. However, this does not allow us to examine whether a system being put in place delivers a net reduction since bioenergy combines the carbon emissions from the land and energy sectors. Therefore, supply chain and life cycle thinking have been applied to concoct other system boundaries, which look at the entire supply chain and evaluate whether the agrochemical use results in a significant offset to the carbon released during combustion, etc. As a simplification, most such systems simply ignore the carbon originally sequestered as it is argued it is 'carbon neutral' having originally come from atmosphere. It would make much more ideological sense to actually account for that carbon when it is sequestered, rather than just ignore it. For short-term crops, it will make no difference to the end-result, but for perennial crops and forestry systems, it would provide a much more concrete evidence base and accurate representation of the real carbon intensity associated with bioenergy systems and would address the temporal argument, particularly around foregone sequestration and long-term carbon stocks.

18.6 SUMMARY OF POSSIBLE ALTERNATIVE SYSTEM SCOPE

Summary of Suggested Alternative Bioenergy System Scope

	Scope of system (existing legislation)	Scope of system (proposed)
Included	I1. Transport emissions I2. Agrochemical emissions I3. Land-use change emissions	I1. Transport emissions I2. Agrochemcial emissions I3. Land-use change emissions I4. Carbon sequestered from start of accounting period (1990 or another relevant period e.g. establishment point for a new crop or the period of the relevant carbon budget) I5. Emissions from biomass combustion

Summary of Suggested Alternative Bioenergy System Scope—cont'd

	Scope of system (existing legislation)	Scope of system (proposed)
Sometimes included	S1. Indirect land-use change emissions S2. Avoided landfill emissions	
Not included	N1. Carbon sequestered N2. Emissions from biomass combustion	N1. Avoided landfill emissions (these are not real if they are actually being avoided). However, if the material is of recent biogenic origin then the sequestration of the carbon embodied in the fuel will be already included in I4 above and so the same net result will be achieved, but with a tighter focus on whether the emissions occur during a relevant accounting period or not

The very significant advantage of this system scope is much greater transparency about the sequestered carbon and emissions from biomass combustion. It would also provide greater clarity around the treatment of forestry systems as it adjusts the timeframe to focus on what is tangibly sequestered and/or being released. It also has the advantage of focusing on the real emissions that are actually occurring and forcing climate change rather than theoretical offsets and unjustifiable projections about what might have otherwise happened.

References

[1] Commission E. Directive on the promotion of the use of energy from renewable sources (2009/28/EC, RES Directive); 2009.

[2] National renewable energy action plan for the United Kingdom.

[3] DIRECTIVE 2009/30/EC—amendment to Directive 98/70/EC on environmental quality standards for fuel (Fuel Quality Directive). European Commission.

[4] Bentsen NS, Jack MW, Felby C, Thorsen BJ. Allocation of biomass resources for minimising energy system greenhouse gas emissions. Energy 2014;69:506–15.

[5] Fazio S, Monti A. Life cycle assessment of different bioenergy production systems including perennial and annual crops. Biomass Bioenerg 2011;35(12):4868–78.

[6] Haberl H, Sprinz D, Bonazountas M, Cocco P, Desaubies Y, Henze M, et al. Correcting a fundamental error in greenhouse gas accounting related to bioenergy. Energy Policy 2012;45:18–23.

[7] Thornley P, Gilbert P, Shackley S, Hammond J. Maximizing the greenhouse gas reductions from biomass: the role of life cycle assessment. Biomass Bioenerg 2015;81:35–43.

[8] Gallager E. The Gallagher review of the indirect effects of biofuels production. Renewable Fuels Agency; 2008.

[9] Commission E. Directive of the European Parliament and of the Council amending Directive 98/70/EC relating to the quality of petrol and diesel fuels and amending Directive 2009/28/EC on the promotion of the use of energy from renewable sources; 2012.

[10] Thornley P, Cooper D. The effectiveness of policy instruments in promoting bioenergy. Biomass Bioenerg 2008;32(10):903–13.

[11] Sterner M, Fritsche U. Greenhouse gas balances and mitigation costs of 70 modern Germany-focused and 4 traditional biomass pathways including land-use change effects. Biomass Bioenerg 2011;35(12):4797–814.

[12] Searchinger T, Heimlich R, Houghton RA, Dong F, Elobeid A, Fabiosa J, et al. Use of U.S. croplands for biofuels increases greenhouse gases through emissions from land-use change. Science 2008;319(5867):1238–40.
[13] Holtsmark B. Harvesting in boreal forests and the biofuel carbon debt. Climatic Change 2011;112(2):415–28.
[14] Commission E. Directive 2009/28/EC of the European Parliament and of the Council of 23 April 2009 on the promotion of the use of energy from renewable sources and amending and subsequently repealing Directives 2001/77/EC and 2003/30/EC (Text with EEA relevance); 2009.
[15] Upham P, Thornley P, Tomei J, Boucher P. Substitutable biodiesel feedstocks for the UK: a review of sustainability issues with reference to the UK RTFO. J Clean Prod 2009;17(Suppl. 1):S37–45.
[16] Raman S, Mohr A, Helliwell R, Ribeiro B, Shortall O, Smith R, et al. Integrating social and value dimensions into sustainability assessment of lignocellulosic biofuels. Biomass Bioenerg 2015;82:49–62.
[17] Purkus A, Röder M, Gawel E, Thrän D, Thornley P. Handling uncertainty in bioenergy policy design—a case study analysis of UK and German bioelectricity policy instruments. Biomass Bioenerg 2015;79:64–79.
[18] Guidelines for national greenhouse gas inventories. IPCC; 2006.
[19] Decision No 529/2013/EU of the European Parliament and of the Council of 21 May 2013 on accounting rules on greenhouse gas emissions and removals resulting from activities relating to land use, land-use change and forestry and on information concerning actions relating to those activities. European Parliament; 2013.
[20] Sylvester-Bradley R, Thorman R, Kindred D, Wynn S, Smith K, Rees R, et al. Minimising nitrous oxide intensities of arable crop products ADAS; 2015.
[21] Commission E. Communication from the Commission on the practical implementation of the EU biofuels and bioliquids sustainability scheme and on counting rules for biofuels. Brussels: Official Journal of the European Union; 2010; C160/8–C160/16.
[22] Sustainability standards for electricity generation from biomass. In: OFGEM DoEaCCa, editor. 2013.
[23] Livingston BE, Beall R. The soil as direct source of carbon dioxide for ordinary plants. Plant Physiol 1934;9:237–59.
[24] Anderson K, Bows A, Mander S. From long-term targets to cumulative emission pathways: reframing UK climate policy. Energy Policy 2008;36(10):3714–22.
[25] Carnevale E, Lombardi L, Zanchi L. Life Cycle Assessment of solar energy systems: comparison of photovoltaic and water thermal heater at domestic scale. Energy 2014;77:434–46.
[26] Kabakian V, McManus MC, Harajli H. Attributional life cycle assessment of mounted 1.8 kWp monocrystalline photovoltaic system with batteries and comparison with fossil energy production system. Appl Energy 2015;154:428–37.
[27] Thornley P, Upham P, Huang Y, Rezvani S, Brammer J, Rogers J. Integrated assessment of bioelectricity technology options. Energy Policy 2009;37(3):890–903.

CHAPTER

19

Outlook–for Low Carbon Bioenergy

Patricia Thornley, Paul Adams†*

*University of Manchester, Manchester, United Kingdom †University of Bath, Bath, United Kingdom

19.1 LOW CARBON BIOENERGY NEED

The threat of climate change is real, significant, and imminent. In 2007, Nicolas Stern reviewed the evidence and concluded that the scientific evidence was overwhelming and failure to act would result in losing at least 5%–20% of GDP each year, forever [1] and that the benefits of early action are profound, with investment needed in the next 10–20 years to avoid risks of major and irreversible changes (ibid.).

It could be argued that the global response to date has arrested the previous rapid growth of greenhouse gas emissions and there has even been unprecedented sustained economic growth simultaneous with a reduction in GHG emissions. The Paris Conference of Parties (COP) in 2015 did indeed succeed in producing an agreement that bound nations to pursue efforts to limit global temperature rises to <1.5°C and has been ratified by 141 of the 197 parties [2]. However, the contributions declared by individual countries to date would still result in global emissions increasing by 40% above 1990 levels by 2025 and 44% above 1990 levels by 2030, with aggregate emission levels still expected to be 19% above the level that would facilitate climate stabilisation at a temperature increase of 2°C [3]. There is no doubt that we urgently need to reduce global greenhouse gas emissions and that progress to date is not on a trajectory that will avoid dangerous climate change.

It is therefore extremely important to ensure that renewable bioenergy systems are actually delivering real greenhouse gas reductions. This book therefore started in Chapter 1 with a review of climate change and the policy initiatives that have tried to promote carbon reductions via bioenergy in Chapter 2. It then progressed to examine how life cycle assessment approaches can be used to determine the extent to which bioenergy systems are really delivering against the challenging targets that we must meet. This often requires exhaustive data gathering, lengthy calculations, and pragmatic modelling frameworks that accurately reflect complex changes and inform global policy. It is incredibly important that this is carried out rigorously, carefully, diligently, and consistently to ensure that global bioenergy is appropriately developed to support climate objectives because the risks of inappropriate development are huge.

Greenhouse Gas Balances of Bioenergy Systems
https://doi.org/10.1016/B978-0-08-101036-5.00019-7

19.2 LOW CARBON BIOENERGY POTENTIAL

The challenge for development is to mitigate the risks and maximise the benefits. This book contains many examples of bioenergy systems that can make positive contributions to greenhouse gas reduction targets and should be developed as matter of urgency. While there remain uncertainties in some areas that must be carefully dealt with (forest carbon stocks, indirect land-use change, black carbon, and aerosols), there is undeniably a huge potential for well-designed bioenergy systems to use sustainable feedstocks to deliver much-needed greenhouse gas reductions. Examples of this have ranged from waste wood in Chapter 12 through forest residues in Chapter 14, energy crops (Chapter 11) and even biodiesel from soy in Chapter 10. All can achieve substantial greenhouse gas reductions over full life cycle and could make a significant contribution to greenhouse gas reductions if implementation were supported and progressed. In recent years, it seems that detailed assessments of system performance, greenhouse gas balances, and wider sustainability of bioenergy systems have paralysed rather than catalysed action. Sometimes the precautionary principle has been invoked to delay implementation and, while there may be some outstanding concerns in relation to some feedstocks or systems, these need to be balanced against consideration of the catastrophic impacts of not acting in an informed manner. The importance of considering the whole system cannot be over-emphasised.

Fig. 19.1 compares first-generation and second-generation biofuel systems. Each has the same initial land area, and the area of the symbol in the diagram is proportional to the energy content in the energy vector at that stage of the process. It should be noted that the proportion of the biomass crop which is suitable for the conversion technology chosen is very significantly higher for the second-generation system since it can deal with the whole crop, rather than selecting only the sugar/starch/digestible proportion for the first-generation technology. This is the key over-riding factor that dominates the resource efficiency of the system and results in a much larger bioenergy output per unit of land for the second-generation compared to the first-generation system. The nominal conversion efficiency of the key technology step is actually higher for the first-generation system, but this is not acting on the whole of the accessible feedstock and the vector produced requires further upgrading/refining to be totally suitable

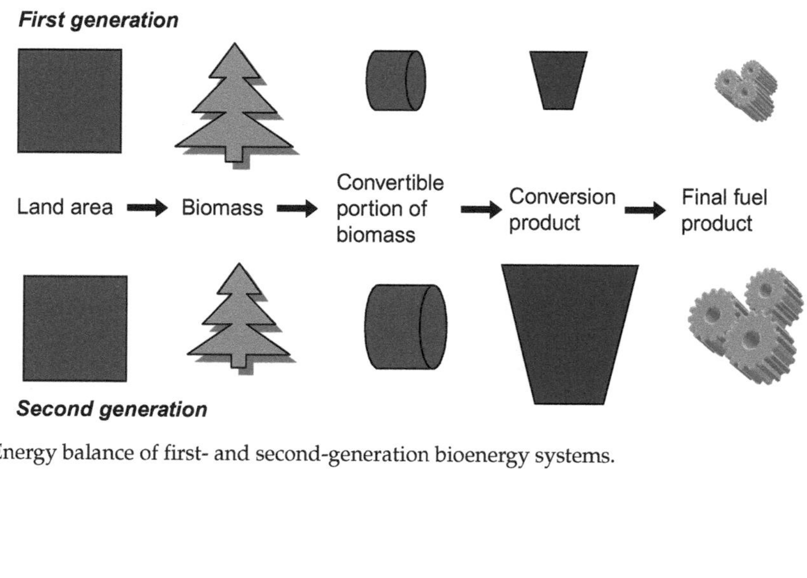

FIG. 19.1 Energy balance of first- and second-generation bioenergy systems.

as a drop-in fuel within the existing infrastructure. These factors combined result in a superior mass and energy balance for the second-generation system. A full LCA would be required to verify the extent to which this improves the greenhouse gas intensity of both systems, but it is clear from the figure that the first-generation system is only likely to be superior if there is a very significant GHG emission (e.g. soil/land-use emissions or emissions associated with agrochemical production) associated with the second-generation system that do not apply to the first-generation system. This may well be the case, e.g. as outlined in Chapter 13, many second-generation ethanol systems require significant pretreatment with associated energy and GHG penalties to release the accessible proportion of the feedstock. However, the energy penalty would have to be very large to offset the supply chain efficiency benefit of being able to make sensible use of the entire biomass feedstock. More detail on this can be found in Thornley [4].

Clearly a whole systems perspective is very important in evaluating the greenhouse gas balances of bioenergy systems. However, this must also be underpinned by *sustainable feedstocks* matched with appropriate *conversion technologies* that maximise benefits and each of these aspects is considered further below.

19.2.1 Sustainable Feedstocks

Sustainable feedstocks are the basic starting point for any sustainable bioenergy system. It should be obvious from the examples in this book that energy crops or biomass utilisation that have large agrochemical (particularly fertiliser) consumption, upset biodiversity levels, aggravate sensitive water balances, or reduce long-term carbon stocks should be avoided. This is as true of bioenergy as it is of other land-use options.

Fortunately, there are many examples of sustainable biomass production that has positive impacts, but this cannot be assured simply by feedstock category/description. Within each biomass category there are many drivers that affect sustainability: for the purposes of ensuring a good greenhouse gas balance, fertiliser use is particularly important for first-generation biofuels; land-use change is important for energy crops; assessment of the long-term carbon stock is key for forestry systems; consideration of composition, alternative uses, and material recovery prospects is important for wastes.

Even when appropriate sustainability constraints on feedstocks are taken into account, there is a very high global bioenergy potential, although the range of that potential is extremely sensitive to modelling assumptions and approaches, with wide ranging estimates [5–7]. It is impossible to be confident simply from the type or label of a feedstock that it is sustainable. There is a need to carefully assess the actual feedstocks being considered and this is challenging when taking into account global supply markets. Certification frameworks (discussed below) become key.

Nevertheless, it becomes abundantly clear that a focus on sustainability of feedstocks will inevitably result in consideration of more wastes and residues [8–10] and this has important implications for conversion technologies.

19.2.2 Appropriate Conversion Technologies

Fig. 19.1 emphasises the importance of being able to access as large a proportion of the crop as possible and Fig. 19.2 shows the broad routes for different feedstock types matched

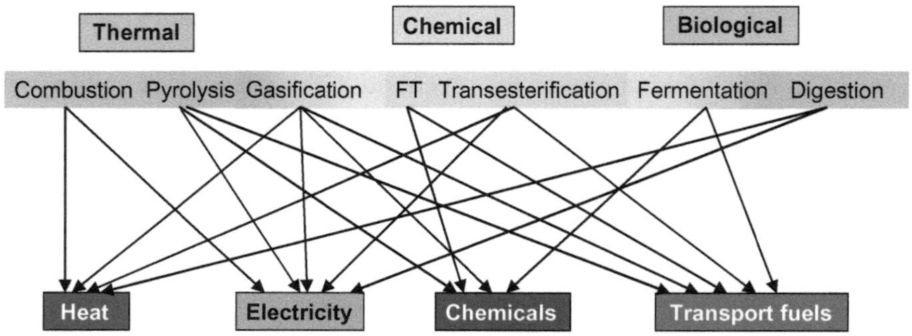

FIG. 19.2 Matching bioenergy conversion technologies to energy vectors.

to different conversion technologies and end-use vectors. Clearly from Fig. 19.2, there is a vast choice of options for configuring processes, and from Fig. 19.1, the appropriate matching of feedstock to conversion technology is critical. There is often a temptation, particularly at lower TRL's or when carrying out challenging laboratory work to focus on a particular feedstock with a fairly narrow spread of properties. This is essential when trying to understand reaction mechanisms, improve performance, and enhance efficiency. However, once that basis has been established, it is critical to evaluate the range of parameters that are tolerable and the extent to which conversion efficiency is affected by variations in feedstock properties.

It may be possible to construct a theoretical system with a very high efficiency and correspondingly good greenhouse gas balance from laboratory trials and to stretch and optimise that efficiency with advanced technologies, exotic materials, and complex process configurations. However, this will only deliver real greenhouse gas reductions if it operates reliably with the feedstocks that are presented on a daily basis in the real world. This means that a plant that can continue to operate smoothly with non-ideal feedstocks is preferred over one that is more selective and that achieving a reasonable efficiency across a range of feedstock parameters is more useful than achieving an excellent efficiency across a narrow range of parameters. So, if we think of process efficiency plotted against any feedstock parameter in which we are interested, we need to take into account that there will be some feedstock with optimal characteristic and some falling to either side of that. For waste-derived feedstocks, the width of that curve will be greater and the peak lower than for purpose grown material which in turn will be worse than for laboratory samples. Critically the greenhouse gas reductions are not related to the output or efficiency peak reached by either of those curves, but to the integral of that output/efficiency with respect to the parameter that is varying, e.g. moisture content, sugar content, and the time for which the facility can operate.

The corollary is that omnivorous technologies that can tolerate wide variations in feedstock parameters are very valuable and could also increase sustainable feedstock quantity.

19.2.3 Maximising Benefits

The imperative need to reduce greenhouse gas emissions and switch to a low carbon economy means that it is key that we make the most of the limited resources (biomass and other) that we have. It does not make environmental sense to deliver low levels of carbon reduction from a unit of biomass resource when we could have delivered a higher level of reduction by using an alternative technology or delivering a different energy vector/servicing a different need. Evaluation of this is challenging since it involves looking at different sectors, different counterfactuals, and the impact in related systems, e.g. is more carbon saved by using a quantity of wood to produce second generation bioethanol, bioelectricity, or a construction beam, given that the counterfactuals for all of those uses are also co-evolving in the transition to a low carbon economy? As discussed in Chapter 18, extracting ideal policy frameworks that will incentivise real carbon reductions is a huge challenge given the inter-related nature of the various co-evolving systems. This needs to be considered in a context beyond bioenergy. However, one point worth noting is that, to date, bioenergy policy and sustainability frameworks have focused on excluding the worst systems and demanding minimum carbon savings for scheme eligibility. That has some merit in avoiding environmental damage, but it falls a long way short of addressing the real challenge of maximising the carbon reduction potential of biomass.

The first step in maximising potential is to actually measure the carbon reductions along the whole system and this is being done in a number of different ways at present. That is not unreasonable, since this book has noted repeatedly that different LCA questions demand different methodological approaches. However, some review and rationalisation could help avoid loopholes that are resulting in certain feedstocks being routed into specific applications, e.g. because they are considered zero carbon at point of collection in that framework, but are required to attribute emissions in another.

The second requirement is that we value the carbon reductions achieved rather than accepting minimum threshold levels. In principle, some form of reward per unit of carbon saved or reward graduated with carbon intensity would make sense and could be linked to the broader emissions trading framework and carbon prices. However, this would only have a significant impact with a carbon price very much higher than todays and there does not seem to be a political appetite for that radical step at present. The UNFCCC provides an appropriate framework for encouraging countries to manage their own emission levels, but the existing framework does not encourage resource efficiency or GHG reductions in the global trade context in which biomass is traded and low carbon solutions need to be delivered.

19.3 LOW CARBON BIOENERGY CONFIDENCE

The role of trade is a very important one for bioenergy. The fact that resources grown and incurring impacts in one country can be used and accrue impacts in another country is notionally valued in the price paid at the point of export/import. So, when a company purchases wood pellets from an overseas forestry company, biodiesel from an overseas producer, or lingo-cellulosic waste from a multinational manufacturing industry, they are not just buying the material, but also buying the associated history and carbon credits. It is therefore essential to have some assurance via certification frameworks and these should go beyond greenhouse gases and indeed beyond environmental impacts to consider wider socio-economic impacts. By importing low carbon fuel, a country is effectively exporting its greenhouse gas reduction commitments and so some responsibility for demonstrating or proving that real GHG reductions have actually been achieved is appropriate. It seems that three components are essential for this:

1. Independent verification of feedstock production standards
2. Transparency of supply chain impacts
3. Commitment to holistic maximisation of benefits and minimisation of impacts

19.3.1 Independent Verification of Feedstock Production Standards (Certification Frameworks)

This book has detailed just a few of the hugely diverse range of feedstocks available for bioenergy systems. Clearly, there is a huge range and it is hugely challenging for bioenergy stakeholders to gather sufficient information on production and supply chain impacts. Already, there are sector-specific organisations set up for soy, palm, etc. that attempt to focus on the steps necessary to improve sustainability of the cultivation of those particular feedstocks. This is a fine balancing act: there is a need to ensure that specific feedstocks are cultivated in a way that is sustainable and this can only really be done by those with detailed knowledge of the impacts of the sector. Separately, there is a need to ensure that the cumulative impact along the supply chain is that greenhouse gas reduction is achieved or low carbon energy is produced.

19.3.2 Transparency of Supply Chain Impacts

Since incentive systems are in place to encourage bioenergy systems for their greenhouse gas, environmental and other benefits it is extremely important that there is transparency about the supply chain impacts of bioenergy systems. This should extend beyond greenhouse gases to consider other environmental impacts since there will inevitably be some trade-offs associated with these balances.

Nevertheless caution needs to be exercised as impacts of supply chains are not necessarily linear and so great care must be taken when using evidence from an existing chain to extrapolate and infer wider impacts. It is also important to consider causality and the extent to which observed impacts have been caused by bioenergy implementation or by other related or unrelated changes.

19.3.3 Commitment to Holistic Maximisation of Benefits and Minimisation of Impacts

As discussed previously there are trade-offs between different environmental and other impacts of bioenergy systems. Delivering energy has a multitude of different impacts, one of which is greenhouse gas emissions. So maximising greenhouse gas reductions may result in other impacts increasing. It is possible to choose and design energy systems with different impacts, but it is not possible to remove them completely. The threshold of what is considered sustainable or unsustainable depends on more than greenhouse gases and should take into account implementation context and stakeholder priorities as well as technical performance and system impacts. In some contexts rural job creation may be valued more than small increases in airborne pollutants; in others carbon soil gains or biodiversity benefits may justify increased energy costs. Reduced greenhouse gas emissions are generally considered as a global benefit and so there may be trade-offs whereby local visual impacts are considered justifiable in pursuit of wider global targets. There are no definitive, 'black and white' thresholds in such assessments: simply multiple shades of grey.

19.4 LOW CARBON BIOENERGY STAKEHOLDERS

While this book has by design focused on the greenhouse gas balances of bioenergy systems, it has also acknowledged the wider environmental impacts of bioenergy systems. However, these are only part of the (positive and negative) impacts that a bioenergy system may have. Social and economic impacts must also be taken into account and the interaction between people and bioenergy systems is particularly strong because of the involvement of land-use and physical labour in biomass production and processing.

In addition it is important to take into account stakeholder priorities when considering bioenergy system design and the trade-offs discussed in Section 19.3.3, since different stakeholders will have different priorities, resulting in preferences for different types of bioenergy system. This provides significant opportunities for development of bioenergy systems to contribute to delivery of sustainable development goals. Bioenergy systems can facilitate access to energy, particularly for rural communities where other options may not be practical and also improve soil quality, fertility and ecosystems with nutritional and health benefits.

19.5 LOW CARBON BIOENERGY SOLUTIONS

The need to decarbonise our energy system is urgent and energy demand is increasing in parts of the world where abundant biomass resources are available. Matching sustainable resources with appropriate conversion technologies can deliver low carbon energy and a host of other environmental benefits. There is a need to manage other impacts and to ensure appropriate design to maximize benefits, but the need for greenhouse gas reductions and energy access are both urgent and reducing greenhouse gas emissions sooner rather than later means that final targets do not have to be as severe and reduction trajectories can be less radical and more manageable. It therefore behaves us to accelerate development of sustainable low carbon bioenergy solutions that can deliver significant greenhouse gas reductions rather than delaying in hope of improved crops, more efficient technologies or alternative options which may not materialize. Bioenergy has a crucial role to play in global decarbonisation and engineers have a crucial role to play in implementing appropriate bioenergy systems with due regard for the full spectrum of impacts, including greenhouse gas balances.

References

[1] Stern N. The economics of climate change. London: HMSO; 2009.
[2] Paris Agreement. In: Nations U, editor. 2015.
[3] Aggregate effect of the intended nationally determined contributions: an update. UNFCCC; 2016.

[4] Thornley P. Biofuels review: report for government office for science. Manchester: University of Manchester; 2012.
[5] Smeets EMW, Faaij APC, Lewandowski IM, Turkenburg WC. A bottom-up assessment and review of global bio-energy potentials to 2050. Prog Energy Combust Sci 2007;33(1):56–106.
[6] Hennig C, Brosowski A, Majer S. Sustainable feedstock potential—a limitation for the bio-based economy? J Clean Prod 2016;123:200–2.
[7] Slade R, Bauen A, Gross R. Global bioenergy resources. Nat Clim Chang 2014;4:99–105.
[8] Searle SY, Malins CJ. Waste and residue availability for advanced biofuel production in EU Member States. Biomass Bioenerg 2016;89:2–10.
[9] Welfle A, Gilbert P, Thornley P. Increasing biomass resource availability through supply chain analysis. Biomass Bioenerg 2014;70:249–66.
[10] Welfle A, Gilbert P, Thornley P. Securing a bioenergy future without imports. Energy Policy 2014;68:1–14.

Index

Note: Page numbers followed by *f* indicate figures and *t* indicate tables.

M

FSC®
www.fsc.org
MIX
Papier | Fördert
gute Waldnutzung
FSC® C083411

Kolibri 360 GmbH
Ferdinand-Jühlke-Straße 7
99095 Erfurt, Deutschland
produktsicherheit@kolibri360.de